Forest Planning and Management

Forest Planning and Management

Edited by Pierre Simone

SYRAWOOD
PUBLISHING HOUSE
New York

Published by Syrawood Publishing House,
750 Third Avenue, 9th Floor,
New York, NY 10017, USA
www.syrawoodpublishinghouse.com

Forest Planning and Management
Edited by Pierre Simone

International Standard Book Number: 978-1-68286-654-2 (Hardback)

Cataloging-in-Publication Data

Forest planning and management / edited by Pierre Simone.
p. cm.
Includes bibliographical references and index.
ISBN 978-1-68286-654-2
1. Forests and forestry--Planning. 2. Forest management. I. Simone, Pierre.

SD131 .F67 2019
634.9--dc23

Printed in China.

TABLE OF CONTENTS

PREFACE

I am honored to present to you this unique book which encompasses the most up-to-date data in the field. I was extremely pleased to get this opportunity of editing the work of experts from across the globe. I have also written papers in this field and researched the various aspects revolving around the progress of the discipline. I have tried to unify my knowledge along with that of stalwarts from every corner of the world, to produce a text which not only benefits the readers but also facilitates the growth of the field.

Forest management is concerned with the administration of forests and development of effective methods of forest conservation. It is a sub-discipline of forestry. All the diverse aspects related to the maintenance and conservation of forests such as aesthetics, silviculture, cutting roads, timber extraction, forest regulation, etc. fall under this field. This book elucidates the concepts and innovative models around prospective developments with respect to forest planning and management. It aims to shed light on some of the unexplored aspects of this field, as well as highlight the modern research studies and trends. While understanding the long-term perspectives of the topics, the book makes an effort in highlighting their impact as a modern tool for the growth of the discipline. Scientists and students actively engaged in this field will find this book full of crucial and unexplored concepts.

Finally, I would like to thank all the contributing authors for their valuable time and contributions. This book would not have been possible without their efforts. I would also like to thank my friends and family for their constant support.

Editor

Extensive Operating Experiments on the Conversion of Fuel-Bound Nitrogen into Nitrogen Oxides in the Combustion of Wood Fuel

Ladislav Dzurenda [1], Emília Hroncová [2,*] and Juraj Ladomerský [2]

[1] Department of Woodworking, Faculty of Wood Sciences and Technology, Technical University in Zvolen, T.G. Masaryka 24, 960 53 Zvolen, Slovakia; dzurenda@tuzvo.sk

[2] Department of Environmental Management, Faculty of Natural Sciences, Matej Bel University, Tajovského 40, 974 01 Banská Bystrica, Slovakia; juraj.ladomersky@umb.sk

* Correspondence: emilia.hroncova@gmail.com or emilia.hroncova@umb.sk

Abstract: This paper presents a wide, targeted, operational experiment on the conversion of fuel-bound nitrogen to nitrogen oxides. The experiment was implemented by studying the production of energy in 16-grate boilers. The conversion factor was determined based on the nitrogen content in individual types of dendromass and on nitrogen oxide emissions in flue gases. In measuring emissions, 103 average values for the concentration of nitrous oxide were found. Every average value was set using three consecutive half-hour emission measurements. The specific thermal output of the boilers was 300 to 2500 kW. The nitrogen content in the deliberately selected dendromass was within a wide interval of N^{daf} values between 0.04% and 1.26%. The dependence of the conversion factor, X_{NOx}, upon the content of fuel-bound nitrogen is very accurately expressed by the power function.

Keywords: combustion; dendromass; conversion factor; fuel-bound nitrogen; nitrogen oxides

---

1. Introduction

Chemically bound nitrogen in organic matter is converted during combustion into a variety of inorganic forms, mainly molecular nitrogen (N_2) and the nitrogen oxides, NO and NO_2. The proportion of nitrogen oxides (NO_x) formed is not clearly predictable because it depends on numerous factors such as the proportion of nitrogen in organic matter, combustion technology, and the specific conditions in the combustion chamber.

As the amount of nitrogen oxides formed during the combustion of fossil fuels and biofuels is very important in terms of air pollution, this issue has been given due attention [1]. In terms of air pollution, it is important to design and operate combustion equipment so that it produces the lowest possible volume of nitrogen oxides. Large combustion plants, designed and operated according to the best available techniques (BAT), operate using effective techniques and measures for the removal of pollutants. The BAT recommendations for the reduction of NO_x emissions and the associated emission levels for various fuels are summarized in a reference document that covers combustion installations with a rated thermal input exceeding 50 MW [2].

The production of biomass-based energy mainly uses grate furnaces with a thermal input of less than 50 MW. The level of development of these medium and small combustion plants, therefore, is not explicitly defined. There is marked variability in the chemical composition of the biomass, which is also reflected in the marked dispersion of concentrations of nitrogen oxides in flue gases from biomass combustion. Because the current emissions limits for NO_x from combusting biomass are not sufficiently strict, researchers have not yet dealt with what minimal values of concentration could be achieved in practice.

Experimental research and modelling of the fuel combustion processes can be used to clarify the formation of nitrogen oxides [3,4]. The combustion of forest biomass (dendromass) should be regulated so that nitrogen oxides emissions are minimized. Pilot experiments and measurements of nitrogen oxide emissions in specific combustion devices have also aimed at minimizing emissions. Statistical evaluation of older available data on types of combusted wood fuel, combustion plants, and pollutant concentrations using factor analysis have shown that combustion technique has a decisive influence on basic pollutant emissions except for NO_x [5]. Nussbaumer (1991) [6] measured conversion factors for NO_x/Fuel-N from 8.4% to 67.3% in laboratory boilers. Stubenberger et al. (2007) [7] investigated the differences of various solid biomass fuels in releasing N and forming N flue gas species using a lab-scale reactor for biomass fuel and one pilot-scale plant. The lab-scale reactor used for the experimental work was a discontinuously operated cylindrical fixed bed reactor, from which it was not possible to get results for the conversion of fuel N into NO_x. In the pilot-scale plant with a reciprocating horizontal grate (the nominal boiler capacity being 440 kWth) and primary and secondary combustion zones, the following values of conversion were measured: −88% for sawdust (fuel concentration N = 0.06%), 22% for bark (fuel concentration N = 0.27%) and 5% for MDF board (fuel concentration N = 6.87%). Pleckaitine and Buinevicius [8] used laboratory experiments to define the dependence of the conversion coefficient on the nitrogen content in fuel. A prototype of a biomass grate-fired unit with a nominal useful output of 250 kWth was used by Díaz-Ramírez et al. (2014) [9] to evaluate the applicability of the three pelletized energy crops. They found conversion factors of NO_x/Fuel-N from 14.5% ± 0.8% to 71.4% ± 1.9%.

To date, studies have consisted of model experiments under laboratory conditions or measurements of the conversion of fuel-bound nitrogen in individual boilers under varying conditions. The actual value of the results of these model experiments is limited and difficult to generalize. Dzurenda et al. (2015) [10] carried out an extensive study combusting spruce wood and monitoring the conversion of fuel-bound nitrogen into nitrogen oxides under real conditions in 16 grate boilers. Their emission measurements were mostly carried out in modern boilers with grate furnaces developed in the early 21st century, for which the conversion factor was 65.4% ± 8.3%.

The existing research therefore has not established representative conversion factor values for the conversion of fuel-bound nitrogen into nitrogen oxides. Representative values must be gathered from a wide spectrum of emission measurements based on operationally combusting dendromass with varying fuel-bound nitrogen contents. This is the aim of this paper. No similar extensive experiment to find the representative values of conversion factor fuel-bound nitrogen into nitrogen oxides has been carried out so far. Establishing representative conversion factors for converting fuel-bound nitrogen into nitrogen oxides can be used to estimate the lowest achievable concentrations of nitrogen oxides.

The dendromass used for our study was debarked deciduous and coniferous trees, bark, and energy chips from plantation-grown, fast-growing tree species. The functional dependency for dendromass with nitrogen in the dry and ash free matter (N^{daf}) was determined to be in the range of 0.04%–1.26%. This dependency was based on extensive targeted operating experiments on energy generation in grate furnace boilers with a rated thermal input of 300–2500 kW.

2. Materials and Methods

2.1. Dendromass for Combustion Tests

The following was specially prepared for the combustion test:

- Moist wood chips and sawdust were prepared from spruce (*Picea exelsa* (L.)), Scots pine (*Pinus sylvestris* (L.)), common beech (*Fagus sylvatica* (L.)), and black locust (*Robinia pseudoacacia*). Wood chips and sawdust were produced from debarked wood from forest stands that had not been chemically treated during their cultivation.

- Crushed tree bark was obtained from the debarking of spruce (*Picea exelsa* (L.)) and English oak (*Quercus robur* (L.)).
- Moist wood chips were also produced from plantation-grown, fast-growing tree species with a short rotation coppice (SRC), including common osier (*Salix viminalis*) and black locust (*Robinia pseudoacacia*); the forest stands had not been treated by any chemical means.

For the collected representative samples of dendromass chips from individual operators, humidity was determined and elementary element analysis was conducted to quantify the C, H, and N contents (though in this paper we indicate only the content of N). In order to determine the nitrogen content in dry ash and ash-free matter (N^{daf}), we also determined the ash content of the investigated dendromass. Differences in ash content between different types of dendromass were relatively small and did not affect concentration of NO_x emissions. For this reason, we have not indicated the value of ash content. For SRC species, the proportion of bark in the chips was determined as well as wood and bark moisture content, after which an elemental analysis of wood and bark was performed. At all operators we studied, the fuel was prepared and stored in the same way in the winter months, so the differences in moisture between the individual plants were small. Given the non-significant differences in the nitrogen content of the same types of samples among the operators, calculations were performed using only the average values.

2.2. *Industrial Wood Boiler*

Emission measurements were carried out on modern, fully automated boilers intended for biofuel combustion (Table 1). The furnaces in Multimiser boilers (Danstoker, Plc., Herning, Denmark) were designed to burn fuel at a low heat load, thereby creating conditions for optimum biofuel combustion and low nitrogen oxide emissions. Similarly, Firematic FR 300 boilers (Herz Energietechnik GmbH, Pinkafeld, Austria) ensure an optimum furnace temperature for biomass combustion by a water-cooled grate and an electronically-controlled BioControl 3000 (Herz Armaturen GmbH Sebersdorf, Austria) for fuel combustion. Other combustion plants use the hot water boiler TPS (ELBH, Ltd., Týn nad Vltavou, Czech Republic).

Table 1. Types of boilers and their boiler input.

Boiler Type	Number of Boilers (pcs)	Boiler Input (kW)
MULTIMISER 20	1	2200
MULTIMISER 19	1	1750
MULTIMISER 16	1	880
MULTIMISER 15	4	725
MULTIMISER 14	1	560
MULTIMISER 13	2	425
MULTIMISER 12	1	350
FIREMATIC FR 300	3	300
TPS 100	1	1163
TPS 35	1	440

2.3. *Modelling the Conversion of Fuel-Bound Nitrogen into Nitrogen Oxides*

The conversion factor for converting fuel-bound nitrogen into emissions of nitrogen oxides X_{NOx} is defined as the percentage ratio of the concentration of nitrogen oxide emissions in the combustion gases, as determined by emission measurements, to the theoretical concentration of nitrogen oxides calculated by Equation (1). Emissions from wood combustion are converted to 11% oxygen content in the resulting dry combustion gases; therefore, the calculated concentration is converted to the same oxygen content in dry combustion gases.

$$X_{NO_x} = \frac{C_{NO_x}}{C_{maxNO_x}} \times 100 \quad (1)$$

where X_{NOx} (%) is the coefficient of the conversion of fuel-bound nitrogen into emissions of nitrogen oxides; C_{NOx} (mg/m^3) is the concentration of NO_x in the combustion gases, determined by emission measurements; and C_{maxNOx} (mg/m^3) is the theoretical (maximum) concentration of NO_x in the combustion gases, formed by the complete oxidation of nitrogen present in the fuel, determined by a technical calculation.

The maximum concentration of NO_x in the combustion gases C_{maxNOx} (kg/m^3) that would be formed by 100% conversion is described by the equation Equations (2)–(4):

$$c_{maxNO_x} = \frac{{}_{max}m_{NO_x}}{{}_{min}V_c} \tag{2}$$

$$c_{maxNO_x} = \frac{\frac{M_{NO_2}}{A_N} \times N^{daf}}{\frac{V_m}{A_C} \times C^{daf} + X_{N_2} \times V_{air} + (\lambda - 1) \times V_{air}} \tag{3}$$

$$c_{maxNO_x} = \frac{3.2857 \times N^{daf}}{1.8555 \times C^{daf} + 0.7905 \times V_{air} + (\lambda - 1) \times V_{air}} \tag{4}$$

where ${}_{max}m_{NOx}$ (kg) is the maximum mass of nitrogen oxides formed by the combustion of 1 kg of combustible matter; ${}_{min}V_c$ (m_n^3) is the minimum volume of combustion gases formed by the combustion of 1 kg of combustible matter according to a stoichiometric calculation; M_{NO_2}(g/mol) is the molar mass; A_N (g/mol) is the standard atomic weight of nitrogen; A_C (g/mol) is the standard atomic weight of carbon; V_m (dm^3/mol) is the molar volume of an ideal gas at 1 atmosphere of pressure is 22.414 dm^3/mol at 0 °C; $X_{N_2}(-)$ is the proportion of nitrogen in the air; N^{daf} (kg/kg) is the nitrogen content in the combustible matter; C^{daf} (kg/kg) is the carbon content in the combustible matter; V_{air} (m_n^3/kg) is the stoichiometric amount of combustion air needed for the combustion of 1 kg of combustible matter; and λ (−) is the excess combustion air.

2.4. Analytical and Measurement Methods

2.4.1. Determination of Water in Dendromass

From the samples collected, three analytical samples were used to determine the water content according to EN 14774-2 [11], i.e., a gravimetric method with drying at atmospheric pressure and a temperature of 105 ± 2 °C.

2.4.2. Determination of C, H, and N in Dendromass

The carbon, hydrogen, and nitrogen contents in the dendromass were determined using an NCS-FLASH EA 1112 analyzer (Thermo Finnigan, San Jose, CA, USA) (though in this paper we indicate only the content of N). The nitrogen contents of SRC were determined by a calculation based on the proportion of bark in the wood chips and on the measured values of nitrogen in the wood and bark using Equation (5):

$$N_{SRC}^{daf} = \left[\frac{100 - X_B}{100}\right] \times N_W^{daf} + \frac{X_B}{100} \times N_B^{daf} \tag{5}$$

where N_{SRC}^{daf} (%) is the nitrogen content in combustible matter of SRC; N_W^{daf} (%) is the nitrogen content in combustible matter of wood; N_B^{daf} (%) is the nitrogen content in combustible matter of bark; and X_B (%) is the proportion of bark in chips.

2.4.3. Emissions Measurement

For this study, we measured emissions from 16 different grate-hearth boilers. In the studied boilers, different types of woody biomass were burnt: *Picea excelsa* L.—wood, bark; *Pinus sylvestris* L.—wood; *Quercus robur* L.—wood, bark; *Fagus sylvatica* L.—wood; *Salix viminalis*—component wood;

Salix viminalis—component bark; *Robinia pseudoacacia*—component wood; *Robinia pseudoacacia*—component bark. Under stable operation of the boilers with minimum output power of 90%, the emission values of NO_x, CO, SO_2, and O_2 were determined according to ISO 10849 [12], EN 15058 [13], ISO 7935 [14], and EN 14789 [15], respectively. In measuring emissions, 103 average values for concentration of nitrogen oxide were found. Every average value was determined using three consecutive half-hour emission measurements.

Emission gas concentrations were measured using an ENDA 6000 portable gas analyzer (Horiba, Kyoto, Japan). The measurement methods used were NDIR for NO_x, CO, and paramagnetism for oxygen. The inert sampling probe and sampling tube were heated to 200 °C. Calibration gases (NO_x in N_2) for the gas analyzers were supplied by Linde Gas (Bratislava, Slovakia) and used at a concentration of 100–250 ppm.

To determine the conversion factor, only the concentrations of nitrogen oxides are important and indicated. To demonstrate the quality of combustion, the concentrations of carbon monoxide are indicated. The relative standard uncertainty of the measurement of nitrogen oxides mass in the combustion gases during the emission measurements with the ENDA 6000 analyzer was $u_{NOx} = 5\%$. The pollutant concentrations measurements correlate to dry gas at 11% oxygen content in the combustion gases.

3. Results

Dendromass for the operating combustion experiments was represented by specially prepared samples of debarked wood from coniferous and deciduous trees, crushed bark, and wood chips from the dendromass of forests stands with plantation-grown, fast-growing tree species. Hence, fuel dendromass was prepared with a wide range of nitrogen content, from 0.04% to 1.26% (dry ash free). The theoretical values of maximum concentrations of NO_x in the combustion gases (according to a stoichiometric calculation) ranged from 136 mg/m_n^3 to 4139 mg/m_n^3 (converted to 11% O_2).

The average values of nitrogen content in the dendromass, including the values of the uncertainty of determining the average value, and the calculated maximum values of the concentration of nitrogen oxides (NO_x) in dry combustion gases (converted to 11% O_2 in the combustion gases) according to Equation (4) are given in Table 2.

Table 2. Average nitrogen content in the combustible matter of dendromass and theoretical values of maximum concentrations of NO_x in the combustion gases (converted to 11% O_2).

Dendromass	N^{daf} (%)		C_{maxNOx} (mg/m_n^3)
Picea excelsa L.—wood	0.04 ± 0.002		136
Pinus sylvestris L.—wood	0.07 ± 0.002		230
Quercus robur L.—wood	0.14 ± 0.01		427
Fagus sylvatica L.—wood	0.15 ± 0.01		493
Chips from SRC			
Salix viminalis—component wood	0.35 ± 0.02	0.58 ± 0.03 *	1906
Salix viminalis—component bark	1.66 ± 0.08		
Robinia pseudoacacia—component wood	0.58 ± 0.02	1.26 ± 0.08 *	4139
Robinia pseudoacacia—component bark	3.52 ± 0.08		
Crushed bark of trees			
Picea excelsa L.—bark	0.38 ± 0.02		1248
Quercus robur L.—bark	0.69 ± 0.03		2267

* weighted average.

The moisture content of the combusted dendromass, as well as the average values and standard deviations of the concentration of nitrogen oxides (NO_x) and carbon monoxide (CO) in the combustion gases, are given in Table 3.

Figure 1 depicts the dependence of the conversion factor of fuel nitrogen in the emitted nitrogen oxides on the nitrogen content in the woody biomass (NO_x/Fuel-N) in the combustion. The process of combustion took place in grate-hearth boilers and the moisture content of the woody biomass was W^r = 33.3%–54.1%.

Table 3. Average concentrations of nitrogen oxides NO_x and carbon monoxide CO in dry combustion gases (converted to 11% O_2).

Dendromass	Amount of Mean Values *	Moisture W^r (%)	Average Half-Hour Emission Concentrations (mg/m_n^3)	
			NO_x	CO
Wood chips and sawdust from debarked wood species				
Picea excelsa L.	14	42.3 ± 2.3	89 ± 12	542 ± 136
Pinus sylvestris L.	12	44.2 ± 2.9	141 ± 19	527 ± 98
Quercus robur L.	13	41.4 ± 3.9	170 ± 28	574 ± 56
Fagus syvatica L.	16	43.1 ± 4.8	192 ± 36	559 ± 74
Chips from SRC				
Salix viminalis	12	48.9 ± 3.8	364 ± 46	534 ± 86
Robinia pseudoacacia	12	36.1 ± 2.8	574 ± 64	526 ± 116
Crushed bark of trees				
Picea excelsa L.	8	49.3 ± 4.8	309 ± 36	532 ± 64
Quercus robur L.	16	46.8 ± 4.1	415 ± 58	558 ± 81

* the mean value is the result of three half-hour measurements at each operator.

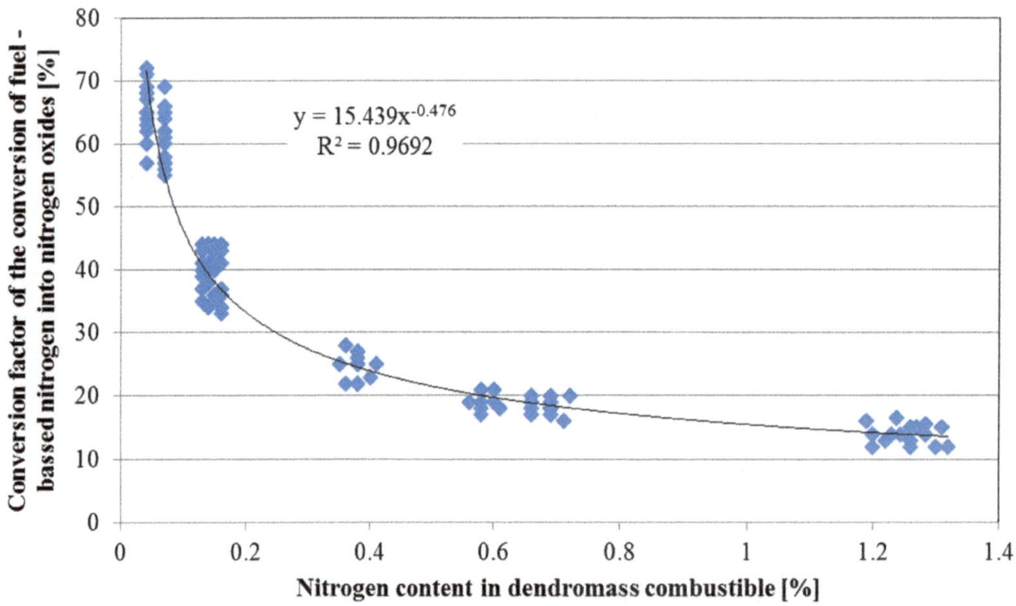

Figure 1. Dependence of the conversion factor of fuel-bound nitrogen into nitrogen oxide emissions on the content of nitrogen in dendromass in the process of combustion of moist dendromass in grate furnaces of heat generators.

The functional dependence relationship of the conversion factor of fuel-bound nitrogen into emission oxides (X_{NOx}), depending on the proportion of nitrogen in the dendromass combustible N^{daf}, is described by Equation (6):

$$X_{NO_x} = 15.439 \times \left(N^{daf}\right)^{-0.476} \quad (6)$$

with the index of determination being R^2 = 0.9692. A necessary condition for the validity of this dependence is that the temperature does not exceed 1124 °C anywhere in the combustion chamber.

4. Discussion

In our operating experiments the concentrations of CO in the flue gases (Table 3) confirmed that the dendromass combustion process was correctly regulated. The values of CO concentrations were very stable, and the differences between the individual types of dendromass were negligible because the experiments took place under operational conditions. Boilers worked in real conditions and emission measurement was carried out during continuous operation of at least 90% of the boiler output. In all cases, the maximum concentrations of CO were significantly below the acceptable emissions limit values (850 mg/m_n^3). In contrast, the concentrations of nitrogen oxides differed significantly, which is certainly related to the differences in the content of fuel-bound nitrogen.

The calculated values of the conversion coefficient for fuel nitrogen in wood and bark to emissions of NO_x, and the dependence of the conversion coefficient of fuel nitrogen in NO_x oxides on fuel nitrogen content in woody biomass, can be considered as representative. This is justified by the conditions of combustion, emissions measurements performed and the technical calculation itself, and confirmed by further knowledge and findings.

According to Marutzky and Seeger [16] and Lin and Murase [17], the concentration of nitrogen oxides in combustion gases from combusting dendromass is also affected by nitrogen-containing adhesive and chemical protectants contaminating the wood. The emission of NO_x is higher during particleboard and glued veneer combustion than it is during the combustion of wood without additives [18]. For this reason, we avoided possible uncertainties of this nature in our experiments. In areas of timber harvesting, no chemical interventions or treatment of forest stands had occurred in the past or during timber harvesting. The chips, sawdust, and bark prepared for the experiment may be considered standard biofuels [19].

All measurements of the concentrations of nitrogen oxides in combustion gases for determining the conversion coefficient were performed during stable operation of the combustion devices and at a minimum power of 90% of the rated power. This minimizes uncertainty in establishing concentrations of NO_x in combustion gases and the values of conversion coefficients.

Although NO_x was found to be formed mainly by the conversion of fuel-bound nitrogen, an important precondition for determining the correct value of the conversion coefficient of fuel-bound nitrogen into nitrogen oxides was making sure that only fuel-bound nitrogen oxides would be formed. Biomass feedstock can burn with water contents up to 60% [20]. Hence we combusted only dendromass with a humidity over $W^r = 23\%$, in which the adiabatic temperature of the flame of the combusted biofuel, in excess air $\lambda = 2.1$, was below the value tad = 1124 °C [21]. Similar data is given, for example, by Nussbaumer (2003) [22]. Maintaining an excess of oxygen in the combustion gases at values $\lambda > 2.0$ was strictly adhered to and thoroughly controlled by emission measurements. By combusting only dendromass with moisture content between $W^r = 33.3\%–54.1\%$, we avoided creating conditions for the formation of high temperature and immediate nitrogen oxides in the boiler furnaces. Based on the mechanism of oxidation of fuel C and N, it follows that with improved oxidation conditions, the concentration of CO falls and the concentration of NO_x increases in combustion gases. This was confirmed in practice [5].

Nitrogen content was the lowest in coniferous wood and approximately two to three times higher in deciduous trees. Nitrogen content was higher in the bark and chips, and was highest in SRC. The difference between the lowest nitrogen content, in the wood of *Picea excelsa* (L.) (0.04%), and that in the bark of *Robinia pseudoacacia* (3.52%) was almost 100-fold. This is consistent with the theoretical values of the maximum concentrations of nitrogen oxides if all the fuel-bound nitrogen of the dendromass was converted to nitrogen oxides during combustion.

However, the real measured values of NO_x concentration in the combustion gases, converted to 11% O_2, were significantly lower than the theoretical values, ranging from 77 mg/m_n^3 to about 600 mg/m_n^3. The relations between nitrogen oxide emissions of coniferous and deciduous trees are therefore similar to the values published in the literature [6,23].

Figure 1 shows a significant decline in the values of the conversion coefficient with an increasing proportion of nitrogen in the dendromass. The declining values of the conversion coefficient of fuel-bound nitrogen in the combustion process of wood is attributed to increased concentration of nitrogen oxide (NO) in the combustion chamber. A similar trend was observed in Pleckaitine and Buinevicius's model (2011) [8], according to which conversion was expressed by the equation $X_{NOx} = 11.312 \times (N_K)^{-0.6196}$ (where N_K is the nitrogen content in fuel, expressed as a percentage). Another graphically similar shape was produced by laboratory measurements of biomass combustion in an electrically heated reactor, where obtained values were considerably lower than the generally recognized values [24]. It follows from this that the results of laboratory modeled experiments cannot always be used for accurate calculations in real practice. Moreover, for combustion of various woods, the change of the design of the secondary air inlets might lower the emissions of NO_x [25].

Our generalized results make it possible to determine the minimum concentrations of nitrogen oxides from combusting woody biomass in the grate furnaces of boilers, depending on the content of fuel-bound nitrogen. If, in one of the grate furnaces, higher concentrations of nitrogen oxides are created than those corresponding with Equation (4), which was derived by our experiments, then the operator can seek the causes of increased creation of nitrogen oxides and modify the parameters of combusting dendromass in grate furnaces in order to minimize nitrogen oxide emissions.

5. Conclusions

This paper describes extensive experiments on the operational combustion of specially prepared samples of eight kinds of dendromass with nitrogen content in the range 0.04%–1.26%, moisture content of W^r = 33.3%–54.1%, and excess air $\lambda = 2.1$. Under these conditions, only fuel-based nitrogen oxides are released. Average concentrations of NO_x in dry combustion gases (converted to 11% O_2) ranged from 89 ± 12 mg/m_n^3 (in chips and sawdust from debarked wood *Picea excelsa* L.) to 574 ± 64 mg/m_n^3 (in chips from SCR *Salix viminalis*). With an increasing proportion of nitrogen in the dendromass (combustible N^{daf} in the range 0.04%–1.26%), the conversion factor X_{NOx} decreased according to the relationship $X_{NOx} = 15.439 \times (N^{daf})^{-0.476}$.

Our analysis shows that approximately 2/3 of the fuel-bound nitrogen was converted into NO_x in the combustion of coniferous wood with a low nitrogen content (up to 0.1%). In the combustion of wood from deciduous trees with a nitrogen content above 0.1%, less than half of the fuel-bound nitrogen is converted into NO_x.

Acknowledgments: This research was supported by the Slovak Grant Agency VEGA under contract no. VEGA 1/0547/15 "Experimental measurement and modelling of fugitive emissions". This work was also supported by the Slovak Research and Development Agency under the contract no. APVV-0353-11 "A proposal and realization of a pilot retort with reduced emissions for charcoal production in marginal zone and verification application".

Author Contributions: Ladislav Dzurenda, Juraj Ladomerský and Emília Hroncová designed the experiments, prepared and carried out the experiments. Ladislav Dzurenda, Emília Hroncová and Juraj Ladomerský prepared the manuscript.

Conflicts of Interest: The authors declare no conflict of interest.

References

1. Chen, W.; Liu, G. Numerical investigation on the flow, combustion and NO_x emission characteristics in a 500 MW_e tangentially fired pulverized-coal boiler. *Open Fuels Energy Sci. J.* **2015**, *8*, 8–13.
2. European Commission (EC). *Large Combustion Plants, Integrated Pollution Prevention and Control*; EC: Brussels, Belgium, 2016.
3. Salzmann, R.; Nussbaumer, T. Fuel staging for NO_x reduction in biomass combustion: Experiments and modelling. *Energy Fuels* **2001**, *15*, 575–582. [CrossRef]
4. Fournel, S.; Marcos, B.; Godbout, S.; Heitz, M. Predicting gaseous emissions from small-scale combustion of agricultural biomass fuels. *Bioresour. Technol.* **2015**, *179*, 165–172. [CrossRef] [PubMed]
5. Ladomerský, J. Emission analysis and minimization from the wood waste combustion. *Wood Res.* **2000**, *45*, 33–44.
6. Nussbaumer, T. Requirements for low-emission wood firings. *Eur. J. Wood Wood Ind.* **1991**, *49*, 445–450. [CrossRef]

7. Stubenberger, G.; Scharler, R.; Obernberger, I. Nitrogen Release Behavior of Different Biomass Fuels Under Lab-Scale and Pilot-Scale Conditions. In Proceedings of the 15th European Biomass Conference & Exhibition, Berlin, Germany, 7–11 May 2007.
8. Pleckaitine, R.; Buinevicius, K. The factors which have influence on nitrogen conversion formation. In *Environmental Engineering*; Vilnius Gediminas Technical University: Vilnius, Lithuania, 2011; pp. 263–269.
9. Díaz-Ramírez, M.; Sebastián, F.; Royo, J.; Rezeau, A. Influencing factors on NOx emission level during grate conversion of three pelletized energy crops. *Appl. Energy* **2014**, *115*, 360–373. [CrossRef]
10. Dzurenda, L.; Ladomersky, J.; Hroncova, E. Conversion Factor of Fuel-Bound Nitrogen to Oxides in the Process of Spruce Wood Combustion in Boiler Grate Furnaces. *Pol. J. Environ. Stud.* **2015**, *24*, 505–509. [CrossRef]
11. European Committee for Standardization. *EN 14774-2 Solid Biofuels—Determination of Moisture Content—Oven Dry Method—Part 2: Total Moisture—Simplified Method*; European Committee for Standardization: Brussels, Belgium, 2010.
12. International Organization for Standardization (ISO). *ISO 10849 Stationary Source Emissions—Determination of the Mass Concentration of Nitrogen Oxides—Performance Characteristics of Automated Measuring Systems*; ISO: Geneva, Switzerland, 2013.
13. European Committee for Standardization. *EN 15058 Air Quality. Stationary Source Emissions. Determination of Mass Concentration of Carbon Monoxide (CO). Reference Method: Non-Dispersive Infrared Spectrometry*; European Committee for Standardization: Brussels, Belgium, 2007; Volume 15058.
14. International Organization for Standardization (ISO). *ISO 7935 Stationary Source Emissions—Determination of the Mass Concentration of Sulfur Dioxide—Performance Characteristics of Automated Measuring Methods*; ISO: Geneva, Switzerland, 2013.
15. European Committee for Standardization. *Stationary Source Emissions. Determination of Volume Concentration of Oxygen (O_2). Reference Method Paramagnetism*; European Committee for Standardization: Brussels, Belgium, 2014.
16. Marutzky, R.; Seeger, K.; Marutzky, R.; Seeger, K. *Energy from Wood and other Biomass*; DRW—Verlag Weinbrenner GmbH &Co.: Leinfelden-Echterdingen, Germany, 1999.
17. Lin, H.C.; Murase, Y. Estimation of the thermal decomposition and analysis of evolved species and char of wood preservatives using thermogravimetric analysis combined with spectrometric techniques. *J. Fac. Agric. Kyushu Univ.* **2007**, *52*, 371–380.
18. Risholm-Sundman, M.; Vestin, E. Emissions during combustion of particleboard and glued veneer. *Eur. J. Wood Wood Ind.* **2005**, *63*, 179–185. [CrossRef]
19. Belbo, H.; Talbot, B. Systems analysis of ten supply chains for whole tree chips. *Forests* **2014**, *5*, 2084–2105. [CrossRef]
20. Nussbaumer, T. *Overview on Technologies for Biomass Combustion and Emission Levels of Particulate Matter*; Federal Office for the Environment (FOEN): Zürich, Switzerland, 2010.
21. Dzurenda, L. Mathematic model for the calculation of the emission value of cmaxNOx for combustion of wet dendromass in grate fire chamber. *Acta Fac. Ecol.* **2008**, *19*, 49–58.
22. Nussbaumer, T. Combustion and co-combustion of biomass: Fundamentals, technologies, and primary measures for emission reduction. *Energy Fuels* **2003**, *17*, 1510–1521. [CrossRef]
23. Mandl, C.; Obernberger, I.; Benesch, C.; Scharler, R. Release and Conversion of Fuel-Bound Nitrogen During Fixed-Bed Gasification and Subsequent Staged Combustion. In Proceedings of the 19th European Biomass Conference & Exhibition, Berlin, Germany, 6–10 June 2011.
24. Houshfar, E.; Løvås, T.; Skreiberg, Ø. Experimental investigation on NO_x reduction by primary measures in biomass combustion: Straw, peat, sewage sludge, forest residues and wood pellets. *Energies* **2012**, *5*, 270–290. [CrossRef]
25. Stålnacke, O.; Zethræus, B. Emissions performance of a 40 MW pulverised wood fired boiler. *IFRF Combust. J.* **2009**, *1*, 1–27.

Influence of Heartwood on Wood Density and Pulp Properties Explained by Machine Learning Techniques

Carla Iglesias [1], António José Alves Santos [2], Javier Martínez [3], Helena Pereira [2] and Ofélia Anjos [2,4,*]

1 Department of Natural Resources and Environmental Engineering, University of Vigo, Vigo 36310, Spain; carlaiglesias@uvigo.es
2 Centro de Estudos Florestais, Instituto Superior de Agronomia, Universidade de Lisboa, Tapada da Ajuda, Lisboa 1349-017, Portugal; antonioalvesantos@gmail.com (A.J.A.S.); hpereira@isa.ulisboa.pt (H.P.)
3 Department of Mathematics and Statistics, Centro Universitario de la Defensa, Escuela Naval Militar, Marín 36920, Spain; javier.martinez@cud.uvigo.es
4 Instituto Politécnico de Castelo Branco, Castelo Branco 6001-909, Portugal
* Correspondence: ofelia@ipcb.pt

Abstract: The aim of this work is to develop a tool to predict some pulp properties e.g., pulp yield, Kappa number, ISO brightness (ISO 2470:2008), fiber length and fiber width, using the sapwood and heartwood proportion in the raw-material. For this purpose, *Acacia melanoxylon* trees were collected from four sites in Portugal. Percentage of sapwood and heartwood, area and the stem eccentricity (in N-S and E-W directions) were measured on transversal stem sections of *A. melanoxylon* R. Br. The relative position of the samples with respect to the total tree height was also considered as an input variable. Different configurations were tested until the maximum correlation coefficient was achieved. A classical mathematical technique (multiple linear regression) and machine learning methods (classification and regression trees, multi-layer perceptron and support vector machines) were tested. Classification and regression trees (CART) was the most accurate model for the prediction of pulp ISO brightness ($R = 0.85$). The other parameters could be predicted with fair results ($R = 0.64$–0.75) by CART. Hence, the proportion of heartwood and sapwood is a relevant parameter for pulping and pulp properties, and should be taken as a quality trait when assessing a pulpwood resource.

Keywords: *Acacia melanoxylon*; heartwood; pulp properties; Multiple Linear Regression; CART; Multi-Layer Perceptron (MLP); Support Vector Machines (SVM)

---

1. Introduction

Blackwood and other acacia were introduced in Portugal at the beginning of the twentieth century in order to colonize dry and poor sandy soils along the coast. Some of the species are considered as invasive as they are characterized by vigorous tree growth, root sprouting and fire stimulated seed germination. Given the high timber value of *Acacia melanoxylon* in its region of origin, a research effort was made to study the wood quality in view of its valorization as a timber species in Europe. For instance, studies were made on heartwood development [1], density and mechanical properties [2] as well as on its utilization as raw-material for pulp and paper [3,4].

Blackwood is a medium density hardwood presenting basic density ranging between 450 and 850 kg/m^3 [2]. The density that is usually higher for heartwood will influence the paper properties [5,6].

The wood quality of blackwood trees is primarily given by the amount of heartwood, which generally shows a rich brown color and a high natural durability [7,8]. Heartwood, which has a high content of extractives, mostly dark-colored and phenolic, impacts negatively on wood chemical pulping [9], pulp yields, pulp brightness [10–13] and affects paper manufacturing [14–16]. However, the presence of heartwood color can be a positive detail in furniture. The variation of heartwood and sapwood in *Acacia melanoxylon* growth in Portugal showed no significant variation between trees and site [1].

Knowledge about the influence of heartwood percentage on the properties of the final product is important for stand management and raw material selection in the industry.

Machine learning focuses on the biological learning process and tries to emulate it through algorithms that are able to learn from given data and provide new results. Machine learning techniques have shown to be an effective tool for modelling and predicting natural parameters in different fields. Among the examples in the literature, these techniques were used to evaluate manufactured paper [17], to obtain the mechanical properties of wood and cork from physical properties [18–20], to predict paper properties from its density [5], to discriminate wood types [21] and to evaluate tree growth [22], to name a few.

The objective of this work is to develop a tool to predict pulp properties (pulp yield, Kappa number, ISO brightness (ISO 2470:2008), fiber length and fiber width) from characteristics of the raw material such as sapwood and heartwood proportion, stem eccentricity in N-S and E-W directions and the relative position of samples with respect to the total tree height. For this purpose, we first applied classical mathematical techniques (multiple linear regression), and then used different machine learning techniques, namely classification and regression trees (CART), Multi-Layer Perceptron (MLP), which is a particular case of artificial neural networks, and support vector machines (SVM).

2. Materials and Methods

2.1. Dataset

Transversal stem sections of *A. melanoxylon* belonging to 20 trees from four sites in Portugal were used—the Camarido National Forest (MNC), at the mouth of Minho River, in the littoral north close to Caminha; the Forest Perimeter of Ovar Dunes (PFDOVM), in the littoral north close to Ovar; the Forest Perimeter of Rebordões de Santa Maria (PFRSM), in the north mid-interior close to Ponte de Lima; and the Forest Perimeter of Crasto Mountain (PFC), in the centre interior close to Viseu [23]. Selective harvesting was done for a sawn timber with diameter at breast height (*dbh*) above 40 cm over-bark, corresponding to a revolution age of about 50 years. Six samples were taken from each tree at different heights: one from the top, one from the bottom, and one from the 65%, 35%, 15% and 5% of the total height of the tree.

The marking out of heartwood was done visually by its color, its area being estimated by the calculation of a circle area, using the mean radius of direct measurements on four geographic exposures of each disc. The tree and heartwood volumes were calculated per stump comprised between the different height levels of the sampling. Conical sections for 0%–5%, 5%–15%, 15%–35%, 35%–65% and 65%–top were considered. The sapwood volume was calculated by the difference of heartwood with respect to total wood, according Rucha et al. [24].

The samples were subsequently chipped to particles with approximately 50 mm length, 10 mm width and 5 mm thickness, taking care to avoid any knotwood—since their extractive content may influence pulping [25]—and submitted to conventional Kraft pulping in a forced circulation digester under the following reaction conditions: active alkali charge 21.3% (as NaOH); sulfidity index 30%; liquor/wood ratio 4/1; time to temperature 90 min; time at temperature (160 °C) 90 min. Experiments were carried out with 25 g (oven dry) of wood, as reported in Santos et al. [6].

For each sample, the following parameters were gathered: position of the samples with respect to the relative height (%), total tree height (m), absolute height of the sample (m), tree diameter at different heights (cm), stem eccentricity in N-S and E-W directions (cm), sapwood and heartwood proportion and total area (cm^2) according to [23]. Also, density (kg/m^3) of the wood chips was measured [26]. The pulp was characterized by measuring pulp yield (%), Kappa number, ISO brightness (%), fiber length (mm) and fiber width (μm) according Santos et al. [6].

2.2. Classical Techniques: Multiple Linear Regression

In a multiple linear regression model, two or more explanatory variables relate to a single response variable by means of an adjusted linear function,

$$\hat{y}_i = \beta_0 + \sum_{k=1}^{K} \beta_k f_k\left(X_{i_1}, X_{i_2}, \dots, X_{i_p}\right) + \varepsilon_i \tag{1}$$

where $\hat{y}_i$ represents the estimated response of y_i [27,28]. The similarity of the prediction method is maximized with the least squares method and the correlation coefficient is defined as

$$R^2 = 1 - \frac{\frac{1}{n}\sum_{i=1}^{n} (y_i - \hat{y}_i)^2}{S_Y^2} \tag{2}$$

being S_Y^2 the variance of Y.

2.3. Machine Learning Techniques: CART

Classification and regression trees (CART) are defined as a non-parametric model that is able to explain the response of a dependent variable given a set of independent variables [29]. The process is based on the creation of subgroups from the initial input data, then analyzing their variance and minimizing this parameter until they are homogeneous.

The prediction of pulp properties was tackled as a regression problem, so the regression trees defined were optimized using mean-squared error (MSE).

Depending on the type of problem (regression or classification), different algorithms can be used to minimize the variance of the subgroups: (MSE) for regression trees, towing rule, cross entropy or Gini's diversity index for classification trees [30].

2.4. Machine Learning Techniques: MLP

Artificial neural networks (ANNs) emulate the natural connections between our neurons and how they process information. The structure of multilayer perceptron (MLP) is based on layers (input, intermediate or hidden and output) formed by perceptions that are not back propagated, connected by activation functions and weights that are adjusted by training the algorithm until the best output is obtained [31].

Hence, an ANN can be defined by (1) how its layers are interconnected; (2) how its weights are updated, i.e., how it learns; (3) how the activation function converts input neurons into their output [31,32].

The training of the network was completed with a 10-fold cross-validation process that guaranteed that all the data were used as training data and as testing data to optimize the weights.

2.5. Machine Learning Techniques: SVM

Since Cortes and Vapnik introduced a new methodology to approach binary classification problems [33], support vector machines (SVM) have become an important machine learning technique for pattern recognition problems due to the global optimum character, parsimony and flexibility [34]. This method is based on finding an optimal separating hyperplane able to classify the observations of an initial input space into a certain number of classes, which are at maximum distance from each other. However, SVM have also been applied to regression problems, usually renamed as support vector regression (SVR) [35,36].

2.6. Predictive Model: Input and Output Data

The predictive model defined in this paper consists of some input variables used to obtain other output variables (the desired prediction results) by means of the selected machine learning technique (Figure 1).

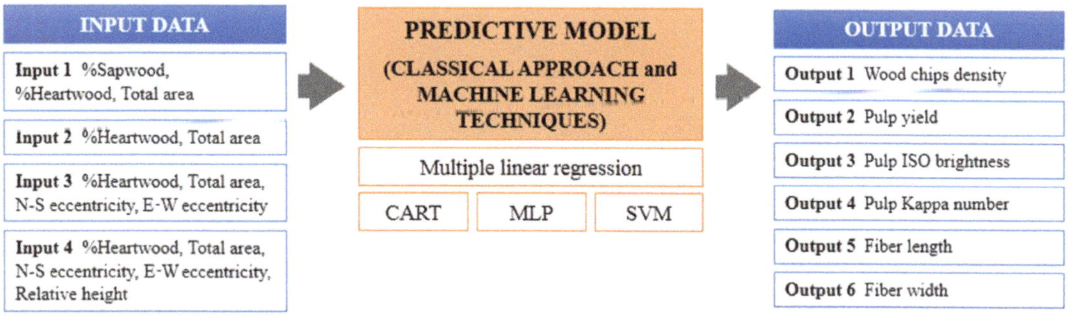

Figure 1. Predictive model defined in this paper. CART, Classification and Regression Trees; MLP, Multi-Layer Perceptron; SVM, Support Vector Machines; N, North; S, South; E, East; W, West.

The problem was approached as a regression problem, where a numeric output is given. Several input sets were tested in order to find the most suitable input data for the model:

- Input 1: sapwood in the wood discs (%), heartwood in the wood discs (%), total area of the wood discs (cm^2);
- Input 2: heartwood in the wood discs (%), total area of the wood discs (cm^2);
- Input 3: heartwood in the wood discs (%), total area of the wood discs (cm^2), stem eccentricity in N-S direction (cm), stem eccentricity in E-W direction (cm);
- Input 4: heartwood in the wood discs (%), total area of the wood discs (cm^2), stem eccentricity in N-S direction (cm), stem eccentricity in E-W direction (cm), relative position of the samples with respect to the total tree height (m).

The relevance of input variables was studied by means of PCA, whose results will be discussed in the corresponding section.

Regarding the output variables, the prediction of the following properties was assessed:

- Output 1: density of wood chips (kg/m^3);
- Output 2: pulp yield (%);
- Output 3: pulp ISO brightness (%);
- Output 4: Kappa number of the pulp;
- Output 5: fiber length (mm);
- Output 6: fiber width (μm).

3. Results

The basic statistic parameters of the selected samples used as input variables to obtain other output variables with the machine learning technique are reported in Table 1, while Table 2 contains a summary of the results.

Table 1. Basic descriptive statistics for the parameters that compose our dataset.

		Min	Max	Average	Median	Standard Deviation	*N* Samples
Discs sampling characteristics	Relative height (%)	0.10	91.6	34.0	25.0	31.1	120
	Total tree height (m)	24.8	36.3	30.2	29.4	3.2	120
	Absolute sample height (m)	0.03	32.2	10.2	7.0	9.5	120
	Tree diameter (cm)	0.06	64.3	20.4	14.0	19.0	120
Stem samples characteristics	N-S stem eccentricity (cm)	0.00	11.2	1.7	1.0	1.9	120
	E-W stem eccentricity (cm)	0.00	10.6	1.6	1.1	1.9	120
	% Sapwood	11.9	100	47.5	37.6	24.9	115
	% Heartwood	0	88.1	52.5	62.4	24.9	115
	Total area (cm^2)	17.4	2355.3	723.3	715.0	534.0	115
Wood chips characteristics	Basic density (kg/m^3)	449	649	524	523	36	100
Pulp characteristics	Yield (%)	47.0	58.2	50.7	50.4	2.0	100
	Kappa number	10.9	18.4	13.7	13.6	1.5	100
	ISO Brightness (%)	14.9	45.6	27.4	27.2	5.4	100
	Fiber length (mm)	0.66	0.94	0.74	0.74	0.05	100
	Fiber width (μm)	16.2	22.9	18.9	18.6	1.5	100

Table 2. Summary of the best results obtained for multiple linear regression, Classification and Regression Tress (CART), Multi-Layer perceptron (MLP) and Support Vector Machines (SVM) for each Output.

		Mult. Linear Regression	CART	MLP	SVM
Output 1	*R* Input	$R^2 = 0.14$ Input 1	0.68 Input 3	0.31 Input 2	0.23 Input 2
Output 2	*R* Input	$R^2 = 0.03$ Input 1	0.68 Input 4	0.20 Input 3 or 4	0.23 Input 3
Output 3	*R* Input	$R^2 = 0.46$ Input 1	0.85 Input 4	0.48 Input 4	0.24 Input 3
Output 4	*R* Input	$R^2 = 0.01$ Input 1	0.73 Input 4	0.24 Input 1	0.32 Input 3
Output 5	*R* Input	$R^2 = 0.05$ Input 1	0.63 Any input	0.26 Input 1 or 4	- -
Output 6	*R* Input	$R^2 = 0.21$ Input 1	0.67 Any input	0.36 Input 4	0.29 Input 4

PCA analysis was performed in order to study the relevance of the variables of each input set. Since these variables belong to different properties and magnitudes, they were standardized before performing PCA. The results regarding the total variance explained by each principal component are included in Table 3.

Table 3. Principal Component Analysis (PCA) results: total variance explained for Input 1, Input 2, Input 3 and Input 4 sets.

	Input 1	Input 2	Input 3	Input 4
Princ. Comp. 1	886,540,996	869,919,835	577,608,239	621,724,612
Princ. Comp. 2	113,459,004	130,080,165	237,653,129	216,512,849
Princ. Comp. 3	1,98E-30		131,618,457	105,298,331
Princ. Comp. 4			531,201,749	426,028,046
Princ. Comp. 5				138,614,039

The performance of the predictive models was assessed by means of their Pearson's correlation coefficient (*R*). Furthermore, root mean square error (RMSE) was also considered to assess CART, MLP and SVM predictive models. It gives the standard deviation of the model prediction error, so the smaller its value, the better the model performance is. Regarding multiple linear regression results, R^2 was obtained.

Table 4 shows in more detail the performance of CART models for the different inputs and outputs. Figures 2 and 3 display the real and predicted values for Output 3 and Output 4, respectively, when obtained by means of the trained CART from Input 4.

Table 4. Results for the CART model. Best results are displayed in bold.

		Output 1	Output 2	Output 3	Output 4	Output 5	Output 6
Input 1	*R*	0.63	0.51	0.69	0.58	0.63	0.67
	RMSE	24.077	1.427	3.471	0.957	0.033	0.852
Input 2	*R*	0.63	0.53	0.70	0.59	0.64	0.68
	RMSE	24.250	1.404	3.485	0.946	0.033	0.855
Input 3	*R*	0.68	0.62	0.81	0.66	0.64	0.67
	RMSE	22.599	1.263	2.727	0.868	0.033	0.865
Input 4	*R*	0.66	0.68	0.85	0.74	0.64	0.67
	RMSE	23.409	1.156	2.425	0.758	0.033	0.859

RMSE, root mean square error.

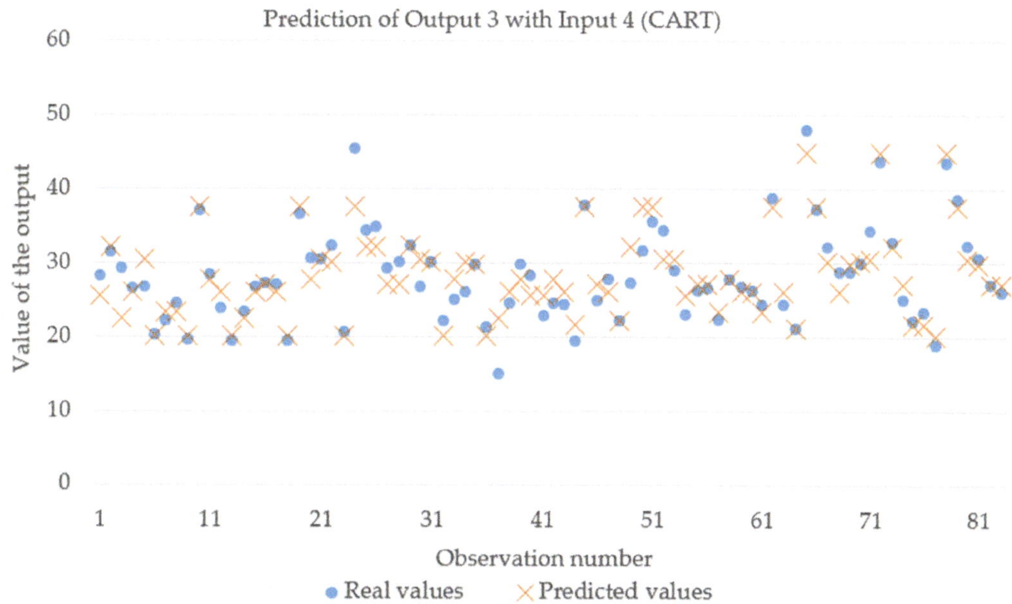

Figure 2. Prediction of Output 3 with the trained CART model from Input 4.

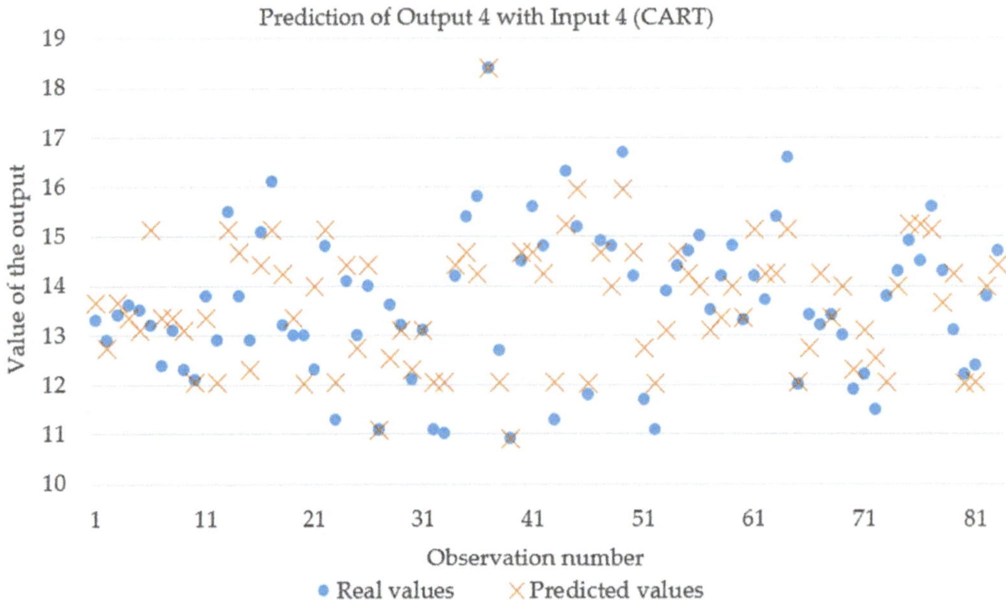

Figure 3. Prediction of Output 4 with the trained CART model from Input 4.

4. Discussion

PCA analysis provides valuable information regarding the role played by each variable in the predictive model (Table 3). The four input sets were studied and their principal components show the following: regarding Input 1, there is a linear dependence between two of the variables (quite obvious, since % Sapwood and % Heartwood make 100%). The remaining Input sets do not show a linear dependence between their variables and none of them could be removed without losing a considerable amount of information.

According to the results obtained, multiple linear regression turned out not to be appropriate for the given data when the input was Input 2, Input 3 or Input 4 since the resulting R^2 statistic was negative. When Input 1 was used, multiple linear regression performed poorly whatever the output was, with Output 3 showing the highest $R^2 = 0.46$. These results are coherent with PCA analysis. Furthermore, given the complexity of the data, results of multiple linear regression with Inputs 2, 3 and 4 are proof of the non-linear relationship among the variables considered. Bearing this in mind, machine learning techniques were applied to the problem.

Among the machine learning techniques tested, MLP and SVM performed similarly: poor results were obtained regardless of the input and output, as shown in Table 2. This fact could be explained by the limited number of input variables and observations available, since they usually perform well with considerable databases [37].

Based on our results, CART is the technique that best predicts wood pulp properties, better than multiple linear regression, MLP and SVM. All the output variables tested showed higher correlation coefficients and lower RMSE values when they were obtained through CART models, regardless of the input parameters (Table 4). Pulp ISO brightness gave the best results ($R = 0.85$, RMSE = 2.425), followed by Kappa number ($R = 0.74$, RMSE = 0.758), whereas the remaining properties (wood chips density, pulp yield, fiber length and fiber width) had similar results ($R = 0.64$–0.68 and RMSE consistent with the dispersion of data). Furthermore, the results show that Input 3 and Input 4 perform better in general terms than Input 1 and 2, which is in accordance with PCA results that indicated the distributed importance of the variables.

No models were found in the literature to predict these output variables with these input variables. However, other models with similar Pearson's correlation coefficients were found for wood quality prediction. Fritz et al. [38] used structure-from-motion with multi-view stereo-photogrammetry and obtained a $R = 0.70$, to estimate the stem radius of oak (*Quercus robur*), hornbeam (*Carpinus betulus*), and maple (*Acer pseudoplatanus*) trees. Knapic and Pereira [39] found significant positive correlations for heartwood diameter with tree diameter at breast height ($R = 0.63$).

Furthermore, CART provides valuable information regarding the influence of the different input parameters, since the structure of the tree represents the regression process until the result is reached [29]. Stem eccentricity in both N-S and E-W directions showed up as an important input parameter for an accurate prediction of all the output variables. It is also the case of the relative position of the samples with respect to the total tree height, which is included in the Input 4 configuration that gave the most accurate results for five of the six outputs predicted.

Stem eccentricity is frequently linked to the presence of reaction wood that is produced on inclined trunks or branches. Depending on reaction wood proportion, the wood properties could be affected to a greater or lesser extent. The eccentricity of a log strongly influences timber processing and the yield of wood components. Tension wood i.e., reaction wood in hardwoods, produces paper with inferior strength properties (namely tensile index, brust index and tear index) in comparison to normal wood, given the differences in the morphologic ultra-structure [40].

5. Conclusions

This research work focuses on predicting pulp properties from raw-material characteristics, testing the suitability and applicability of different mathematical techniques to correlate the available data and obtain the density of wood chips, pulp yield, Kappa number of the pulp, pulp ISO brightness and fiber length and width.

A classical mathematical approach, namely multiple linear regression, as well as machine learning techniques, namely CART, MLP and SVM, were used for this purpose. CART has been shown as the most appropriate technique to predict pulp properties among the four tested. The poor results of the remaining techniques can be due to the limited database and the heterogeneity of the data, thus being difficult to model. The pulp properties best predicted were ISO brightness and Kappa number, which could be modelled reasonably well by means of CART.

Taking advantage of CART structure, the influence of the different input parameters could be analyzed. Thereby, stem eccentricity, an indicator of the presence of reaction wood, was shown to influence the strength of the final paper products due to the resultant differences in the morphologic ultra-structure.

This kind of study provides valuable information to correct or prevent undesirable properties in the final products, as well as highlighting the existing relationships between the different variables of the process.

Acknowledgments: The first author acknowledges the Spanish Ministry of Education, Culture and Sport [grant number FPU 12/02283 to C.I.]. The second author acknowledges a scholarship (SFRH/BD/42073/2007) given by Fundação para a Ciência e a Tecnologia (FCT), Portugal. Centro de Estudos Florestais is a research unit funded by FCT (UID/AGR/UI0239/2013).

Author Contributions: A.J.A.S., H.P. and O.A. conceived and designed the experiments; A.J.A.S., and O.A. performed the experiments; C.I. and J.M. analyzed the data; A.J.A.S., C.I., O.A. and H.P. wrote the paper.

Conflicts of Interest: The authors declare no conflict of interest. The founding sponsors had no role in the design of the study; in the collection, analyses, or interpretation of data; in the writing of the manuscript, and in the decision to publish the results.

References

1. Knapic, S.; Tavares, M.; Pereira, H. Heartwood and sapwood variation in *Acacia melanoxylon* R. Br. trees in Portugal. *Forestry* **2006**, *79*, 371–380. [CrossRef]
2. Machado, J.S.; Louzada, J.L.; Santos, A.J.A.; Nunes, L.; Anjos, O.; Rodrigues, J.; Simões, R.M.S.; Pereira, H. Variation of wood density and mechanical properties of blackwood (*Acacia melanoxylon* R. Br.). *Mater. Des.* **2014**, *56*, 975–980. [CrossRef]

3. Anjos, O.M.S.; Santos, A.J.A.; Simões, R.M.S. Effect of *Acacia melanoxylon* fibre morphology on papermaking potential. *Appita J.* **2011**, *64*, 185–191.
4. Santos, A.J.A.; Anjos, O.; Simões, R.M.S. Paper making potencial of Acacia. *Appita J.* **2006**, *59*, 58–64.
5. Anjos, O.; García-Gonzalo, E.; Santos, A.J.A.; Simões, R.; Martínez-Torres, J.; Pereira, H.; García-Nieto, P. Using Apparent Density of Paper from Hardwood Kraft Pulps to Predict Sheet Properties, based on Unsupervised Classification and Multivariable Regression Techniques. *BioResources* **2015**, *10*, 5920–5931. [CrossRef]
6. Santos, A.J.A.; Anjos, O.; Amaral, M.E.; Gil, N.; Pereira, H.; Simões, R. Influence on pulping yield and pulp properties of wood density of *Acacia melanoxylon*. *J. Wood Sci.* **2012**, *58*, 479–486. [CrossRef]
7. Harrison, C.M. Heartwood content patterns in *Acacia melanoxylon* in the southern cape. *For. S. Afr.* **1974**, *15*, 31–34.
8. Villaverde, J.J.; De Vega, A.; Ligero, P.; Freire, C.S.R.; Pascoal Neto, C.; Silvestre, A.J.D. Miscanthus x giganteus bark organosolv fractionation: Fate of lipophilic components and formation of valuable phenolic byproducts. *J. Agric. Food Chem.* **2010**, *58*, 8279–8285. [CrossRef] [PubMed]
9. Miranda, I.; Gominho, J.; Lourenço, A.; Pereira, H. Heartwood, extractives and pulp yield of three *Eucalyptus globulus* clones grown in two sites. *Appita J.* **2007**, *60*, 485–500.
10. Higgins, G. Pulp and paper. In *Eucalyptus for Wood Production*; Hillis, W.E., Brown, A.G., Eds.; CSIRO/Academic Press: Melbourne, Australia, 1984; pp. 290–316.
11. Santos, A.J.A.; Anjos, O.; Pereira, H. Estimation of *Acacia melanoxylon* unbleached Kraft pulp brightness by NIR spectroscopy. *For. Syst.* **2015**, *24*, eRC03. [CrossRef]
12. Santos, A.J.A.; Anjos, O.; Pereira, H. Prediction of blackwood Kraft pulps yields with wood NIR–PLSR models. *Wood Sci. Technol.* **2016**, *50*, 1307–1322. [CrossRef]
13. Domingues, R.M.A.; Patinha, D.J.S.; Sousa, G.D.A.; Villaverde, J.J.; Silva, C.M.; Freire, C.S.R.; Silvestre, A.J.D.; Pascoal Neto, C. *Eucalyptus* biomass residues from agro-forest and pulping industries as sources of high-value triterpenic compounds. *Cellul. Chem. Technol.* **2011**, *45*, 475–481.
14. Hillis, W.E. *Wood Extractives and Their Significances to the Pulp and Paper Industries*; Academic Press: London, UK; New York, NY, USA, 1962.
15. Pereira, H.; Santos, A.J.A.; Anjos, O. Fibre Morphological Characteristics of Kraft Pulps of *Acacia melanoxylon* Estimated by NIR-PLS-R Models. *Materials* **2016**, *9*, 8. [CrossRef]
16. Santos, A.J.A.; Anjos, O.; Simões, R.; Rodrigues, J.; Pereira, H. Kappa number prediction of *Acacia melanoxylon* unbleached kraft pulps using NIR-PLSR models with a narrow interval of variation. *BioResources* **2014**, *9*, 6735–6744. [CrossRef]
17. García Nieto, P.J.; Martínez Torres, J.; de Cos Juez, F.J.; Sánchez Lasheras, F. Using multivariate adaptive regression splines and multilayer perceptron networks to evaluate paper manufactured using *Eucalyptus globulus*. *Appl. Math. Comput.* **2012**, *219*, 755–763. [CrossRef]
18. García, Á.; Anjos, O.; Iglesias, C.; Pereira, H.; Martínez, J.; Taboada, J. Prediction of mechanical strength of cork under compression using machine learning techniques. *Mater. Des.* **2015**, *82*, 304–311. [CrossRef]
19. Iglesias, C.; Anjos, O.; Martínez, J.; Pereira, H.; Taboada, J. Prediction of tension properties of cork from its physical properties using neural networks. *Eur. J. Wood Wood Prod.* **2015**, *73*, 347–356. [CrossRef]
20. Tiryaki, S.; Özşahin, Ş.; Yıldırım, İ. Comparison of artificial neural network and multiple linear regression models to predict optimum bonding strength of heat treated woods. *Int. J. Adhes. Adhes.* **2014**, *55*, 29–36. [CrossRef]
21. Turhan, K.; Serdar, B. Support vector machines in wood identification: The case of three Salix species from Turkey. *Turk. J. Agric. For.* **2013**, *37*, 249–256.
22. Görgens, E.B.; Montaghi, A.; Rodriguez, L.C.E. A performance comparison of machine learning methods to estimate the fast-growing forest plantation yield based on laser scanning metrics. *Comput. Electron. Agric.* **2015**, *116*, 221–227. [CrossRef]
23. Santos, A.J.A.; Simões, R.; Tavares, M. Variation of some wood macroscopic properties along the stem of *Acacia melanoxylon* R. Br. adult trees in Portugal. *For. Syst.* **2013**, *22*, 463–470. [CrossRef]
24. Rucha, A.; Santos, A.J.A.; Campos, J.; Anjos, O.; Tavares, M. Two methods for tree volume estimation of *Acacia melanoxylon* in Portugal. *Floresta* **2011**, *41*, 169–178. [CrossRef]
25. Pietarinen, S.P.; Willför, S.M.; Ahotupa, M.O.; Hemming, J.E.; Holmbom, B.R. Knotwood and bark extracts: Strong antioxidants from waste materials. *J. Wood Sci.* **2006**, *52*, 436–444. [CrossRef]

26. Santos, A.J.A.; Alves, A.; Rogério, S.; Pereira, H.; Rodrigues, J.; Schwanninger, M. Estimation of wood basic density of *Acacia melanoxylon* (R. Br.) by near infrared spectroscopy. *J. Near Infrared Spectrosc.* **2012**, *2*, 267–274. [CrossRef]
27. Neter, J.; Kutner, M.; Nachtsheim, C.; Wasserman, W. *Applied Linear Statistical Models*; The McGraw-Hill Companies, Inc.: Boston, MA, USA, 1996.
28. Seber, G. *Linear Regression Analysis*; John Wiley and Sons, Inc.: New York, NY, USA, 1977.
29. Breiman, L.; Friedman, J.; Stone, C.J.; Olshen, R.A. *Classification and Regression Trees*; Wadsworth Inc.: Belmont, Australia, 1984.
30. Brown, S.D.; Myles, A.J. Decision Tree Modeling in Classification. In *Reference Module in Chemistry, Molecular Sciences and Chemical Engineering*; Elsevier: Oxford, UK, 2013.
31. Bishop, C.M. *Neural Networks for Pattern Recognition*; Oxford University Press: New York, NY, USA, 2008.
32. Heaton, J. *Introduction to the Math of Neural Networks*; Heaton Research: New York, NY, USA, 2012.
33. Cortes, C.; Vapnik, V. Support-vector networks. *Mach. Learn.* **1995**, *20*, 273–297. [CrossRef]
34. Burges, C.J.C. A tutorial on support vector machines for pattern recognition. *Data Min. Knowl. Discov.* **1998**, *2*, 121–167. [CrossRef]
35. Smola, A.J.; Schölkopf, B. A tutorial on support vector regression. *Stat. Comput.* **2004**, *14*, 199–222. [CrossRef]
36. Fletcher, R. *Practical Methods of Optimization*; John Wiley & Sons, Inc.: New York, NY, USA, 1989.
37. Iglesias, C.; Martínez Torres, J.; García Nieto, P.J.; Alonso Fernández, J.R.; Díaz Muñiz, C.; Piñeiro, J.I.; Taboada, J. Turbidity Prediction in a River Basin by Using Artificial Neural Networks: A Case Study in Northern Spain. *Water Resour. Manag.* **2014**, *28*, 319–331. [CrossRef]
38. Fritz, A.; Kattenborn, T.; Koch, B. UAV-based photogrammetric point clouds—Tree stem mapping in open stands in comparison to terrestrial laser scannerpoint clouds. *Int. Arch. Photogramm. Remote Sens. Spat. Inf. Sci. ISPRS* **2013**, *XL-1/W2*, 141–146. [CrossRef]
39. Knapic, S.; Pereira, H. Within-tree variation of heartwood and ring width in maritime pine (*Pinus pinaster* Ait.). *For. Ecol. Manag.* **2005**, *210*, 81–89. [CrossRef]
40. Parham, R.A.; Robinson, K.W.; Isebrands, J.G. Effects of tension wood on kraft paper from a short-rotation hardwood (*Populus Tristis* No. 1). *Wood Sci. Technol.* **1977**, *11*, 291–303. [CrossRef]

3

Influence of Fuel Load Dynamics on Carbon Emission by Wildfires in the Clay Belt Boreal Landscape

Aurélie Terrier [1,*], Mathieu Paquette [1], Sylvie Gauthier [2], Martin P. Girardin [2], Sylvain Pelletier-Bergeron [3] and Yves Bergeron [1,4]

1 Chaire Industrielle en Aménagement Forestier Durable (NSERC-UQAT-UQAM), Université du Québec à Montréal; P.O. Box 8888, Stn. Centre-ville, Montréal, QC H3C 3P8, Canada; paqmat@gmail.com

2 Natural Resources Canada, Canadian Forest Service, Laurentian Forestry Centre, 1055 du PEPS, P.O. Box 10380, Stn. Sainte-Foy, Québec, QC G1V 4C7, Canada; sylvie.gauthier2@canada.ca (S.G.); martin.girardin@canada.ca (M.P.G.)

3 Département des Sciences du Bois et de la Forêt, Université Laval, 2405 de la Terrasse, Québec, QC G1V 0A6, Canada; Sylvain.Pelletier-Bergeron@mffp.gouv.qc.ca

4 Forest Research Institute, Université du Québec en Abitibi-Temiscamingue, 445 blvd de l'Université, Rouyn-Noranda, QC J9X 5E4, Canada; yves.bergeron@uqat.ca

* Correspondence: terrier.aurelie@courrier.uqam.ca

Abstract: Old-growth forests play a decisive role in preserving biodiversity and ecological functions. In an environment frequently disturbed by fire, the importance of old-growth forests as both a carbon stock as well as a source of emissions when burnt is not fully understood. Here, we report on carbon accumulation with time since the last fire (TSF) in the dominant forest types of the Clay Belt region in eastern North America. To do so, we performed a fuel inventory (tree biomass, herbs and shrubs, dead woody debris, and duff loads) along four chronosequences. Carbon emissions by fire through successional stages were simulated using the Canadian Fire Effects Model. Our results show that fuel accumulates with TSF, especially in coniferous forests. Potential carbon emissions were on average 11.9 $t{\cdot}ha^{-1}$ and 29.5 $t{\cdot}ha^{-1}$ for old-growth and young forests, respectively. In conclusion, maintaining old-growth forests in the Clay Belt landscape not only ensures a sustainable management of the boreal forest, but it also optimizes the carbon storage.

Keywords: boreal forest; fuel load dynamics; fire behavior; carbon emission modelling; sustainable management; mitigation management

---

1. Introduction

Forest management that optimizes carbon storage in the boreal forest could be an effective tool to mitigate anthropogenic carbon emissions [1]. By storing more than one third of the global terrestrial carbon [2,3], the boreal biome is the world's largest carbon stock. However, limited by the cold temperature and short growing season, boreal forests' annual carbon uptake (0.004 $Pg{\cdot}C/m^2{\cdot}year$) is low in comparison with the tropical forests (0.008 $Pg{\cdot}C/m^2{\cdot}year$) or the temperate forests (0.009 $Pg{\cdot}C/m^2{\cdot}year$) [4]. Fortunately, large amounts of carbon can accumulate on the boreal forest's floor since the decomposition of dead material is also limited by the cold climate conditions that prevail therein [2]. Although disturbances emit large quantities of carbon, the boreal forest is a carbon sink since regeneration and time between disturbances allows total recovery of emitted carbon [5].

In Canada, the boreal forest represents approximately 60% of the economic resources for the forest industry [6]. In addition, boreal ecosystems provide a diverse range of habitats for wildlife [6]. During the last few decades, the expansion of intensive harvesting has prompted the need to find a compromise between preserving biodiversity, maintaining forest ecosystems resilience, and providing resources for the wood and paper industries [6]. Facing these needs, interests for a sustainable forest management approach based on natural fire dynamics has increased [7–10]. Natural fires shape landscape structures and composition by creating forests of various ages and by favoring the conservation of shade-intolerant species (e.g., jack pine, trembling aspen) in areas with high fire activity and of tolerant species (e.g., balsam fir) in landscapes rarely influenced by fire [6–9]. Forest management that recreates landscapes similar to those generated by natural fire preserves biodiversity and long-term ecosystem functionality [10–13].

A major challenge in sustainable management lies in maintaining forest structures the way they occur under natural conditions. Several studies have highlighted old-growth forests' importance in preserving biodiversity and ecological functions [10,14,15], for example, by supporting the presence of specialized plant and animal species or by favoring forest resilience. While forest harvesting has resulted in the creation of younger landscapes in the eastern part of North America [14,16,17], sustainable forest management aims to promote silvicultural practices that maintain late-successional forests, or forests that have the same characteristics [10,16–18]. These strategies aim to ensure forest sustainability, but their relevance can be brought into question in a mitigation management context under climate change. In old boreal forests, carbon has had time to accumulate for several decades [17]; when disturbed by fire, 40% [19] to 60% [20] of the carbon accumulated in boreal forests is likely to be emitted back into the atmosphere [21]. This phenomenon is likely to be amplified over the next century under global warming owing to the more frequent and more severe drought conditions [22], which may directly increase wildfire activity (e.g., [23–25]) and reduce tree growth rates [26]. These changes will affect the time needed for forests to recover the carbon released [5]. Given their large carbon pools and the anticipated risks posed to these pools under climate change, maintaining old-growth forests, as opposed to restoring productivity through harvesting, could have adverse effects on the global climate. However, old-growth forests are characterized by a particular fuel structure that tends to maintain higher moisture levels in comparison with younger forest stands [27–29]. This fuel structure may provide old-growth forests with some resistance to fire regime changes.

This study estimates the potential amount of carbon released into the atmosphere by fire activity along successional stages for the dominant forests (trembling aspen (TA), jack pine (JP), and black spruce (BS) forests) of the Clay Belt boreal forest in eastern North America. The Clay Belt region constitutes one of the world's largest terrestrial carbon stocks, estimated at 201 to 250 $t \cdot ha^{-1}$, with a large proportion of forests reaching values of up to 1050 $t \cdot ha^{-1}$ [30]. Therefore, it is justified to assess carbon emissions caused by fire disturbance specifically in the Clay Belt forest. The notion of the impact of fuel accumulation with the time since the last fire (TSF) on fire behavior was first mentioned by Brown [31] for *Pinus ponderosa* ecosystems in the Interior West of the United States. The policy of fire exclusion has created a situation in which fuel accumulates, resulting into catastrophic and severe fire events [32], and induced higher carbon emissions. In 2001, Johnson et al. [33] showed that this notion is not valid for boreal ecosystems. Fire cycles (time required to burn an area equal in size to the study area) in these ecosystems are longer than those in *Pinus ponderosa* ecosystems. Fuel accumulates and does not constitute a limiting factor for fire events: large severe fires in these ecosystems occur under extreme climatic conditions whatever the forest. Studies conducted in Clay Belt BS forests show that peat mosses accumulate with the TSF, thereby inducing a decrease in tree productivity (tree fuel load) [34]. Organic layer accumulation leads to a rise in the water table [35] and creates high soil moisture conditions [34]. Depth of burn (i.e., the depth of the soil organic layer consumed during a fire) is consequently low [36–39] and the fire only burns surface fuel. However, these studies did not consider all fuel material and forest types. First, we wanted to address the hypothesis that fuel accumulates with TSF in the dominant Clay Belt boreal forests, considering all fuel material. Second, it can be postulated that the high soil moisture content and low depth of burn in BS forests may induce less carbon emissions, while the higher fuel availability in TA and JP forests could induce higher carbon emissions. From this hypothesis, we verified whether or not fuel accumulation could induce higher carbon emission by fires in late-successional TA and JP forests, and not in black spruce forests.

To do so, we performed a complete fuel inventory (tree biomass, herbs and shrubs, dead woody debris (DWD), and duff loads) in 61 sites differing in their forest type and TSF. Empirical fuel loads were then used to calculate the carbon emitted by one single simulated fire at each sampling site using the Canadian Fire Effects Model (CanFIRE) [40,41].

2. Materials and Methods

2.1. Study Area

The study area (49°00′–50°00′ N, 78°30′–80°00′ W) is located in the Clay Belt boreal forest in eastern North America, stretching across the Quebec and Ontario border (Figure 1).

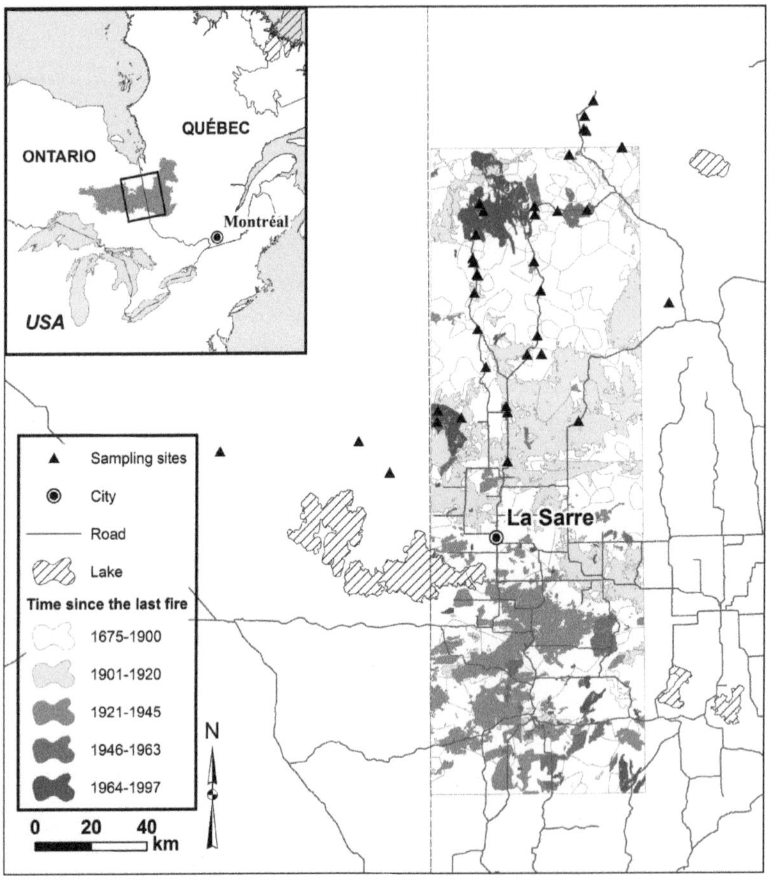

Figure 1. Geographic location of the sampling sites in the Clay Belt boreal forest, eastern North America.

A former proglacial lake (Lake Barlow-Ojibway) left a thick deposit of clay, forming the physiographic unit known today as the Clay Belt, which covers an area of approximately 145,470 km^2 [42]. The current level of fire activity is low, with a fire cycle estimated at 398 years from 1959 to 1999 [43]. The climate is subpolar and subhumid continental, characterized by long, harsh, and dry winters and, short, hot, and humid summers [44]. Average annual temperature from 1971 to 2000 recorded at the closest weather station to the north (Matagami, 49°46′ N, 77°49′ W) and to the south (La Sarre, 48°46′ N, 79°13′ W) was −0.7 °C and 0.7 °C, respectively. Mean total annual precipitation was 906 and 890 mm, respectively [44]. The poor drainage conditions induced by the presence of an impermeable clay substrate, flat topography, historical low fire activity, and cold climate facilitated the accumulation of thick layers of organic soil, a process often described as paludification [45,46]. In parts of the region, peat mosses accumulate on initially mesic soils, independently of topography or drainage, and are related to forest succession [34]. Therefore, in the prolonged absence of fire, these forests tend to convert into less productive spruce-*Sphagnum* opened forests regardless of the initial species composition [34,47–50]. Burned area and residual organic layers (i.e., layers that are not consumed by fire) jointly control forest structure and composition [51]. Shallow residual organic layers on the ground lead to the establishment of dense forests composed of black spruce (*Picea mariana* (Mill.) BSP), trembling aspen (*Populus tremuloïdes* Michx.), or jack pine (*Pinus banksiana* Lamb.) on mesic sites [48,51–53]. In contrast, thick residual organic layers favor black spruce self-replacement [53,54] and accelerate the process of paludification [34,51].

2.2. Fuel Load Dynamics with TSF

Data were collected to characterize trees, herbs, shrubs, DWD, and duff loads (t·ha^{-1}) along the successional stages of the dominant Clay Belt forests (TA, JP, BS; see Figure 5 in [48]). Black spruce forests were separated into black spruce forests originating from severe fire (BS-S) and black spruce forests originating from non-severe fire (BS-NS), since fire severity influences tree and duff load dynamics in BS forests [34]. Successional stages were determined with TSF. We used information (tree basal area, organic layer depth, TSF) that was already available for 37 sites to calculate tree and duff loads [34,48,51]. We visited these sites to complete the inventory with information on DWD, shrubs, and herbs. We selected 24 additional sites to expand TSF variability for each forest type (Table 1). TSF was determined for these additional sites by overlaying a fire reconstruction map previously published for the study area [43]. A total of 61 sites were sampled. At each site, a 30-m-sided equilateral triangle was defined in which the fuel sampling protocol was performed.

Table 1. Distribution of study sites in different successional stages and forest types.

TSF (Years)	Forest Type			
	TA	**JP**	**BS-S**	**BS-NS**
<90	6	7	5	5
90–150	4	1	5	5
>150	2	7	8	6
Total	12	15	18	16

Values express the number of sites. TSF: time since the last fire; TA: trembling aspen; JP: jack pine; BS-S: black spruce originating from severe fire; BS-NS: black spruce originating from non-severe fire.

Basal area for live trees larger than 3 cm in diameter at breast height (DBH) were sampled by species for the 24 additional sampling sites using the prism method (factor 2, metric). Tree loads were calculated using stand-level equations [55]. Shrub load was sampled using the method described in Brown et al. [56]. Nine 1-m^2 plots were established evenly along the triangle transect. Shrub species were identified and stem basal diameters (cm) were measured. Loads were calculated using equations linking shrub species weight in grams with the stem diameter. Previously determined equations [57] were used for *Lonicera canadensis* Bartr., *Ribes* sp., *Rosa acicularis* Lindl., and *Viburnum edule* Raf. New equations were determined with additional shrub samples for *Chamaedaphne calyculata* (L.) Moench, *Kalmia angustifolia* L., *Rhododendron groenlandicum* (Oeder) K.A. Kron & Judd, and *Vaccinium myrtilloid*es Michx. (see details in Appendix A, Table A1). We measured herb loads (all surface vegetation that is not woody (e.g., forbs and graminoids)) using a weighted-estimate approach [56]. We established four plot rectangles (0.5 m^2), three at the center of each side and one in the middle of the triangle. We selected the plot with the most weight, which was identified as the base plot. Herbaceous vegetation was collected in this base plot and oven dried at 95 °C for 24 h for dry weight measurements. The weight of the three other plots was estimated as a fraction of the base plot and fuel load was calculated using Brown et al.'s equation [56]. Fuel load of DWD was sampled using the line intersect method [58] along the triangle sides. Pieces smaller than 7 cm were measured with a "go-no-go" gauge, and counted according to five diameter sizes (class I: 0–0.49 cm; class II: 0.5–0.99 cm, class III: 1–2.99 cm, class IV: 3–4.99 cm; class V: 5–6.99 cm). For pieces bigger than 7 cm, we measured the diameter size. DWD fuel loads were calculated for each size classes using McRae et al.'s equations [58]. Size-class specific coefficients values were extracted from Hély et al. [57] for white cedar, and from McRae et al. [58] for the other species (BS, JP, TA, WB). Finally, duff loads were calculated by multiplying total organic layer depth by bulk density. We used organic layer depths data from Simard et al. [34] for the 37 previously sampled sites. Additional organic layer depth was measured for the other 24 sites in parallel with shrub sampling in each of the nine 1-m^2 plots and averaged by sites. We used previously published mean bulk density ($kg{\cdot}m^{-3}$) information [36] to convert organic layer depth into duff load ($kg{\cdot}m^{-2}$). Duff data were missing for old BS-NS sites because of technical problems.

The relationship between TSF and (i) the fuel load of each material separately and (ii) total fuel load was investigated by forest type. We performed polynomial regressions using the R freeware [59]. Herb and shrub fuel loads were combined into fine aerial vegetation. DWD corresponded to the sum of all classes. Loads were logarithmically transformed (ln + 1) to linearize the relationship. First- to third-order polynomial regressions were tested, and we promoted the significant minimum-order polynomial regressions (p-values ≤ 0.05) in accordance with the parsimony principle.

2.3. *Simulation of Carbon Emissions by Fire*

Potential carbon emission (tons/ha) was investigated using the Canadian Fire Effects Model (CanFIRE, formerly the Boreal Fire Effects Model, BORFIRE) [40,41], including the modified depth of burn equations for BS forests dominated by *Sphagnum* spp. in the Quebec Clay Belt boreal landscape [36,60]. CanFIRE is a collection of Canadian fire behavior models that are used to estimate first-order fire effects on physical characteristics, and to estimate ecological effects at the stand level. For any given fire, CanFIRE calculates fire behavior information (e.g., fire intensity, rate of spread) based on the pre-fire amount of fuel and components of the Canadian Forest Fire Weather Index (FWI) System [61]. Tree fuel consumption takes place only when CanFIRE predicts a crown fire. Values are estimated as the sum of foliage and bark using tree biomass algorithms [62] and are comparable to overstory fuel consumption data recorded on experimental burns in the Canadian Forest Fire Behavior Prediction (FBP) System database [63]. DWD and duff consumption follows McRae et al.'s [64] and de Groot et al.'s [65] equations, respectively. These algorithms are driven by the Buildup Index (BUI) and Drought Code (DC) values of the FWI System for DWD and duff consumption, respectively, and were built from empirical pre-burn and post-burn fuel data collected in prescribed fires.

Since simulation objectives were to analyze differences in carbon emission with fuel structure variations under similar climatic conditions, we simulated the potential carbon emitted by one single simulated fire, using the same fire weather conditions at each sampling site. Sensitivity analyses were made to ensure that fire weather values above average did not modify our conclusions (see details in Appendix B). We determined FWI System component values by averaging daily FWI System components for each natural fire start point encountered during the interval 1971–2000 period in the study area (FWI = 18) [60]. Forest cover was determined based on species' empirical basal area as described in the previous section. Since basal area was sampled for trees with DBH larger than 3 cm, basal area, and, consequently, cover were 0 for all species in young sites. In this case, we attributed a value of 100% to the dominant species of the forest type (e.g., cover: 100% of jack pine in JP). Tree fuel load was directly estimated by CanFIRE using species' basal area. To do so, species' site index and age were used to calculate basal area with Plonski's yield tables [65]. We used TSF for species age and the same site indexes as those used by Terrier et al. [60]. Prior to running simulations, we ensured that the use of Plonski's yield tables instead of our empirical basal area data would not bring a bias to our conclusions (see details in Figure C1). DWD and duff loads were provided by our empirical data. Total duff was separated into duff-fibric and duff-humus using the fibric layer equations described in Terrier et al. [60]. Since the database was not complete for all sites (technical problems in duff sampling in old BS-NS sites), a subset of 55 sites was used for the simulations. Simulation results were averaged by forest types and successional stage classes (TSF < 90 years; TSF between 90 and 150 years; TSF > 150 years).

Some additional changes were also made to the model. In our study area, high soil moisture conditions in BS forests dominated by *Sphagnum* spp. resulted in lower depth of burn, and, consequently, in lower forest floor fuel consumption (FFFC) in comparison with other boreal forest types [36]. In the CanFIRE model, FFFC is a function of duff load and climate conditions [37], and does not consider the lower depth of burn for calculations. Therefore, FFFC equations had to be modified to reflect specific carbon emission in BS forests dominated by *Sphagnum* spp. We defined FFFC as a function of the potential depth of burn as Equation (1):

$$FFFC\left(\frac{\text{kg}}{\text{m}^2}\right) = \text{depth of burn (m)} \times \text{bulk density} \left(\frac{\text{kg}}{\text{m}^3}\right) \tag{1}$$

Since bulk density varies with the organic layer depth [37], we came up with a model that linked measured bulk density with organic layer depth. To do so, we used a published dataset that comprised 103 peat measurements of bulk density at different organic layer depths sampled in 11 sites [34] (details are presented in Appendix D).

3. Results

3.1. Fuel Load Dynamics with TSF

TSF was significantly related to all fuel loads materials (trees, fine aerial vegetation, DWD, and duff) and total fuel loads for TA, JP, BS-S, and BS-NS boreal forests in the Clay Belt. Exceptions were observed for fine aerial vegetation in coniferous forests, duff load in TA forests, and DWD of JP forests (Figure 2, Table 2).

Tree fuel loads (Table 2; Figure 2A) were predicted to approximate 0 (t·ha^{-1}) just after fire (exponential of the polynomial regression intercept ~0), except for BS-NS, where the intercept equals 0.79. Tree fuel loads for coniferous forests increased gradually before decreasing after 150 and 200 years in JP and BS forests, respectively (Figure 2A). Maxima reached 5.47, 4.75, and 4.39 for JP, BS-S, and BS-NS, respectively. In the case of TA forests, tree fuel loads increased particularly rapidly at early successional stages until around 75 years (logarithmic maximum = 6.46), followed by a slight decrease. Finally, tree fuel loads increased again at late successional stages.

Fine aerial vegetation load did not vary significantly with TSF in coniferous forests (Figure 2B); logarithmic values varied near 0 along the successional stages (Table E1). In contrast, the amount of fine aerial vegetation was particularly high at early successional stages in TA forests. A value of 570 tons/ha was observed for the younger site (TSF = 11 years) (Table E1). Fuel loads decreased significantly, reached 0 at 120 years old (Table E1), and increased rapidly at late-successional stages.

DWD load dynamics for BS forests, represented in Figure 2C, followed the same trend as those observed for the tree load: an increase was predicted at early successional stages, maxima logarithmic values of 2.79 and 1.86 for BS-S and BS-NS, respectively, were reached at 200 years, and were followed by a decrease in values. DWD load trend in TA forests was the opposite of tree load: a value of 90 t·ha^{-1} was estimated in the youngest site (TSF = 11 years) (Table E1). Values decreased for 75 years, then increased slightly, and finally decreased gradually at late successional stages. DWD load for JP forests showed no significant relationship with TSF. Values ranged from 1 to 86 t·ha^{-1} (Table E1).

Duff load showed a significant gradual increase with TSF in coniferous forests. Higher values were predicted for BS-NS (Figure 2D). Values were on average 70, 125, 292 t·ha^{-1} for JP, BS-S, and BS-NS, respectively, in sites younger than 50 years old (Table E1). The increase in duff load was faster in BS forests in comparison with JP forests (higher equation slope in Table 2). Variation in duff load was not significantly related to TSF in TA forests. Values averaged 150 all along the chronosequence.

Finally, variation in total fuel load with TSF was significant for all forest types (Figure 2E). Total fuel load gradually increased along successional stages in coniferous forests. Values varied from 57 t·ha^{-1} to 382 t·ha^{-1} in JP, from 77 t·ha^{-1} to 767 t·ha^{-1} in BS-S, and from 160 t·ha^{-1} 1084 t·ha^{-1} in BS-NS (Table E1). Total fuel in TA forests decreased slightly at early successional stages, reaching a minimum at 140 years, and increased at late successional stages (Figure 2E). Values of total fuel ranged from 241 t·ha^{-1} to 730 t·ha^{-1} in TA forests (Table E1).

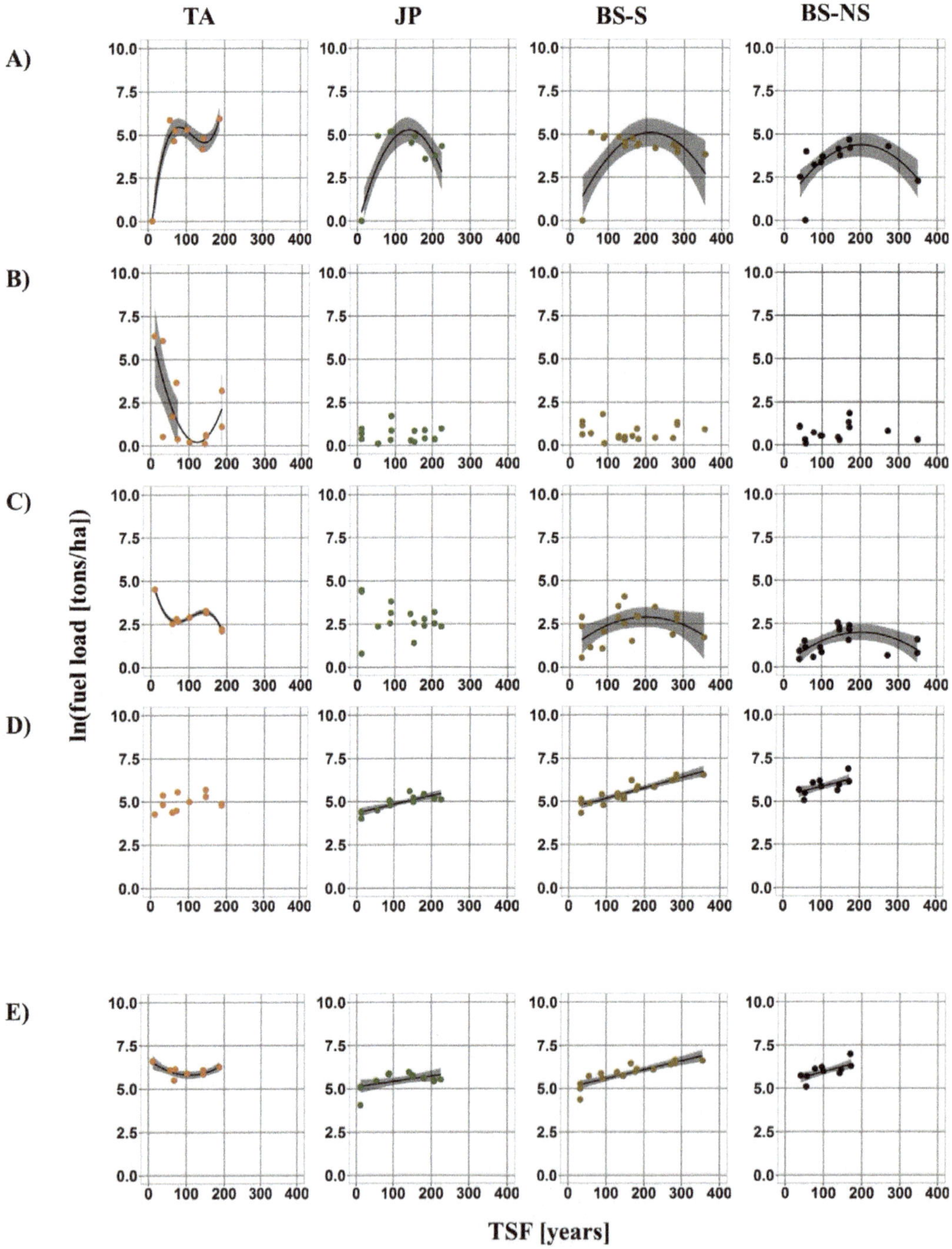

Figure 2. Changes in loads of (**A**) tree; (**B**) fine aerial vegetation (herbs and shrubs); (**C**) dead woody debris (DWD); (**D**) duff; and (**E**) total fuel (**A** + **B** + **C** + **D**) with TSF for TA, JP, BS-S, and BS-NS dominated forests. Fuel loads were logarithmically transformed and are represented by color points. The black line expresses the significant relationship (p-value ≤ 0.05) in the polynomial regression analysis and the grey shape corresponds to the regression standard error.

Table 2. p-value, adjusted R^2 and equations extracted from the regression polynomial analysis of tree, fine aerial vegetation (shrubs + herbs), DWD, duff, and total fuel loads with TSF for TA, JP, BS-S, and BS-NS dominated forests. NS stands for a non-significant relationship.

	Fuel	p-Value	R^2	Model
TA	Trees	$\leq$0.005	0.92	$\ln(load) = -2.21 + 0.23 \times TSF - 0.002 \times TSF^2 + 0.000006 \times TSF^3$
	Fine aerial	$\leq$0.05	0.47	$\ln(load) = 6.91 - 0.11 \times TSF + 0.00045 \times TSF^2$
	DWD	$\leq$0.001	0.84	$\ln(load) = 5.31 - 0.093 \times TSF + 0.00099 \times TSF^2 - 0.000003 \times TSF^3$
	Duff	NS	NS	NS
	Total	$\leq$0.05	0.49	$\ln(load) = 6.72 - 0.017 \times TSF + 0.00008 \times TSF^2$
JP	Trees	$\leq$0.005 [1]	0.78	$\ln(load) = -0.38 + 0.084 \times TSF - 0.0003 \times TSF^2$
	Fine aerial	NS	NS	NS
	DWD	NS	NS	NS
	Duff	$\leq$0.001	0.65	$\ln(load) = 4.35 + 0.0049 \times TSF$
	Total	$\leq$0.05	0.19	$\ln(load) = 5.1 + 0.003 \times TSF$
BS-S	Trees	$\leq$0.001 [1]	0.52	$\ln(load) = -0.049 + 0.048 \times TSF - 0.00012 \times TSF^2$
	Fine aerial	NS	NS	NS
	DWD	$\leq$0.05 [1]	0.16	$\ln(load) = 1.03 + 0.018 \times TSF - 0.000046 \times TSF^2$
	Duff	$\leq$0.001	0.81	$\ln(load) = 4.59 - 0.006 \times TSF$
	Total	$\leq$0.001	0.71	$\ln(load) = 5.06 + 0.0052 \times TSF$
BS-NS	Trees	$\leq$0.001 [1]	0.48	$\ln(Fuel\ load) = 0.79 + 0.036 \times TSF - 0.00009 \times TSF^2$
	Fine aerial	NS	NS	NS
	DWD	$\leq$0.05 [1]	0.34	$\ln(Fuel\ load) = 0.06 + 0.019 \times TSF - 0.00005 \times TSF^2$
	Duff	$\leq$0.05	0.38	$\ln(Fuel\ load) = 5.3 + 0.0055 \times TSF$
	Total	$\leq$0.005	0.48	$\ln(Fuel\ load) = 5.34 + 0.006 \times TSF$

[1] Non-significant intercept.

3.2. *Simulation of Carbon Emission by Fire*

The CanFIRE model was used to simulate carbon emissions during a single fire along the different successional stages of the four dominant forest types (TA, JP, BS-N, and BS-NS) of the Clay Belt boreal forest. Simulation results are presented in Figure 3.

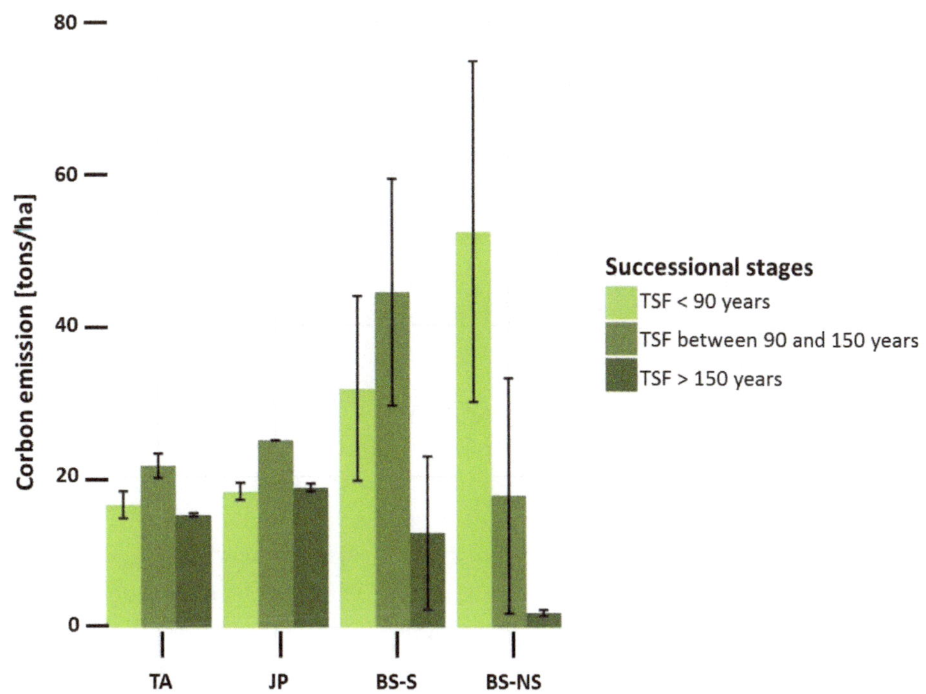

Figure 3. CanFIRE potential carbon emissions averaged by early (TSF < 90 years), intermediate (TSF between 90 and 150 years), and late (TSF > 150 years) successional stages, and by dominant forest types of the Clay Belt boreal landscape (trembling aspen (TA), jack pine (JP), black spruce originating from severe fire (BS-S), and black spruce originating from non-severe fire (BS-NS) forests). Error bars represent the mean standard error.

Mean carbon emissions by fire were estimated to range from 14.83 to 21.37 $t{\cdot}ha^{-1}$ in TA, from 17.97 to 24.7 $t{\cdot}ha^{-1}$ in JP, from 12.48 to 44.3 $t{\cdot}ha^{-1}$ in BS-S, and from 1.92 to 52.31 $t{\cdot}ha^{-1}$ in BS-NS. Simulated carbon emissions were on average lower for sites older than 150 years, except for JP forests, where carbon emission estimates averaged 16 $t{\cdot}ha^{-1}$ for forests younger than 90 years and 19 $t{\cdot}ha^{-1}$ for forests older than 150 years. A maximum average of 53 $t{\cdot}ha^{-1}$ of carbon emitted was simulated for BS-NS with TSF < 90 years, while a particularly low value (2 tons/ha) was simulated for older BS-NS. Aside from BS old-growth forests, potential carbon emission values simulated were lower for TA. Table 3 presents the contribution in percentage of each fuel material to carbon emission by fire. Trees barely contribute to emission, which is mostly generated by duff and DWD combustion.

Table 3. Values of total potential carbon emission by forest type and successional stage, and contribution of each fuel category to the total emission.

Forest Type	Successional Stage	Total (t/ha)	Contribution of Fuel Category		
			Tree (%)	DWD (%)	Duff (%)
TA	TSF < 90 years	16.2	0	20	80
	TSF between 90 and 150 years	21.4	0	11	89
	TSF > 150 years	14.8	0	10	90
JP	TSF < 90 years	17.9	14	21	65
	TSF between 90 and 150 years	24.7	0	9	91
	TSF > 150 years	18.5	0	6	94
BS-S	TSF < 90 years	31.6	6	3	91
	TSF between 90 and 150 years	44.3	6	5	89
	TSF > 150 years	12.5	2	12	86
BS-NS	TSF < 90 years	52.3	3	1	96
	TSF between 90 and 150 years	17.4	3	6	91
	TSF > 150 years	1.9	0	53	47

4. Discussion

4.1. Fuel Load Dynamics with TSF

Our results confirmed our first hypothesis that fuel accumulates with TSF in Clay Belt coniferous forests, despite variations in the dynamics of the different fuel materials. These variations were confirmed and explained by previous studies on fuel load dynamics.

Tree dynamics strongly affect variations in fuel material with TSF [66–69]. It is therefore important to first understand Clay Belt tree dynamics to discuss fine aerial, DWD, and duff dynamics. Light, nutrients, and space are abundant just after a fire event, and vegetation can rapidly colonize disturbed sites [70]. Trees grow more or less rapidly depending on the composition [65] or depth of the residual organic layer [51,71]. Consequently, it was not surprising to predict in our study an increase in tree biomass at early successional stages for all forest types. This increase was faster for TA and JP forests, because growth rates are faster for these trees [65,72] and the low residual organic layers required for TA and JP postfire recruitment lead to the establishment of productive and dense forests [49,52–54]. In contrast, BS-NS forests showed the slowest increase due to the thick residual organic layers left after fire which favor unproductive open BS self-replacement [34,48,52–54]. Following the growth influx, a decrease in tree loads [34,73] was predicted at 80, 110, and 200 years for TA, JP, and BS forests, respectively. Over time, tree mortality starts, gaps are created, and shade-tolerant species appear in the canopy [74,75]. Our predictions concur with previous observations: Pothier et al. [76] estimated tree productivity loss after about 60 years for TA forest in Quebec; Belleau et al. [50] estimated composition changes to occur in the Clay Belt region at around 70 to 120 years for JP and TA forests, respectively, and at around 100 to 200 years for BS forests. In our analysis, TA experienced a recovery in tree biomass after 150 years. Low organic layers observed at late successional stages in TA forests probably made this productivity recovery possible.

Fine aerial vegetation establishment and growth rates are primarily limited by the understory light availability that is created by the overstory canopy [67–69]. The shady understory environment provided by coniferous species [77] results in less abundant fine aerial vegetation in coniferous forests [68,70]. Forest openings induced by the establishment of the second cohort was not associated with the development of fine aerial vegetation in old-growth coniferous forests because the cold soil temperature and nitrogen limitation induced by organic layer accumulation at the late successional stage [34] probably acted to prevent herbs and shrubs development. Herbs and shrubs dynamics in TA forests reflect higher light transmission by the deciduous canopy. At early successional stages, light availability decreases in response to tree growth, and fine aerial vegetation declines in abundance and growth [64]. Tree mortality at intermediate successional stages increases light availability, allowing understory recolonization by herbs and shrubs [70,78,79]. Late successional stages were consequently characterized by an increase in fine aerial vegetation.

DWD variations with TSF can be explained by coarse woody debris (CWD) dynamics (pieces > 7 cm) since CWD constitute most of the DWD loads [29,66]. CWD dynamics in TA forests followed the traditional U-shaped distribution summarized by Brassard and Chen [68]. Pre-disturbance and disturbance-generated woody debris that were not consumed generated a high CWD quantity at early successional stages. As TSF increased this material decomposed and a decrease in CWD occurred. Tree mortality at intermediate successional stages provided high loads of woody debris. Old-growth forests were characterized by a low quantity of CWD, as new cohort growth did not generate dead woody supply [68]. DWD debris dynamics in BS forests corresponded closely to tree fuel load dynamics, with an absence of DWD high value at early successional stages, compared with what was observed in TA forests. Our data showed an absence or low quantity of CWD in young BS forests.Such a feature in coniferous forests has already been mentioned in previous studies [66,80,81]. It may be induced by a faster decay rate after fire and the absence of pre-fire large-diameter trees [66]. CWD could also be buried in the moss layer in BS-NS due to the rapid growth of *Sphagnum* species [31]. No trend in DWD was observed in JP forests, although Brais et al. [82] found that DWD dynamics followed a 'U-shaped' temporal pattern in pure Clay Belt JP forests. We explain these differences in observations by the facts that our data cover a larger variability of TSF and our forests were not pure JP forests. Few studies have compared CWD dynamics in different cover types [69]. Differences in DWD dynamics between coniferous forests and TA forests may be induced by the lower pre-fire tree productivity (Figure 2) and the presence of woody debris with higher flammability [83–86] in coniferous forests.

Decomposition rates were higher in TA than in coniferous forests [80,86]; consequently, organic layer accumulation did not occur in TA forests. Conversely, duff loads increased gradually in coniferous choronosequences, since the lower decomposition rates associated with the paludification process favor organic layer accumulation [34,45,46]. The most interesting fact is that despite a tree and DWD load loss with TSF in coniferous forests, total fuel loads increased with TSF.

Scharlemann et al. [30] provided a global map representing carbon stocks in tree biomass and soils. As mentioned in the introduction, they estimated a higher carbon amount in the Clay Belt boreal forests, with values ranging from 201 to 250 $t{\cdot}ha^{-1}$ (402–500 $t{\cdot}ha^{-1}$ of fuel load), and forests reaching values up to 1050 $t{\cdot}ha^{-1}$ (2100 $t{\cdot}ha^{-1}$ of fuel load). Our values are in accordance with the authors: we observed fuel loads ranging from around 60 to 1080 $t{\cdot}ha^{-1}$, with the majority of sites containing 200 to 500 $t{\cdot}ha^{-1}$. This study confirmed the importance of the region for global carbon stocks.

4.2. Simulation of Carbon Emission by Fire

This research is the first that we are aware of to use empirical data that include both spatial and temporal variations in fuel loads to quantify fire carbon emission in boreal ecosystems. Our approach differs from other studies [20,27,28,87,88], as fuel loads were approximated by fuel type and did not consider that the temporal changes [20,27] or amount of carbon emitted were a function of the burned area [28,88] or of the landscape's total tree loads [87]. We estimated that mean carbon emission by fire ranged from 2 to 53 $t \cdot ha^{-1}$, depending on the forest type and age category. These results are consistent with values provided by previous studies. Harden et al. [87] estimated long-term average for C exchange in boreal forests near Thompson, Manitoba, Canada. The values ranged from 11 to 25 $t \cdot ha^{-1}$ for JP forests and from 14 to 70 $t \cdot ha^{-1}$ for BS-feathermoss forests. No observed data were offered for BS-*Sphagnum* forests, but the authors simulated an average of 20 $t \cdot ha^{-1}$. Simulated carbon emissions using the CanFIRE model in forests of Alberta, Canada, ranged from 11.3 to 42.6 $t \cdot ha^{-1}$, depending on the month during which fire occurred [19]. Amiro et al. [28] estimated a mean of 20.1 $t \cdot ha^{-1}$ of fuel consumed by individual fire (carbon emissions of around 100 ($t \cdot ha^{-1}$)) in the boreal east shield from 1959 to 1999.

The most interesting point to emerge from our simulations is that despite fuel accumulation with TSF, old-growth forests emitted less carbon by fire on average, whatever the forest type. Pre-fire fuel structure determined the amount of carbon emitted by fire [27–29] since total fuel consumption in CanFIRE was a function of the amount of each fuel material. More specifically, as previously observed [27], more than 85% of the carbon emissions were generated by duff and DWD combustion. It was not surprising to simulate a low tree contribution to carbon emission, since fire does not consume trees, it kills them [89]. Emissions are consequently generated after the fire, through dead wood decomposition [90]. However, this particularity does not influence the conclusions of this study since tree fuel load is as low in late-successional than in early-successional forests. Variations in carbon emission through BS successional stages were mainly influenced by duff dynamics. As mentioned previously, organic layer accumulation leads to a rise in the water table [35] and creates high soil moisture conditions [35]. Depth of burn is consequently low [36–39]. Duff loads in TA and JP forests present little variation across successional stages, as carbon emission was determined by DWD dynamics. Old-growth forests emitted less carbon because DWD loads were lower (even if DWD variations with TSF were not significant in JP forests, values were lower in old-growth forests).

These results have important implications for forest management in the context of climate change mitigation. Managers should consider practices that favor mature forests for harvesting but also increase potential forest vulnerability to higher carbon emission and decrease long-term carbon storage [36]. In fact, although the climatic conditions expected by the end of the 21st century may induce an increase in fire activity in the Clay Belt boreal region (e.g., [22–24]), expansion of old-growth BS forests in the landscape is projected [60]. If fire activity over the next three to four decades remains similar to current levels, an increase in the proportion of forests characterized by high soil moisture conditions will occur [60]. These moisture conditions may provide landscape resilience to increased fires. Fuel should consequently continue to accumulate without generating more emissions during fires. Peatland protection (such as reducing or preventing peatland drainage) could be an alternative in order to increase forest resistance to fire and reduce future fire carbon emissions in eastern Canadian forests.

Finally, it is important to note that the CanFIRE model does not include fine aerial consumption in its calculations. However, this omission would not change the findings of this study. Fine aerial load is low for coniferous forests all along the successional stages. High fine aerial load in TA forests at early and late successional stages may amplify our results as moisture contents of this fuel reduce fire intensity and, therefore, carbon emissions.

5. Conclusions

Empirical fuel structure dynamics were examined and potential carbon emissions by wildfire were simulated along four dominant chronosequences of the Clay Belt boreal forest in eastern North America. Fuel structure was an important factor influencing carbon emission by the simulated fire, while fuel availability was not a determining factor. Given our results, we argue that maintaining old-growth forests in the Clay Belt landscape not only promotes a sustainable management of the boreal forest, it also optimizes carbon sink.

A considerable effort was made in this study to sample covering a wide variability in TSF, and our results showed a low value for TA in the TSF > 150 years category and for JP in the TSF between 90 and 150 years category. However, these conclusions may apply only to the region studied, since the paludification process brings specific high moisture dynamics in old-growth forests [34]. Direct impacts of climatic change on fuel loads should also be investigated. This includes the potential increase in tree growth with the lengthening of the growing season [91,92] or decomposition [93]. However, the particularly high organic layer thickness and high soil moisture content may prevent such impacts by limiting tree growth and decomposition.

Acknowledgments: This project was financially supported by the Natural Sciences and Engineering Research Council of Canada (NSERC; Strategic Project; Discovery Grants Program), the Canada Chair in Forest Ecology and Management (Yves Bergeron), the Canadian Forest Service Tembec Inc. Danielle Charron and Marie-Hélène Longpré contributed to the field work logistics. We are particularly grateful to Mélanie Desrochers for her help with ARCGIS maps, to Rémi St-Amant for providing climate data and help with the BioSIM software, to Martin Simard and Nicolas Lecomte for providing their data and, finally, to Christelle Hély and Samira Ouarmim for their availability and helpful advice on fuel load fieldwork and calculations. We extend our thanks to Samira Ouarmim for reviewing the paper and providing helpful comments and to Isabelle Lamarre for technical editing. Finally, we are grateful to two anonymous reviewers for their helpful comments on an earlier version of the manuscript.

Author Contributions: S.G. and Y.B. conceived and coordinated the research project. M.P. and S.G. designed data collection. M.P. and S.P.-B. did the field work. M.P., A.T., and S.P.-B. performed fuel load calculations. A.T. and M.P.G. conceived modeling and simulations analyses. A.T. and M.P. analyzed the data. A.T. wrote the paper and all the authors revised it.

Conflicts of Interest: The authors declare no conflict of interest. The founding sponsors had no role in the design of the study; in the collection, analyses, or interpretation of data; in the writing of the manuscript; and in the decision to publish the results.

Abbreviations

The following abbreviations are used in this manuscript:

BS	Black spruce
BS-S	Black spruce forests originating from severe fire
BS-NS	Black spruce forests originating from non-severe fire
CWD	Coarse woody debris
DBH	Diameter at breast height
DWD	Dead woody debris
FFFC	Forest floor fuel consumption
FWI	Fire weather index
JP	Jack pine
TA	Trembling aspen
TSF	Time since the last fire

Appendix A. Shrub Load Equations

Previously determined equations were used for load calculations of *Lonicera canadensis* Bartr., *Ribes* sp., *Rosa acicularis* Lindl., and *Viburnum edule* Raf [57].

New equations were determined with additional shrub samples for load calculations of *Chamaedaphne calyculata* (L.) Moench, *Kalmia angustifolia* L., *Rhododendron groenlandicum* (Oeder) K.A. Kron & Judd, and *Vaccinium myrtilloid*es Michx. (Table A1). A total of 90 additional shrub samples were randomly collected in various triangle transects to determine equations linking species' weight (g) with stem diameter (cm) for *Chamaedaphne calyculata* (sample size = 19), *K. angustifolia* (sample size = 28), *R. groenlandicum* (sample size = 28), and *V. myrtilloid*es (sample size = 15). Efforts were made to collect a high variability in diameter size. Each sample was weighted and linked with basal diameter using linear regressions. Weight and diameter values were logarithmically transformed to normalize the relationship. Linear regressions were performed using *R* freeware [59].

Table A1. Equations and adjusted R^2 from linear regressions analysis linking shrub weight (g) with stem basal diameter (cm) by species. All equations and coefficients were significant with a *p*-value ≤ 0.05.

Species	Equations	Adjusted R^2
Kalmia angustifolia [2]	$\ln(weight) = -2.13 + 2.5\ \ln(D)$	0.91
Lonicera canadensis [1]	$\ln(weight) = -2.33 + 2.64\ \ln(D)$	0.82
Chamaedaphne calyculata [2]	$\ln(weight) = -3.25 + 3.25\ \ln(D)$	0.93
Rhododendron groenlandicum [2]	$\ln(weight) = -2.4 + 2.52\ \ln(D)$	0.94
Ribes sp. [1]	$\ln(weight) = -2.18 + 2.33\ \ln(D)$	0.79
Rosa acicularis [1]	$\ln(weight) = -2.09 + 2.41\ \ln(D)$	0.82
Vaccinium myrtilloides [1]	$\ln(weight) = -1.93 + 1.92\ \ln(D)$	0.65
Viburnum edule [2]	$\ln(weight) = -2.55 + 2.62\ \ln(D)$	0.84

[1] Hély's equations [57]; [2] New equations from additional shrub sample.

Appendix B. Sensitivity Analysis of the Influence of Fire Weather Variations on Potential Carbon Emissions across Successional Stages

Sensitivity analyses were made to ensure that fire weather values above average did not modify our conclusions. To do so, we performed simulations using daily FWI System components for each natural fire start point during the interval 1971–2000 period in the study area [60] (Table B1). A total of eight fires occurred during this period, burning a total of 56,080 ha.

Table B1. Daily FWI System components for each natural fire start point encountered during the interval 1971–2000 period.

Name	FFMC	DMC	DC	ISI	BUI	FWI
Fire 1	89	22	168	15	33	24
Fire 2	93	42	85	17	41	30
Fire 3	88	22	167	9	33	17
Fire 4	90	50	125	15	49	29
Fire 5	83	26	99	5	32	10
Fire 6	82	24	101	4	30	8
Fire 7	88	30	99	8	34	15
Fire 8	88	31	88	6	33	13

FFMC: Fine Fuel Moisture Code; DMC: Duff Moisture Code; DC: Drought Code; ISI: Initial Spread Index; BUI: Buildup Index; FWI: Fire Weather Index.

Simulation results showed that potential carbon emission was the lowest for forests older than 150 years, regardless of the FWI System components (Figure B1).

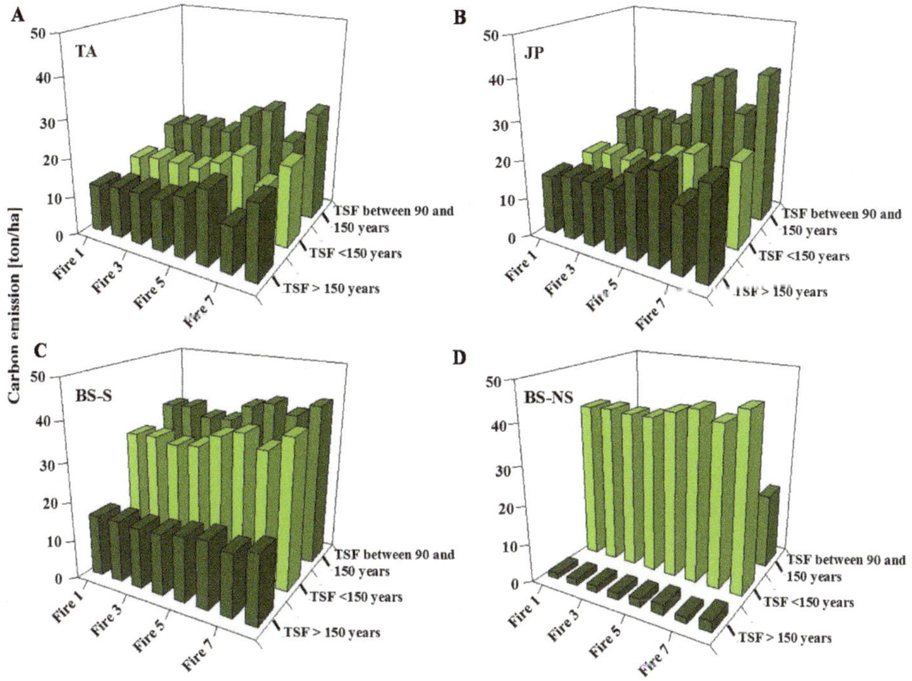

Figure B1. CanFIRE potential carbon emissions averaged by early (TSF < 90 years), intermediate (TSF between 90 and 150 years) and late (TSF > 150 years) successional stages, and by dominant forest types of the Clay Belt boreal landscape ((**A**) trembling aspen (TA), (**B**) jack pine (JP), (**C**) black spruce originating from severe fire (BS-S), and (**D**) black spruce originating from non-severe fire (BS-NS) forests). Simulations were performed using FWI System components for each natural fire start from 1971 to 2000.

Appendix C. Comparison of Empirical Basal Area and Simulated Basal Area with Plonski's Yield Table

Prior to simulations of carbon emission by fire, we ensured that the use of Plonski's yield table [65] instead of our empirical basal area data would not bring a bias to our conclusions. We calculated Pearson's correlation between tree basal areas extracted from Plonski's tables and our empirical basal areas. Comparisons showed good predictive skills, with a Pearson correlation coefficient of 0.72 (Figure C1). Plonski's tables, however, overestimated the basal area, but this difference does not change the conclusions of our study because trees contribute only slightly to the amount of carbon emitted during fire (Table 3) and simulated values are significantly similar to observed values.

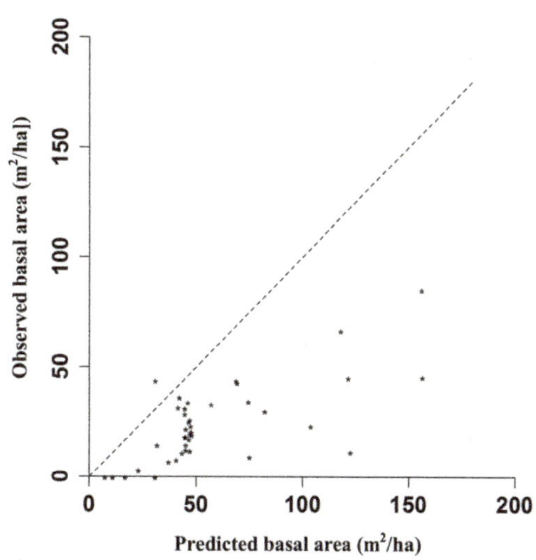

Figure C1. Comparison of empirical basal area and simulated basal area using Plonski's yield table.

Appendix D. Calibration of Bulk Density with Duff Fuel Loads

We calibrated a model linking measured bulk density (kg·m^{-3}) with organic layer depth (m) to consider heterogeneity in organic bulk density with the organic layer depth [37]. We made use of peat datasets comprising 103 measurements of bulk density at different organic layer depths selected from 11 sites in the Quebec Clay Belt boreal forest [34]. Bulk density and organic layer depths were logarithmically transformed to linearize the relationship. Linear regressions were performed using *R* freeware [59].

The resulting model explained 45% of the variance in bulk density (adjusted $R^2 = 0.45$, p-value ≤ 0.001). The model took on the following form:

$$\ln(Bulk\ Density + 1) = 2.72 + 0.57 \times \ln(Organic\ Layer\ Depth + 1) \quad \text{(D1)}$$

Therein bulk density progressively increases with organic layer depth (Figure D1A). Comparison of observed and predicted values indicated relatively good predictive skills, with a Pearson correlation coefficient of 0.73 (Figure D1B).

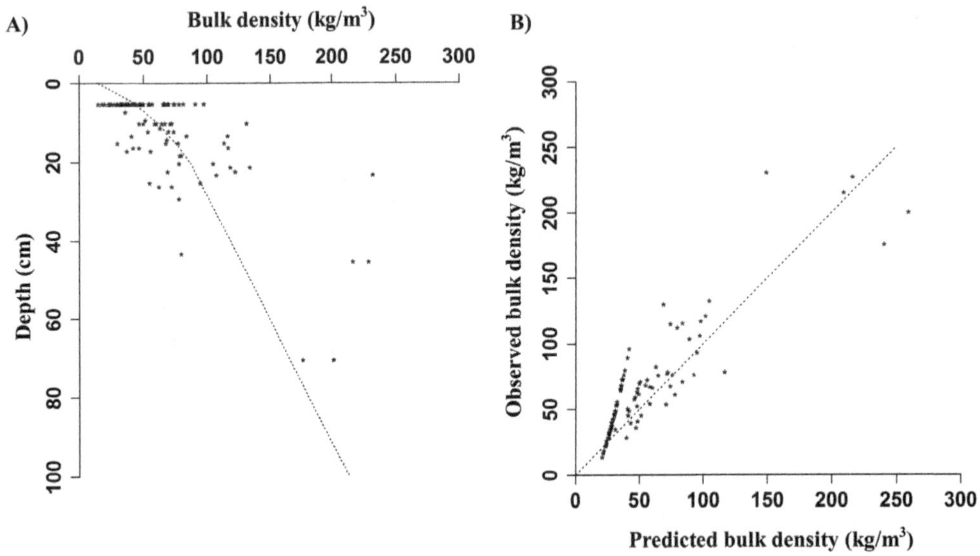

Figure D1. (**A**) Bulk density (kg/m^3) variation with organic layer depth for forested peats in the Clay Belt boreal forests. Stars represent sampling data collected by Simard et al. [34]. The dotted line represents the regression relationship; (**B**) Predicted bulk density with regression analysis versus observed values.

Appendix E. Calibration of Bulk Density with Duff Fuel Loads

Table E1. Empirical TSF and fuel loads (tree, fine aerial vegetation, DWD, duff, and total) for TA, JP, BS-S, and BS-NS dominated forests.

			Fuel Material			
Forest Type	TSF (Years)	Tree (t/ha)	Fine Aerial (t/ha)	DWD (t/ha)	Duff (t/ha)	Total (t/ha)
	11	0	569	90	70	730
	57	346	4	12	78	440
	68	100	37	16	88	241
	71	190	0	14	257	460
TA	102	200	0	17	145	362
	145	120	1	22	292	435
	145	120	1	26	197	343
	187	375	23	8	117	524
	187	375	2	7	129	514

Table E1. *Cont.*

			Fuel Material			
Forest Type	**TSF (Years)**	**Tree (t/ha)**	**Fine Aerial (t/ha)**	**DWD (t/ha)**	**Duff (t/ha)**	**Total (t/ha)**
JP	11	0	1	1	54	57
	11	0	0	86	81	168
	11	0	2	79	77	158
	54	137	0	10	87	234
	87	173	0	12	154	339
	89	175	4	22	152	353
	89	175	1	44	115	335
	142	91	0	21	270	382
	152	132	0	3	182	317
	152	132	1	12	140	285
	180	34	1	15	225	276
	180	34	0	10	225	270
	207	42	0	12	171	225
	207	42	0	23	173	238
	225	74	2	10	163	248
BS-S	32	0	1	17	170	188
	32	0	3	10	131	144
	32	0	2	1	74	77
	55	164	1	2	131	298
	87	118	5	2	223	348
	91	134	0	7	117	258
	129	127	1	33	190	350
	129	127	1	16	232	376
	145	72	1	58	169	299
	145	94	0	11	204	310
	165	115	1	3	506	625
	177	74	2	18	289	383
	181	85	0	17	350	453
	225	66	1	31	338	436
	272	79	0	6	496	581
	283	52	2	14	698	767
	283	66	3	19	541	629
	356	45	1	4	682	733
BS-NS	41	11	2	1	292	306
	41	11	2	2	292	307
	41	11	2	1	292	306
	55	0	0	4	157	160
	57	53	0	2	239	295
	78	24	1	1	435	461
	96	29	1	2	483	515
	100	40	1	1	351	393
	142	63	1	12	281	356
	146	43	0	7	389	440
	146	43	0	9	389	441
	170	106	3	4	972	1084
	172	66	5	8	464	543
	172	66	2	10	464	542

References

1. Lemprière, T.C.; Kurz, W.A.; Hogg, E.H.; Schmoll, C.; Rampley, G.J.; Yemshanov, D.; McKenney, R.; Gilsenan, A.; Beatch, D.; Blain, J.S.; et al. Canadian boreal forests and climate change mitigation 1. *Environ. Rev.* **2013**, *21*, 293–321. [CrossRef]
2. Lorenz, K.; Lal, R. *Carbon Sequestration in Forest Ecosystems*; Springer: Dordrecht, The Netherlands, 2010; pp. 5–11.
3. Dixon, R.K.; Solomon, A.M.; Brown, S.; Houghton, R.A.; Trexler, M.C.; Wisniewski, J. Carbon pools and flux of global forest ecosystems. *Science* **1994**, *263*, 185–190. [CrossRef] [PubMed]

4. Luyssaert, S.; Inglima, I.; Jung, M.; Richardson, A.D.; Reichstein, M.; Papale, D.; Piao, S.L.; Schulze, E.-D.; Wingate, L.; Matteucci, G.; et al. CO_2 balance of boreal, temperate, and tropical forests derived from a global database. *Glob. Chang. Biol.* **2007**, *13*, 2509–2537. [CrossRef]
5. Terrier, A.; Girardin, M.; Bergeron, Y. Les réservoirs de carbone en forêt boréale à l'est du Canada: Acquis et incertitudes dans la modélisation face aux changements climatiques. *VertigO-la Revue Électronique Sci. L'environnement* **2012**, *11*, 3. [CrossRef]
6. Burton, P.J.; Messier, C.; Weetman, G.F.; Prepas, E.E.; Adamowicz, W.L.; Tittler, R. The current state of boreal forestry and the drive for change. In *Towards Sustainable Management of the Boreal Forest*; Burton, P.J., Messier, C., Smith, D.W., Adamowicz, W.L., Eds.; NRC Research Press: Ottawa, ON, Canada, 2003; pp. 1–40.
7. Attiwill, P.M. The disturbance of forest ecosystems: The ecological basis for conservative management. *For. Ecol. Manag.* **1994**, *63*, 247–300. [CrossRef]
8. Bergeron, Y.; Harvey, B. Basing sylviculture on natural ecosystems dynamics: An approach applied to the southern boreal mixedwood forest of Quebec. *For. Ecol. Manag.* **1997**, *92*, 235–242. [CrossRef]
9. Angelstam, P.K. Maintaining and restoring biodiversity in European boreal forests by developing natural disturbance regimes. *J. Veg. Sci.* **1998**, *9*, 593–602. [CrossRef]
10. Gauthier, S.; Vaillancourt, M.-A.; Leduc, A.; de GrandPré, L.; Kneeshaw, D.; Morin, H.; Drapeau, P.; Bergeron, Y. *Ecosystem Management in the Boreal Forest*; Presses de l'Uniersité du Québec: Québec, QC, Canada, 2009.
11. Franklin, J.F. Preserving biodiversity: Species, ecosystems, or landscapes? *Ecol. Appl.* **1993**, *3*, 202–205. [CrossRef] [PubMed]
12. Gauthier, S.; Leduc, A.; Bergeron, Y. Forest dynamics modelling under natural fire cycles: A tool to define natural mosaic diversity for forest management. *Environ. Monit. Assess.* **1996**, *39*, 417–434. [CrossRef] [PubMed]
13. Perera, A.H.; Buse, L.J. Emulating Natural Disturbance in Forest Management. In *Emulating Natural Forest Landscape Disturbances*; Perera, A.H., Buse, L.J., Weber, M.G., Eds.; Columbia University Press: New York, NY, USA, 2004; pp. 3–7.
14. Niemela, È.J. Management in relation to disturbance in the boreal forest. *For. Ecol. Manag.* **1999**, *115*, 127–134. [CrossRef]
15. Bergeron, Y.; Fenton, N.J. Boreal forests of eastern Canada revisited: Old growth, nonfire disturbances, forest succession, and biodiversity. *Botany* **2012**, *90*, 509–523. [CrossRef]
16. Bergeron, Y.; Harvey, B.; Leduc, A.; Gauthier, S. Forest management guidelines based on natural disturbances dynamics: Stand- and forest-level considerations. *For. Chron.* **1999**, *75*, 49–54. [CrossRef]
17. Cyr, D.; Gauthier, S.; Bergeron, Y.; Carcaillet, C. Forest management is driving the eastern North American boreal forest outside its natural range of variability. *Front. Ecol. Environ.* **2009**, *7*, 519–524. [CrossRef]
18. Luyssaert, S.; Schulze, E.D.; Börner, A.; Knohl, A.; Hessenmöller, D.; Law, B.E.; Ciais, P.; Grace, J. Old-growth forests as global carbon sinks. *Nature* **2008**, *455*, 213–215. [CrossRef] [PubMed]
19. Amiro, B.D.; Cantin, A.; Flannigan, M.D.; de Groot, W.J. Future emissions from Canadian boreal forest fires. *Can. J. For. Res.* **2009**, *39*, 383–395. [CrossRef]
20. De Groot, W.J.; Cantin, A.S.; Flannigan, M.D.; Soja, A.J.; Gowman, L.M.; Newbery, A. A comparison of Canadian and Russian boreal forest fire regimes. *For. Ecol. Manag.* **2013**, *294*, 23–34. [CrossRef]
21. Chapin, F.; Woodwell, G.; Randerson, J.; Rastetter, E.; Lovett, G.; Baldocchi, D.; Clark, D.; Harmon, M.; Schimel, D.; Valentini, R.; et al. Reconciling carbon-cycle concepts, terminology, and methods. *Ecosystems* **2006**, *9*, 1041–1050. [CrossRef]
22. Yang, G.; Di, X.-Y.; Guo, Q.-X.; Shu, Z.; Zeng, T.; Yu, H.-Z.; Wang, C. The impact of climate change on forest fire danger rating in China's boreal forest. *J. For. Res.* **2011**, *22*, 249–257. [CrossRef]
23. Flannigan, M.D.; Logan, K.A.; Amiro, B.D.; Skinner, W.R.; Stocks, B.D. Future area burned in Canada. *Clim. Chang.* **2005**, *72*, 1–16. [CrossRef]
24. Bergeron, Y.; Cyr, D.; Girardin, M.P.; Carcaillet, C. Will climate change drive 21st century burn rates in Canadian boreal forest outside of its natural variability: Collating global climate model experiments with sedimentary charcoal data. *Int. J. Wildl. Fire* **2010**, *19*, 1127–1139. [CrossRef]
25. Boulanger, Y.; Gauthier, S.; Burton, P.J.; Vaillancourt, M.A. An alternative fire regime zonation for Canada. *Int. J. Wildl. Fire* **2012**, *21*, 1052–1064. [CrossRef]

26. Girardin, M.P.; Hogg, E.H.; Bernier, P.Y.; Kurz, W.A.; Guo, X.J.; Cyr, G. Negative impacts of high temperatures on growth of black spruce forests intensify with the anticipated climate warming. *Glob. Chang. Boil.* **2016**, *22*, 627–643. [CrossRef] [PubMed]
27. De Groot, W.J. Modeling Canadian wildland fire carbon emissions with the Boreal Fire Effects (BORFIRE) model. *For. Ecol. Manag.* **2006**, *234*, S224. [CrossRef]
28. Amiro, B.D.; Todd, J.B.; Wotton, B.M.; Logan, K.A.; Flannigan, M.D.; Stocks, B.J.; Mason, J.A.; Martell, K.G.; Hirsch, K.G. Direct carbon emissions from Canadian forest fires, 1959–1999. *Can. J. For. Res.* **2001**, *31*, 512–525. [CrossRef]
29. Johnston, D.C.; Turetsky, M.R.; Benscoter, B.W.; Wotton, B.M. Fuel load, structure, and potential fire behaviour in black spruce bogs. *Can. J. For. Res.* **2015**, *45*, 888–899. [CrossRef]
30. Scharlemann, J.P.; Hiederer, R.; Kapos, V.; Ravilious, C. *Updated Global Carbon Map*; UNEP-WCRC & EU-JRC: Cambridge, UK, 2009.
31. Brown, J.K. Unatural Fuel Buildup Issue. In Proceedings of the Symposium and Workshop on Wilderness Fire, Missoula, MT, USA, 15–18 November 1983.
32. Ryan, K.C.; Knapp, E.E.; Varner, J.M. Prescribed fire in North American forests and woodlands: History, current practice, and challenges. *Front. Ecol. Environ.* **2013**, *11*, e15–e24. [CrossRef]
33. Johnson, E.A.; Miyanishi, K.; Bridge, S.R.J. Wildfire regime in the boreal forest and the idea of suppression and fuel buildup. *Conserv. Biol.* **2001**, *15*, 1554–1557. [CrossRef]
34. Simard, M.; Lecomte, N.; Bergeron, Y.; Bernier, P.Y.; Paré, D. Forest productivity decline caused by successional paludification of boreal soils. *Ecol. Appl.* **2007**, *17*, 1619–1637. [CrossRef] [PubMed]
35. Fenton, N.J.; Bergeron, Y. Facilitative succession in a boreal bryophyte community driven by changes in available moisture and light. *J. Veg. Sci.* **2006**, *17*, 65–76. [CrossRef]
36. Terrier, A.; de Groot, W.J.; Girardin, M.P.; Bergeron, Y. Dynamics of moisture content in spruce-feather moss and spruce–*Sphagnum* organic layers during an extreme fire season and implications for future depths of burn in Clay Belt black spruce forests. *Int. J. Wildl. Fire* **2014**, *23*, 490–502. [CrossRef]
37. Benscoter, B.W.; Thompson, D.K.; Waddington, J.M.; Flannigan, M.D.; Wotton, B.M.; de Groot, W.J.; Turetsky, M.R. Interactive effects of vegetation, soil moisture and bulk density on depth of burning of thick organic soils. *Int. J. Wildl. Fire* **2011**, *20*, 418–429. [CrossRef]
38. Benscoter, B.W.; Wieder, R.K. Variability in organic matter lost by combustion in a boreal bog during the 2001 Chisholm fire. *Can. J. For. Res.* **2003**, *33*, 2509–2513. [CrossRef]
39. Shetler, G.; Turetsky, M.R.; Kane, E.; Kasischke, E. *Sphagnum* mosses limit total carbon consumption during fire in Alaskan black spruce forests. *Can. J. For. Res.* **2008**, *38*, 2328–2336. [CrossRef]
40. De Groot, W.J.; Bothwell, P.M.; Carlsson, D.H.; Logan, K.A. Simulating the effects of future fire regimes on western Canadian boreal forests. *J. Veg. Sci.* **2003**, *14*, 355–364. [CrossRef]
41. De Groot, W.J.; Landry, R.; Kurz, W.A.; Anderson, K.R.; Englefield, P.; Fraser, R.H.; Hall, R.J.; Banfield, E.; Raymond, D.A.; Decker, V.; et al. Estimating direct carbon emissions from Canadian wildland fires. *Int. J. Wildl. Fire* **2007**, *16*, 593–606. [CrossRef]
42. Vincent, J.S.; Hardy, L. L'évolution et l'extension des lacs glaciaires Barlow et Ojibway en territoire québécois. *Géogr. Phys. Quat.* **1977**, *31*, 357–372. [CrossRef]
43. Bergeron, Y.; Gauthier, S.; Flannigan, M.D.; Kafka, V. Fire regimes at the transition between mixedwoods and coniferous boreal forest in northwestern Quebec. *Ecology* **2004**, *85*, 1916–1932. [CrossRef]
44. Environment Canada. National Climate Data and Information Archive. 2016. Available online: http://climate.weatheroffice.gc.ca/ (accessed on 7 June 2016).
45. Fenton, N.; Lecomte, N.; Légaré, S.; Bergeron, Y. Paludification in black spruce (*Picea mariana*) forests of eastern Canada: Potential factors and management implications. *For. Ecol. Manag.* **2005**, *213*, 151–159. [CrossRef]
46. Lavoie, M.; Paré, D.; Fenton, N.; Groot, A.; Taylor, K. Paludification and management of forested peatlands in Canada: A literature review. *Environ. Rev.* **2005**, *13*, 21–50. [CrossRef]
47. Harper, K.; Boudreault, C.; de Grandpré, L.; Drapeau, P.; Gauthier, S.; Bergeron, Y. Structure, composition and diversity of old-growth black spruce boreal forest of the Clay Belt region in Québec and Ontario. *Environ. Rev.* **2003**, *11*, S79–S98. [CrossRef]
48. Lecomte, N.; Bergeron, Y. Successional pathways on different surficial deposits in the coniferous boreal forest of the Quebec Clay Belt. *Can. J. For. Res.* **2005**, *35*, 1984–1995. [CrossRef]

49. Fenton, N.J.; Béland, C.; de Blois, S.; Bergeron, Y. *Sphagnum* establishment and expansion in black spruce (*Picea mariana*) boreal forests. *Can. J. Bot.* **2007**, *85*, 43–50. [CrossRef]
50. Belleau, A.; Leduc, A.; Lecomte, N.; Bergeron, Y. Forest succession rate and pathways on different surface deposit types in the boreal forest of northwestern Quebec. *Ecoscience* **2011**, *18*, 329–340. [CrossRef]
51. Lecomte, N.; Simard, M.; Bergeron, Y. Effects of fire severity and initial tree composition on stand structural development in the coniferous boreal forest of northwestern Québec, Canada. *Ecoscience* **2006**, *13*, 152–163. [CrossRef]
52. Greene, D.F.; Macdonald, S.E.; Haeussler, S.; Domenicano, S.; Noël, J.; Jayen, K.; Charron, I.; Gauthier, S.; Hunt, S.; Gielau, E.T.; et al. The reduction of organic-layer depth by wildfire in the North American boreal forest and its effect on tree recruitment by seed. *Can. J. For. Res.* **2007**, *37*, 1012–1023. [CrossRef]
53. Johnstone, J.F.; Chapin, F.S., III; Hollingsworth, T.N.; Mack, M.C.; Romanovsky, V.; Turetsky, M. Fire, climate change, and forest resilience in interior Alaska. *Can. J. For. Res.* **2010**, *40*, 1302–1312. [CrossRef]
54. Van Cleve, K.; Oliver, L.; Schlentner, R.; Viereck, L.A.; Dyrness, C. Productivity and nutrient cycling in taiga forest ecosystems. *Can. J. For. Res.* **1983**, *13*, 747–766. [CrossRef]
55. Paré, D.; Bernier, P.; Lafleur, B.; Titus, B.D.; Thiffault, E.; Maynard, D.G.; Guo, X. Estimating stand-scale biomass, nutrient contents, and associated uncertainties for tree species of Canadian forests. *Can. J. For. Res.* **2013**, *43*, 599–608. [CrossRef]
56. Brown, J.K.; Oberheu, R.D.; Johnston, C.M. *Handbook for Inventorying Surface Fuels and Biomass in the Interior West*; General Technical Report INT-GTR-129; U.S. Department of Agriculture, Forest Service, Intermountain Forest and Range Experimental Station: Ogden, UT, USA, 1982; p. 48.
57. Hély, C.; Université de Montpellier 2, Montpellier, France. Personal communication, 2001.
58. McRae, D.J.; Alexander, M.E.; Stocks, B.J. *Measurement and Description of Fuels and Fire Behavior on Prescribed Burns: A Handbook*; Great Lakes Forest Research Centre: Sault Ste Mary, ON, Canada, 1979.
59. R Development Core Team. *R: A Language and Environment for Statistical Computing*, R Development Core Team: Vienna, Austria, 2010.
60. Terrier, A.; Girardin, M.P.; Cantin, A.; Groot, W.J.; Anyomi, K.A.; Gauthier, S.; Bergeron, Y. Disturbance legacies and paludification mediate the ecological impact of an intensifying wildfire regime in the Clay Belt boreal forest of eastern North America. *J. Veg. Sci.* **2015**, *26*, 588–602. [CrossRef]
61. Van Wagner, C.E. *Development and Structure of the Canadian Forest Fire Weather Index System*; Technical Report No. 35; Canadian Forestry Service: Ottawa, ON, Canada, 1987.
62. Lambert, M.C.; Ung, C.H.; Raulier, F. Canadian national tree aboveground biomass equations. *Can. J. For. Res.* **2005**, *35*, 1996–2018. [CrossRef]
63. Forestry Canada Fire Danger Group. *Development and Structure of the Canadian Forest Fire Behavior Prediction System*; Information Report ST-X-3; Forestry Canada Fire Danger Group: Ottawa, ON, Canada, 1992.
64. De Groot, W.J.; Pritchard, J.M.; Lynham, T.J. Forest floor fuel consumption and carbon emissions in Canadian boreal forest fires. *Can. J. For. Res.* **2009**, *39*, 367–382. [CrossRef]
65. Plonski, W.L. *Normal Yield Tables (Metric) for Major Forest Species of Ontario*; Ontario Ministry of Natural Resources: Toronto, ON, Canada, 1974; p. 40.
66. Hély, C.; Bergeron, Y.; Flannigan, M.D. Coarse woody debris in the southeastern Canadian boreal forest: Composition and load variations in relation to stand replacement. *Can. J. For. Res.* **2000**, *30*, 674–687. [CrossRef]
67. Légaré, S.; Bergeron, Y.; Paré, D. Influence of forest composition on understory cover in boreal mixedwood forests of western Quebec. *Silva Fenn.* **2002**, *36*, 353–366. [CrossRef]
68. Brassard, B.W.; Chen, H.Y. Stand structural dynamics of North American boreal forests. *Crit. Rev. Plant Sci.* **2006**, *25*, 115–137. [CrossRef]
69. Hart, S.A.; Chen, H.Y. Understory vegetation dynamics of North American boreal forests. *Crit. Rev. Plant Sci.* **2006**, *25*, 381–397. [CrossRef]
70. Greene, D.F.; Zasada, J.C.; Sirois, L.; Kneeshaw, D.; Morin, H.; Charron, I.; Simard, M.-J. A review of the regeneration dynamics of North. American boreal forest tree species. *Can. J. For. Res.* **1999**, *29*, 824–839. [CrossRef]
71. Johnstone, J.F.; Chapin, F.S., III. Effects of soil burn severity on post-fire tree recruitment in boreal forest. *Ecosystems* **2006**, *9*, 14–31. [CrossRef]

72. Pothier, D.; Savard, F. *Actualisation des Tables de Production Pour les Principales Espèces du Québec*; RN98-3054; Gouvernement du Québec, Ministère des Ressources Naturelles, Bibliothèque Nationale du Québec: Québec, QC, Canada, 1998.
73. Wardle, D.A.; Walker, L.R.; Bardgett, R.D. Ecosystem properties and forest decline in contrasting long-term chronosequences. *Science* **2004**, *305*, 509–513. [CrossRef] [PubMed]
74. Kneeshaw, D.D.; Bergeron, Y. Canopy gap characteristics and tree replacement in the southeastern boreal forest. *Ecology* **1998**, *79*, 783–794. [CrossRef]
75. Chen, H.Y.; Popadiouk, R.V. Dynamics of North American boreal mixedwoods. *Environ. Rev.* **2002**, *10*, 137–166. [CrossRef]
76. Pothier, D.; Raulier, F.; Riopel, M. Ageing and decline of trembling aspen stands in Quebec. *Can. J. For. Res.* **2004**, *34*, 1251–1258. [CrossRef]
77. Messier, C.; Parent, S.; Bergeron, Y. Effects of overstory and understory vegetation on the understory light environment in mixed boreal forests. *J. Veg. Sci.* **1998**, *9*, 511–520. [CrossRef]
78. Grandpré, L.; Gagnon, D.; Bergeron, Y. Changes in the understory of Canadian southern boreal forest after fire. *J. Veg. Sci.* **1993**, *4*, 803–810. [CrossRef]
79. Pham, A.T.; Grandpré, L.D.; Gauthier, S.; Bergeron, Y. Gap dynamics and replacement patterns in gaps of the northeastern boreal forest of Quebec. *Can. J. For. Res.* **2004**, *34*, 353–364. [CrossRef]
80. Pedlar, J.H.; Pearce, J.L.; Venier, L.A.; McKenney, D.W. Coarse woody debris in relation to disturbance and forest type in boreal Canada. *For. Ecol. Manag.* **2002**, *158*, 189–194. [CrossRef]
81. Bond-Lamberty, B.; Gower, S.T. Decomposition and fragmentation of coarse woody debris: Re-visiting a boreal black spruce chronosequence. *Ecosystems* **2008**, *11*, 831–840. [CrossRef]
82. Brais, S.; Paré, D.; Lierman, C. Tree bole mineralization rates of four species of the Canadian eastern boreal forest: Implications for nutrient dynamics following stand-replacing disturbances. *Can. J. For. Res.* **2006**, *36*, 2331–2340. [CrossRef]
83. Päätalo, M.-L. Factors influencing occurrence and impacts of fires in northern European forests. *Silva Fenn.* **1998**, *32*, 185–202. [CrossRef]
84. Hély, C.; Flannigan, M.; Bergeron, Y.; McRae, D. Role of vegetation and weather on fire behavior in the Canadian mixedwood boreal forest using two fire behavior prediction systems. *Can. J. For. Res.* **2001**, *31*, 430–441. [CrossRef]
85. Arienti, M.C.; Cumming, S.G.; Boutin, S. Empirical models of forest fire initial attack success probabilities: The effects of fuels, anthropogenic linear features, fire weather, and management. *Can. J. For. Res.* **2006**, *36*, 3155–3166. [CrossRef]
86. Krawchuk, M.A.; Cumming, S.G.; Flannigan, M.D.; Wein, R.W. Biotic and abiotic regulation of lightning fire initiation in the mixedwood boreal forest. *Ecology* **2006**, *87*, 458–468. [CrossRef] [PubMed]
87. Harden, J.W.; Trumbore, S.E.; Stocks, B.J.; Hirsch, A.; Gower, S.T.; O'neill, K.P.; Kasischke, E.S. The role of fire in the boreal carbon budget. *Glob. Chang. Biol.* **2000**, *6*, 174–184. [CrossRef]
88. Turetsky, M.R.; Kane, E.S.; Harden, J.W.; Ottmar, R.D.; Manies, K.L.; Hoy, E.; Kasischke, E.S. Recent acceleration of biomass burning and carbon losses in Alaskan forests and peatlands. *Nat. Geosci.* **2011**, *4*, 27–31. [CrossRef]
89. Whelan, R.J. *The Ecology of Fire*; Cambridge University Press: Cambridge, UK, 1995.
90. Fleming, R.A.; Candau, J.-N.; McAlpine, R.S. Landscape-Scale Analysis of Interactions between Insect Defoliation and Forest Fire in Central Canada. *Clim. Chang.* **2002**, *55*, 251–272. [CrossRef]
91. Gignac, L.D.; Vitt, D.H. Responses of northern peatlands to climate change: Effects on bryophytes. *Hattori Shokubutsu Kenkyujo Hokoku* **1994**, *75*, 119–132.
92. Breeuwer, A.; Robroek, B.J.; Limpens, J.; Heijmans, M.M.; Schouten, M.G.; Berendse, F. Decreased summer water table depth affects peatland vegetation. *Basic Appl. Ecol.* **2009**, *10*, 330–339. [CrossRef]
93. Wickland, K.P.; Neff, J.C. Decomposition of soil organic matter from boreal black spruce forest: Environmental and chemical controls. *Biogeochemistry* **2008**, *87*, 29–47. [CrossRef]

4

Pre-logging Treatment of Invasive Glossy Buckthorn (*Frangula alnus* Mill.) Promotes Regeneration of Eastern White Pine (*Pinus strobus* L.)

Thomas D. Lee [1,*], Stephen E. Eisenhaure [2] and Ian P. Gaudreau [3]

1. Department of Natural Resources and the Environment, University of New Hampshire, 56 College Road, Durham, NH 03824, USA
2. Office of Woodlands and Natural Areas, University of New Hampshire, Kingman Farm, Route 155, Madbury, NH 03823, USA; Stephen.Eisenhaure@unh.edu
3. Department of Civil and Environmental Engineering, University of New Hampshire, 33 Academic Way, Durham, NH 03824, USA; iangaudreau@gmail.com

* Correspondence: tom.lee@unh.edu

Abstract: Non-native glossy buckthorn (*Frangula alnus* Mill.) is invasive in forests of the northeastern USA but little is known of its effects on tree regeneration. We tested whether killing buckthorn stems before logging reduces its post-logging abundance and increases the density and height of eastern white pine (*Pinus strobus* L.) seedlings. Three 0.4 ha plots were clearcut, three were thinned, and three were left as controls. Each plot had previously been divided into three subplots that received different buckthorn treatments during the two years before logging. Buckthorn treatments were (1) stems cut at base five times; (2) stems cut once then heat killed four times; (3) untreated control. Three years post-logging, buckthorn density and stem height were unaffected by logging but equally reduced by the two buckthorn treatments. Buckthorn reduction increased density and height of pine seedlings, and seedling height also increased with logging. In the fifth year post-logging, pine height growth and biomass were greater in clearcut than in thinned treatments, greater in areas of buckthorn removal and, within treated subplots, greater in areas with low buckthorn density than in thickets of recovering buckthorn. Thus, although buckthorn inhibited regenerating pine, pre-logging destruction of buckthorn stems reduced such competition for at least four years.

Keywords: basal cutting; biological invasion; competition; exotic shrub; flame gun; glossy buckthorn (*Frangula alnus* Mill.); eastern white pine (*Pinus strobus* L.)

---

1. Introduction

Glossy buckthorn (*Frangula alnus* Mill., hereafter known as 'buckthorn'), a tall shrub or small tree native to Europe, is one of over 20 non-native, woody plant species reported as invasive in forests of eastern North America [1]. Moderately shade tolerant [2–4], buckthorn can form a dense shrub layer that dominates the forest understory especially under canopies of eastern white pine (*Pinus strobus* L.) [2,5–8]. Buckthorn abundance in white pine stands is enhanced by logging, with resulting densities sometimes exceeding 60,000 stems$\cdot$ha^{-1} [4]. Consequently, buckthorn can potentially inhibit native tree regeneration before logging occurs and especially so afterward.

As is true for most eastern North American woody invasive plants, relatively little is known about the competitive effects of buckthorn on other forest species. Correlative evidence suggests that buckthorn is associated with lower herb cover [2,9] and lower density, growth, and survival of tree seedlings and saplings [2,7]. For example, Fagan and Peart [7] found that seedlings of economically important eastern white pine growing under buckthorn had lower growth and survival than those away from buckthorn. Despite the strong association of buckthorn with pine-dominated forests, only two studies [8,10] have reported the effects of experimental buckthorn removal on eastern white pine regeneration and, in both of these, white pine was one of many understory species examined. Neither found significant buckthorn effects, but pine abundance and samples sizes were low.

Buckthorn's ability to alter forest structure has prompted managers to seek control methods. Potential controls include biological agents, mechanical methods, and herbicides. Despite a major effort, no promising buckthorn-specific biocontrol agent has been found [11]. Thus, buckthorn control is presently limited primarily to chemical and mechanical means. One or two applications of herbicide cause buckthorn mortality [12,13], although time of year [14] and soil moisture level [15] may influence herbicide effectiveness.

Few peer-reviewed publications discuss mechanical control of buckthorn and none employ the scheme of timing and repetition we describe in this paper. Mechanical uprooting is possible and has been used to control other invasive shrubs [16,17] but it is labor and time intensive, difficult to apply to large areas, and often requires follow-up control [18]. Single applications of stem girdling [19] and flaming [20–22] are reported to be ineffective for buckthorn, but have been tested under a limited range of conditions. Repeated cutting, mowing, or burning, however, seem to provide buckthorn control [23,24]. Indeed, repeated application of mechanical techniques, such as flaming, have been effective in controlling other invasive shrub species [25,26].

Effectiveness of mechanical control of buckthorn is limited by the production of butt or stump sprouts immediately after cutting [12,19,27]. To minimize such sprouting, some authors emphasize the timing of control. Kays and Canham [28] and Love and Anderson [16] advocate cutting stems of shrubs when root reserves are low, thus reducing sprouting and maximizing the chance of plant death. Cutting buckthorn stems not only promotes sprouting, but is often followed by germination of buried dormant seed [29].

We sought to determine the effect of mechanical removal of buckthorn on the abundance and growth of eastern white pine seedlings. Moreover, we sought to quantify these effects in association with a logging event. Although an obvious approach to minimizing buckthorn effects on pine regeneration would be to treat buckthorn at or just after the time of logging, the novel approach we report here involved treatment beginning two years prior to logging. We expected that repeated cutting or flaming (heat-killing) of buckthorn stems while they remained beneath a forest canopy in a competitive environment would more effectively reduce buckthorn and liberate pine seedlings than in the light- and nutrient-rich post-logging environment.

Our first objective was to test the hypotheses that destruction of buckthorn stems by cutting, or cutting and heat killing, prior to logging would (1) reduce post-logging density and height of buckthorn and (2) increase post-logging density and size of eastern white pine seedlings. After logging of the forest canopy, we found that buckthorn recovered to some degree within most of our treated areas, and sometimes formed dense, multi-individual patches with continuous cover of foliage. We refer to such patches as 'buckthorn thickets'. Our second objective was to test the hypothesis that white pine seedlings in such thickets had reduced height, biomass, and growth relative to seedlings in non-thicket areas.

2. Materials and Methods

2.1. Study Area

The MacDonald Lot, Durham, New Hampshire, USA, is a 31.9 ha parcel owned and managed by the University of New Hampshire. In Durham, the average annual precipitation of 1169 mm, 10% of which occurs as snow, is evenly distributed over the year and mean annual temperature is 8.9 °C [30]. In the area that we studied, soils were Buxton silt loam and Suffield silt loam, with limited Hollis–Charlton very rocky fine sandy loams [31]. White-tailed deer (*Odocoileus virginianus*) are abundant at this site but rarely browse either eastern white pine or buckthorn. Windstorms are occasional, with the February 2010 storm causing some windthrow and windsnap of pines on this property.

A 2006 inventory of the MacDonald Lot identified several stand types, one of which was old-field eastern white pine approximately 60 years old. Typical of old-field pine, many of these trees had either multiple tops, severe deformity in the first log, or unusually large lower branches. In such stands, a reasonable silvicultural approach is to clearcut the area with the intent of regenerating a higher quality pine stand. Although some white pine advanced regeneration was present on site, the understory was dominated by buckthorn (density given below). Clearcutting the stand without first treating the understory would likely have resulted in an increase of buckthorn [4].

2.2. Study Design

We selected nine 0.4 ha plots of forest with eastern white pine canopies and understories dominated by buckthorn. Three plots were randomly chosen to be clearcut, three to be thinned, and three to serve as uncut controls. Logging treatments were applied in the winter of 2009–2010. Each 0.4 ha plot was divided into three 0.13 ha sub-units ("subplots"), each of which received a different buckthorn treatment in 2008–2009, the two years before logging. The buckthorn treatments were (1) manually cutting buckthorn stems at the base ('cut' treatment), (2) initial cutting of stems followed by flaming of stem bases after initial cutting ('flame' treatment), and (3) untreated. Pre-treatment density of buckthorn, estimated using three 4 m^2 circular quadrats per subplot, averaged 7.00 stems m^{-2} (s.d. = 4.07; n = 27; observations were means of three 4 m^2 quadrats).

2.3. Treatments

Each of the understory buckthorn treatments was applied five times over two years, twice in 2008 (June and September) and three times in 2009 (June, July, September). The strategy was to destroy stems immediately after bud break in June when root reserves were low [16,28], and again late in the season to minimize overwintering reserves. Based on 2008 observations indicating greater than expected sprout growth by mid-summer, in 2009 a third application was added in mid-July. The first of the five applications, in June 2008, involved manual cutting of stems regardless of whether subplots were assigned for 'cut' or 'flame' treatments. This initial application consisted of, first, using Stihl FS 110R line trimmers (Stihl Inc., Virginia Beach, VA, USA) to remove the smallest stems (<3.2 mm basal diameter). Second, Stihl FS 450 brush saws and MS 361 chainsaws were used to cut larger stems. Large material, if cut first, would have hidden seedlings and interfered with the cutting of small stems using line trimmers. Larger material was cut into smaller pieces so as to make follow-up treatments as easy as possible. In the four subsequent applications (September 2008 and all three 2009 applications), the targets—newly emerging seedlings and stump sprouts (from surviving individuals)—were reduced with just two tools, Weed Dragon VT 2-23SVC flame guns (Flame Engineering, Inc., LaCrosse, KS, USA) for the flame treatments and Stihl FS 110R line trimmers for cut treatments. In all treated subplots, cutting and flaming was done as close to the buckthorn stem base as physically possible. All cut materials were left in place.

Tree harvest was conducted in the winter of 2009–2010 by a whole-tree harvesting contractor. All marked trees greater than 10.2 cm diameter at breast height were harvested. In clearcuts, all canopy trees were harvested. In thinned plots, less valuable canopy trees (deformed, injured) were harvested along with more valuable trees to create a uniform canopy cover. Thinning was intended to reduce standing live basal area (BA) to the same value in each plot, ca. 30 $m^2 \cdot ha^{-1}$, which required an 18%–29% reduction in BA from pre-treatment levels. Trees were taken off-site whole, using a grapple skidder.

2.4. Intensive Quadrat Sampling: 2012

Vegetation response to logging and buckthorn removal was assessed using quadrat sampling in July–August 2012, the third growing season after logging. In each subplot, we randomly placed six, 2 m $\times$ 4 m (8 m^2) quadrats and in each quadrat we tallied the number of buckthorn stems and the number of eastern white pine seedlings from the 2010 cohort. We also measured the height of the tallest eastern white pine seedling (2010 cohort) and tallest buckthorn stem within each quadrat. Density and height data from the six quadrats in each subplot were averaged to obtain single values to be used in statistical analysis. Independent variables were buckthorn density, height of tallest buckthorn, pine seedling density, and height of tallest pine seedling. We used a split-plot ANOVA in the statistical analysis program JMP Pro 12 (SAS Inc., Cary, NC, USA) to determine the response of each independent variable to the dependent variables: logging treatment, buckthorn treatment, and their interaction. All independent variables were $\log_e$-transformed to homogenize variances. Where appropriate, treatment and cell means were compared using Tukey HSD (honestly significant difference) in JMP Pro 12.

2.5. Buckthorn Origin

Our 2012 sampling revealed that buckthorn had recovered on some subplots where it had been treated with cutting or cutting and flaming. Between June 4 and 11, 2013, we sampled living buckthorn plants in the treated subplots (in which buckthorn was cut or flamed). We wanted to determine if these buckthorn individuals had established before or after the logging event of winter 2009–2010. In each subplot, we excavated 60 randomly selected buckthorn individuals. We counted wood rings at the stem base to estimate age and then examined each plant for evidence of previous cutting or burning. Plants establishing prior to 2010 would have survived at least some of the cutting or flaming treatments and so should have exhibited some evidence of stem damage. Based on plant age and evidence of cutting or flaming, we classified our selected buckthorn plants into those having established prior to and after logging treatments were imposed. The cut and cut-and-flamed subplots in one clearcut had insufficient buckthorn for sampling, so the number of subplots sampled was 16 (i.e., cut and cut-and-flamed subplots in all three thinned treatments, all three uncut treatments, and in two of the three clearcuts). For statistical analysis, the independent variable, proportion of buckthorn stems established after logging, was arcsine square root transformed and then tested against the dependent variables of logging, buckthorn removal, and their interaction using split-plot analysis of variance (ANOVA) in JMP Pro 12.

2.6. Intensive Quadrat Sampling: 2014

The 2012 quadrat study suggested that buckthorn regrowth in some of the treated subplots was sufficient to inhibit growth of regenerating white pine (see Results). Consequently, in July and August, 2014, we examined white pine seedling density and height in relation to buckthorn abundance in the two clearcut plots and the two thinned plots that had abundant recovering buckthorn. In the cut and cut/flamed subplots of these four plots, we delineated two habitats: buckthorn 'thickets,' characterized by buckthorn plants with continuous foliar cover and with foliage from adjacent buckthorn individuals separated by no more than 20 cm (horizontal distance), and 'non-thicket areas', characterized by non-continuous buckthorn cover and with foliage from adjacent buckthorn individuals separated by more than 20 cm. In each subplot, we used a compass and Distance Measuring Equipment (DME, Haglöf Sweden, Längsele, Sweden) to estimate the area occupied by thicket and non-thicket. To estimate buckthorn and pine density and height in thicket and non-thicket areas of each subplot, we randomly located twenty 1 m $\times$ 1 m quadrats in buckthorn thickets and 20 more in non-thicket areas. In addition, we deployed 20 such quadrats to estimate buckthorn and pine performance in the untreated buckthorn subplots of each plot, which were generally dominated by buckthorn thickets.

In each quadrat, we tallied the number of buckthorn stems arising from the ground surface by height class (<1 m and $\geq$1 m tall). We also tallied the number of eastern white pine seedlings from the 2010 cohort, per quadrat, and determined the height of the tallest pine seedling per quadrat. These tallest pine seedlings were cut at the base and dried in an oven at 90 °C to a constant weight and then weighed to determine aboveground biomass. Maximum height of the buckthorn stems above each plot was estimated to the nearest 0.5 m.

Pine density and, for the tallest white pine per plot, stem height, height growth in 2014, and aboveground biomass were designated as independent variables. We used a split-plot ANOVA in JMP Pro 12 to determine the response of each independent variable to the dependent variables: logging treatment, buckthorn treatment, and thicket presence. All independent variables were $\log_e$-transformed to homogenize variances. Treatment and cell means were compared using Tukey HSD in JMP Pro 12.

3. Results

3.1. Glossy Buckthorn Density and Height: 2012

Three years after logging, glossy buckthorn stem density (stems·m^{-2}) varied across buckthorn treatments (split-plot ANOVA on $\log_e$-transformed density, $F_{2,12} = 24.8$, $P < 0.001$; Figure 1), but was unaffected by logging treatment or the *logging*buckthorn* interaction. Compared to areas with buckthorn left intact, buckthorn density was significantly reduced (ca. 80%) where it was cut or flamed, and cutting and flame treatments did not differ in effects (by Tukey HSD). Mean height of the tallest glossy buckthorn stem per 8 m^2 plot was significantly reduced by buckthorn treatment (split-plot ANOVA on $\log_e$-transformed height, $F_{2,12} = 31.5$, $P = <0.001$; Figure 2), whereas logging treatment and the *logging*buckthorn* interaction had no effect. Relative to the untreated plots, buckthorn height was reduced similarly (ca. 60%) by cutting and flaming (by Tukey HSD).

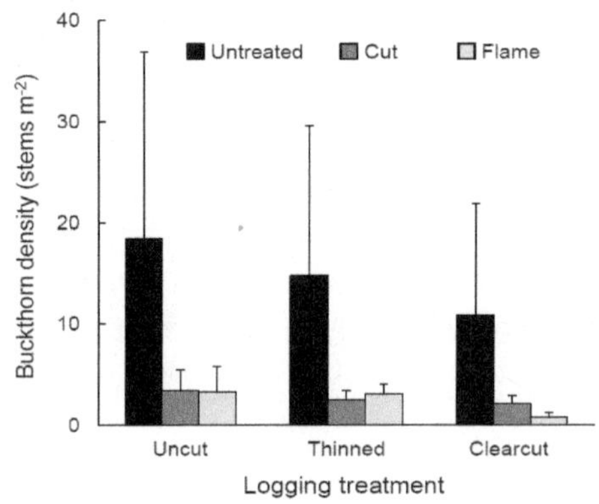

Figure 1. Density of glossy buckthorn (stems m^{-2}) in relation to logging treatment (uncut control, thinned, clearcut) and buckthorn treatment (untreated, cut, flame) at MacDonald Lot, Durham, NH, USA, in 2012, the third year after logging. $n = 27$ subplots.

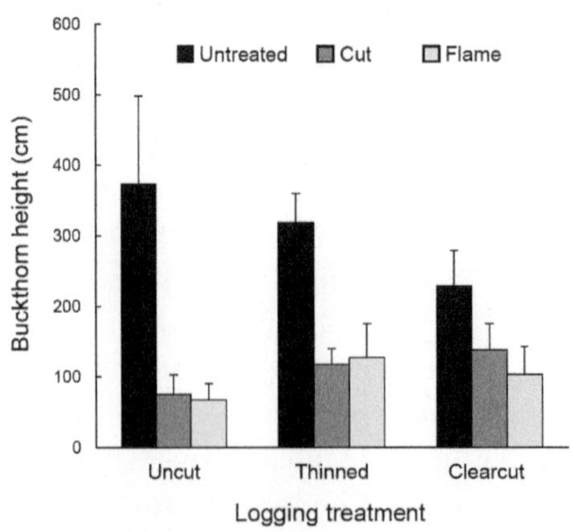

Figure 2. Height (cm) of glossy buckthorn (tallest individual per 8 m^2 plot) in relation to logging treatment (uncut control, thinned, clearcut) and buckthorn treatment (untreated, cut, flame) at MacDonald Lot, Durham, NH, USA, in 2012, the third year after logging. $n = 27$ subplots.

3.2. White Pine Seedling Density and Height: 2012

Based on the 2012 quadrat study (three years after logging), buckthorn removal by cutting or flaming generally increased the density of eastern white pine seedlings, whereas effects of logging treatment were less clear (Figure 3). There was an interaction between buckthorn and logging treatments (split-plot ANOVA on $\log_e$-transformed density, $F_{4,12} = 4.2$, $P = <0.025$). Comparison of treatment means (by Tukey HSD) showed that buckthorn removal in the uncut areas significantly increased pine density, but effects were not significant in thinned and clearcut areas (Figure 3). White pine seedling height was affected by both logging treatment (split-plot ANOVA on $\log_e$-transformed height, $F_{2,6} = 23.5$, $P = <0.005$) and buckthorn treatment (split-plot ANOVA, $F_{2,12} = 29.8$, $P = <0.001$; Figure 4) but not the *logging*buckthorn* interaction. The tallest white pine seedlings occurred in the clearcut plots with buckthorn cut or flamed. Compared to uncut controls, white pine seedlings were 2–3 fold taller in areas where the forest canopy was removed by clearcutting (significantly different by Tukey HSD; differences between cut and flame treatments were not significant). Across all logging treatments, white pine seedlings were ca. 0.5-fold taller in areas where buckthorn was cut or flamed (by Tukey HSD, buckthorn treatments significantly increased height, but cut and flame treatments did not differ).

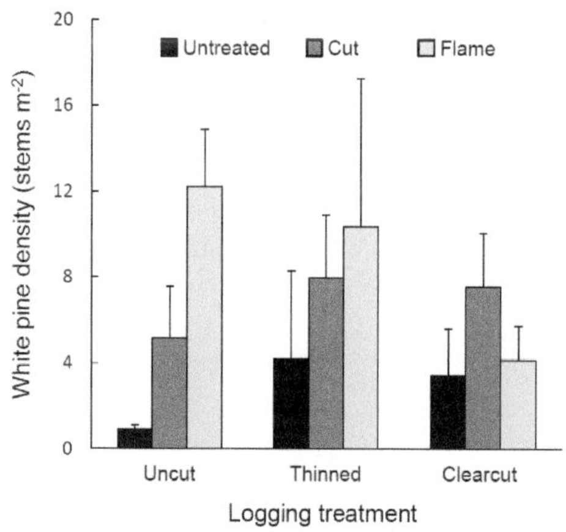

Figure 3. Density (individuals m^{-2}) of eastern white pine seedlings in relation to logging treatment (uncut control, thinned, clearcut) and buckthorn treatment (untreated, cut, flame) at MacDonald Lot, Durham, NH, USA in 2012, the third year after logging. $n = 27$ subplots.

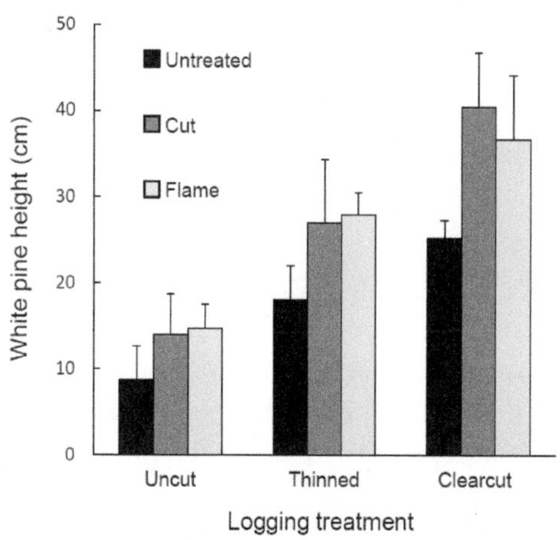

Figure 4. Height (cm) of eastern white pine seedlings (tallest seedling per 8 m^2 plot) in relation to logging treatment (uncut control, thinned, clearcut) and buckthorn treatment (untreated, cut, flame) at MacDonald Lot, Durham, NH, USA, in 2012, the third year after logging. $n = 27$ subplots.

3.3. Buckthorn Origin: 2013

Analysis of the wood rings and stem damage of buckthorn present in 2013 (the fourth year after logging) revealed that on sites with buckthorn removed, 28%–55% of buckthorn stems alive in 2013 had established after logging occurred (Figure 5). Thus, 45%–72% of stems alive in 2013 had survived at least some of the buckthorn treatments. Variances were high, and there was no significant relationship between buckthorn origin and either logging or buckthorn treatments or their interaction.

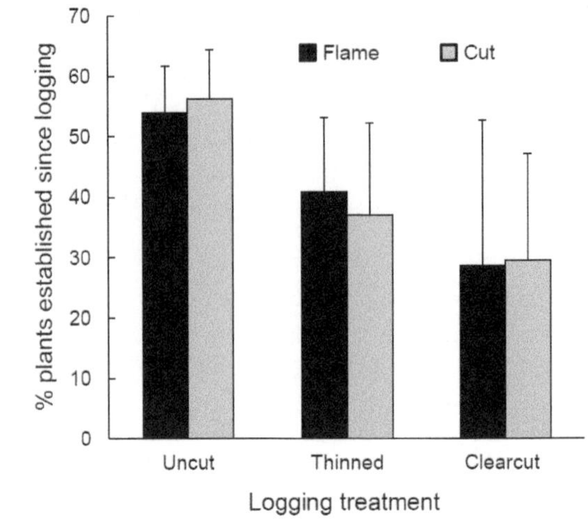

Figure 5. Percent of glossy buckthorn individuals on subplots with buckthorn cut or flamed that established in 2010 or later. Percent of individuals establishing is shown in relation to logging treatment (uncut control, thinned, clearcut). Data collected at MacDonald Lot, Durham, NH, USA, in 2013, the fourth year after logging. $n = 16$ subplots.

3.4. White Pine Response to Recovering Buckthorn: 2014

The 2014 survey (five years after logging) was limited to two clearcut and two thinned plots. Subplots that had not been subject to cutting or flaming had buckthorn thicket covering 82%–95% of the area and buckthorn stem density was estimated at 2.3 to 14.7 stems·m^{-2} (Table 1). In subplots with buckthorn removed by cutting or flaming, new buckthorn thickets (developed since buckthorn treatment) covered between 0 and 33% (Table 1) of subplot area. Average buckthorn stem density in these thickets (9.9–23.8 stems·m^{-2}) was slightly higher than, though comparable to, that in thickets in uncut areas (Table 1). In non-thicket areas within cut or flamed subplots, buckthorn stem density was much lower than in thickets, at 0.8–2.7 stems·m^{-2} (Table 1). Buckthorn was also shorter in non-thicket areas, with stem height up to 3 m less than in thickets (Table 1).

White pine seedling density varied across the untreated and treated subplots (Table 2). Compared to subplots with buckthorn left intact, treated subplots had 50%–170% greater white pine seedling density but there was little difference between thicket and non-thicket habitats in treated subplots (split-plot ANOVA on log$_e$-transformed density, significant buckthorn treatment effect, $F_{1,11} = 8.75$, $P = 0.0128$; Table 2).

Regardless of logging treatments, eastern white pine seedlings were generally taller and had more height growth and biomass where there was less buckthorn (Table 2). Height of the tallest pine per quadrat was greater in subplots with buckthorn removed than in subplots where buckthorn was untreated (split-plot ANOVA on log$_e$-transformed height, significant buckthorn treatment effect, $F_{1,11} = 31.3$, $P = 0.0002$; Table 2). Within the buckthorn removal subplots, pine height was greater in non-thicket than thicket habitat in clearcuts but similar in thinned plots. This difference probably accounts for the weak but significant *logging*thicket* interaction, $F_{1,11} = 5.24$, $P = 0.0427$; Table 2). Annual increase in height and aboveground biomass of the tallest pine seedling per quadrat were also greater in subplots with buckthorn removed (split-plot ANOVA on log$_e$-transformed values, significant buckthorn treatment effect, $F_{1,11} = 14.5$, $P = 0.0029$ and $F_{1,11} = 17.6$, $P = 0.0015$, respectively; Table 2). Moreover, within subplots that had buckthorn removed, both annual increase in height and aboveground biomass of the tallest pine seedling per quadrat were greater in non-thicket than thicket habitats regardless of logging treatment (split-plot ANOVA, significant thicket effect, $F_{1,11} = 22.8$, $P = 0.0006$ and $F_{1,11} = 11.5$, $P = 0.0060$, respectively; Table 2).

Table 1. Description of replicate plots in thinned and clearcut treatments. Percent cover of thicket habitat in each subplot. Mean density and height of buckthorn in thicket and non-thicket habitat in each subplot. Data collected at MacDonald Lot, Durham, NH, USA, in 2014.

Logging Treatment	Replicate	Buckthorn Treatment (Subplot)	Percent of Subplot Area in Thicket	Buckthorn Density (m^{-2}) in Thicket	Buckthorn Density (m^{-2}) in Non-Thicket	Buckthorn Height (m) in Thicket	Buckthorn Height (m) in Non-Thicket
Thinned	1	Cut	9	10.2	1.3	4.5	1.3
		Flame	14	13.9	1.5	3.5	1.4
		Untreated	85	2.3	n/a	2.2	n/a
	2	Cut	33	9.9	2.1	3.0	2.1
		Flame	17	10.6	2.5	3.0	2.2
		Untreated	82	6.9	n/a	2.6	n/a
Clearcut	1	Cut	0	n/a	1.5	n/a	2.3
		Flame	25	8.4	0.8	4.0	2.7
		Untreated	95	13.8	n/a	2.8	n/a
	2	Cut	28	23.8	2.7	2.6	2.0
		Flame	18	15.0	1.4	2.6	2.2
		Untreated	95	14.7	n/a	3.2	n/a

Table 2. Density of eastern white pine seedlings and height, annual increase in height, and aboveground biomass of the tallest white pine seedling per 1 m^2 quadrat in response to logging treatment (clearcut or thinned), buckthorn treatment (untreated or buckthorn removed by cutting or flaming), and presence or absence of buckthorn thicket in treated areas. Values are adjusted means from a split-plot ANOVA with buckthorn treatments nested within logging treatments and dependent variables $\log_e$-transformed. $n = 19$ subplots.

Logging Treatment	Buckthorn Treatment	Thicket or Non-Thicket	Density (Stems $Quadrat^{-1}$)	Height (cm)	Annual Increase in Height (cm)	Above-Ground Biomass (g)
Clearcut	Untreated	Thicket	3.06	44.3	14.3	11.2
	Removed	Thicket	7.38	63.5	17.3	17.5
		Non-thicket	8.23	89.8	33.4	54.9
Thinned	Untreated	Thicket	7.04	28	6.5	3.3
	Removed	Thicket	11.21	47.9	8.4	7.1
		Non-thicket	12.38	45.5	14.7	9.8
Significant effects (split-plot ANOVA; $P < 0.05$):			*Buckthorn treatment*	*Buckthorn treatment, logging*thicket*	*Buckthorn treatment, thicket*	*Buckthorn treatment, thicket*

4. Discussion

4.1. Response of Pine to Buckthorn Removal

As expected given our treatments and study site, glossy buckthorn inhibited eastern white pine seedlings. Post-logging density and especially height of pine seedlings were greater in areas where buckthorn stems had been killed by cutting or burning than in untreated controls. Where recovering buckthorn formed thickets, pine seedlings showed lower annual increase in height and lower aboveground biomass compared to non-thicket areas. Our results thus suggest that pre-logging buckthorn reduction can be effective in releasing an economically important tree species.

Our findings are somewhat consistent with those of Fagan and Peart [7] who found that, in canopy gaps, white pine seedling height growth over one growing season was lower in the presence of buckthorn. Under the closed white pine canopy, however, they found that pine seedling growth did not vary with buckthorn presence. Pine seedling survival in their study showed the reverse pattern, declining under buckthorn only when the pine canopy was intact. Our results differ only in that pine seedling growth and density were higher without buckthorn in both canopy openings and under a closed canopy. Differences in the results of these studies may be due to differences in time span (one vs. five growing seasons) or approach (e.g., correlative vs. experimental). Our results are also consistent with reports showing negative response of various other tree species to buckthorn presence [2,3,8].

4.2. Buckthorn Recovery and Its Consequences for Pine

Our experimental buckthorn treatments neither completely eliminated buckthorn nor prevented it from re-establishing. In the fourth growing season after logging (2013), buckthorn was present in all cut and flamed subplots and, regardless of buckthorn treatment or logging regime, populations were roughly an even mix of plants that established from seed since logging and older plants that survived cutting or flaming and subsequent logging. Despite buckthorn persistence and recovery, buckthorn densities in cut and flamed areas in the fifth growing seasons after application of treatments were generally lower than untreated controls. Thus, treatment effects were not ephemeral.

Buckthorn influence was reduced sufficiently by cutting and flaming so that white pine seedlings were released, a response consistent with an extensive literature indicating that white pine grows better with little competition, e.g., [32,33]. The response of pine seedling density in 2012 to buckthorn removal, however, suggested a complex relationship to logging, with greatest pine response to buckthorn removal occurring in stands that were not logged. The less dramatic response of pine to buckthorn removal in logged stands may have resulted from damage to buckthorn or to potential pine establishment sites from slash deposition and skidder effects.

Although buckthorn removal resulted in taller white pine seedlings three years after logging, the tallest recovering buckthorn (per 8 m^2 quadrat) on these same subplots was already twice the height, on average (>100 cm), of the tallest pine. Given this relationship, one could argue that pine release might be temporary, with pine becoming suppressed by the taller buckthorn. However, in buckthorn removal areas, buckthorn density was generally low, with patches of continuous buckthorn cover (thickets) 33% or less. Indeed, over most of the treated plots, buckthorn remained sparse even in the fifth year after logging.

Our 2014 data also suggest that the buckthorn thickets arising anew on buckthorn removal subplots affected pine differently than the thickets in untreated areas. Thickets on buckthorn removal plots seemed to be less resistant to pine seedling success than were the older, established thickets on the untreated subplots. Although white pine seedling density was similar in both thicket types, white pine biomass and annual increase in height were significantly greater in thickets on treated subplots than in those on untreated ones. New thickets on treated subplots may have allowed more pine growth for two reasons. First, in the untreated subplots, pine had to establish and grow under pre-existing buckthorn plants while, in treated areas, pine and buckthorn both established at the same time and the pine may thus have experienced less competition. Second, thickets in untreated subplots were large and continuous, whereas those in treated plots occurred in small patches whose edges received peripheral light.

There is little information available on the quantitative relationship between buckthorn density and the intensity of competition with white pine seedlings or those of other tree species. Based on buckthorn removal experiments conducted elsewhere in New Hampshire, Koning and Singleton [8] found that buckthorn effects on understory woody plant growth were not detectable until buckthorn density exceeded 1.5–8.25 stems m^{-2}. Within buckthorn thickets in our study, stem densities averaged 15–16 m^{-2} and so, assuming that the Koning and Singleton data apply, we might expect competitive effects on pine in these habitats. Outside of thickets, however, buckthorn density was ca. 3.6 stems m^{-2}, within the Koning and Singleton limit of detectable effects. Given that buckthorn thickets were limited in extent, as noted above, white pine has been able to grow with little competition from buckthorn in well over half the area treated.

Our experiment revealed strong treatment effects on both buckthorn and white pine density and growth despite low replication and high residual variation. Within-treatment (residual) variation may have been due to local heterogeneity in soils, slash deposition, and logging traffic. Although the canopy gaps created by a February, 2010, windstorm likely added to this variation, the large herd of white-tailed deer on site seemed to avoid browsing both buckthorn and pine.

4.3. Potential for Buckthorn Control

The two methods we used for killing buckthorn stems, (1) repeated manual cutting at the stem base and (2) a single manual cut followed by repeated flaming (heat) treatments of the stem, were similarly successful, not only in reducing buckthorn, but also in releasing eastern white pine seedlings. However, our second method was initiated with one round of stem cutting because using flame guns to kill the large buckthorn stems we encountered initially would have been prohibitive. Working with Japanese barberry, Ward et al. [26], showed that the labor cost of initial stem destruction was five times greater for flaming than for cutting with a brush saw. Moreover, cutting and flaming differ greatly in utility, as flaming is limited to field conditions in which there is no chance of initiating wildfire whereas cutting is not so constrained.

The reduction in buckthorn density and height caused by our methods is difficult to compare with the rather limited literature available on buckthorn control. Although one or two mechanical treatments do not reduce buckthorn [19–21], our results concur with reports that repeated cutting, mowing, or burning provide some level of buckthorn control [24–26]. Herbicides (e.g., glyphosate) have been successful with buckthorn [12,13] and offer an alternative that will likely have lower costs than mechanical methods [17,26].

Ultimately, whether or not one chooses to manage buckthorn—and what control method one chooses should they decide to manage—depends on the (1) values identified by the landowner; (2) effectiveness of the method in realizing those values; and (3) cost of the method. Based on current costs and the return on eastern white pine timber in today's market, it is questionable whether mechanical control would be cost effective if only timber values were considered. Inclusion of values associated with wildlife, biodiversity, and aesthetics, however, may influence a landowner's perspective.

4.4. Future Research

Our study was limited to one forest type and one site with particular soil characteristics and climate. To assess the generality of our findings, the study should be repeated in other forest types and at sites with different limiting factors. There are no long-term studies of glossy buckthorn effects on pine and our report provides a limited time frame (five years) for evaluating plant responses to buckthorn and logging treatments. Extending this window will allow more definitive assessment of impacts and we plan to monitor our plots for several more years. Only widely replicated, long-term studies of invasive plant impact and control can inform development of best management practices required by land managers and, increasingly, by government, e.g., [34]. Future research on buckthorn should also quantify the costs of control methods.

Acknowledgments: Partial funding was provided by the New Hampshire Agricultural Experiment Station. This is Scientific Contribution Number 2696. This work was supported by the USDA National Institute of Food and Agriculture McIntire-Stennis Projects 69 and 82. We thank Antoinette Hartgerink, Nick Lanzer, and Mike Simmons for comments on an earlier version of the manuscript. Nathan Roe, Audrey Perkins, Katie Childs, Bryan MacMillan, and Heidi Giguere provided able assistance with field data collection, lab analysis, or both. Mike Simmons, Kelvin Flannery, and Jess Wikle were indefatigable in applying the cutting and flaming treatments.

Author Contributions: S.E. conceived the experiment, which was subsequently designed by S.E. and T.L. Logging and buckthorn removal was directed by S.E. Field data on initial buckthorn densities were collected by S.E. with remaining data (density, height, and biomass) collected by T.L. and I.G. T.L. and I.G. analyzed the data and wrote the first draft of the paper. S.E., I.G., and T.L. edited and revised the paper.

Conflicts of Interest: The authors declare no conflict of interest.

References

1. Webster, C.R.; Jenkins, M.A.; Jose, S. Woody invaders and the challenges they pose to forest ecosystems in the eastern United States. *J. For.* **2006**, *104*, 366–374.
2. Frappier, B.; Eckert, R.T.; Lee, T.D. Potential impacts of the invasive exotic shrub *Rhamnus frangula* L. (glossy buckthorn) on forests of southern New Hampshire. *Northeast. Nat.* **2003**, *10*, 277–296. [CrossRef]
3. Frappier, B.; Lee, T.D.; Olsen, K.F.; Eckert, R.T. Small-scale invasion pattern, spread rate, and lag-phase behavior of *Rhamnus frangula* L. *For. Ecol. Manag.* **2003**, *186*, 1–6. [CrossRef]
4. Lee, T.D.; Thompson, J.H. Effects of logging history on invasion of eastern white pine forests by exotic glossy buckthorn (*Frangula alnus* P. Mill.). *For. Ecol. Manag.* **2012**, *265*, 201–210. [CrossRef]
5. Cunard, C.; Lee, T.D. Is patience a virtue? Succession, light, and the death of invasive glossy buckthorn (*Frangula alnus*). *Biol. Invasions* **2009**, *11*, 577–586. [CrossRef]
6. Catling, P.; Porebski, Z.S. The history of invasion and current status of glossy buckthorn, *Rhamnus frangula*, in southern Ontario. *Can. Field Nat.* **1994**, *108*, 305–310.
7. Fagan, M.E.; Peart, D.R. Impact of the invasive shrub glossy buckthorn (*Rhamnus frangula* L.) on juvenile recruitment by canopy trees. *For. Ecol. Manag.* **2004**, *194*, 95–107. [CrossRef]
8. Koning, C.O.; Singleton, R. Effects of moderate densities of glossy buckthorn on forested plant communities in southwest New Hampshire, USA. *Nat. Area J.* **2013**, *33*, 256–263. [CrossRef]
9. Possessky, S.L.; Williams, C.E.; Moriarity, W.J. Glossy buckthorn, *Rhamnus frangula* L.: A threat to riparian plant communities of the northern Allegheny Plateau (USA). *Nat. Area J.* **2000**, *20*, 290–292.
10. Frappier, B.; Eckert, R.T.; Lee, T.D. Experimental removal of the non-indigenous shrub *Rhamnus frangula* (glossy buckthorn): Effects on native herbs and woody seedlings. *Northeast Nat.* **2004**, *11*, 333–342. [CrossRef]
11. Gassmann, A.; Tosevski, I.; Skinner, L. Use of native range surveys to determine the potential host range of arthropod herbivores for biological control of two related weed species, *Rhamnus cathartica* and *Frangula alnus*. *Biol. Control* **2008**, *45*, 11–20. [CrossRef]
12. Pergams, O.R.W.; Norton, J.E. Treating a single stem can kill the whole shrub: A scientific assessment of buckthorn control methods. *Nat. Area J.* **2006**, *26*, 300–309. [CrossRef]
13. Corace, I.; Gregory, R.; Leister, K.P.; Brosnan, E. Efficacy of different glyphosate concentrations in managing glossy buckthorn (*Frangula alnus*) resprouts (Michigan). *Ecol. Restor.* **2008**, *26*, 111–113. [CrossRef]
14. Reinartz, J.A. Winter season offers many advantages for treating invasive buckthorns and honeysuckles. *Ecol. Restor.* **2002**, *20*, 286–287.
15. Dornbos, D.L., Jr.; Pruim, R. Moist soils reduce the effectiveness of glyphosate on cut stumps of buckthorn. *Nat. Area J.* **2012**, *32*, 240–246. [CrossRef]
16. Love, J.P.; Anderson, J.T. Seasonal effects of four control methods on the invasive Morrow's honeysuckle (*Lonicera morrowii*) and initial responses of understory plants in a southwestern Pennsylvania old field. *Restor. Ecol.* **2009**, *17*, 549–559. [CrossRef]
17. Oneto, S.R.; Kyser, G.B.; DiTomaso, J.M. Efficacy of mechanical and herbicide control methods for Scotch broom (*Cytisus scoparius*) and cost analysis of chemical control options. *Invasive Plant Sci. Manag.* **2010**, *3*, 421–428. [CrossRef]
18. Albrecht, W.D.; Maschinski, J.; Mracna, A.; Murray, S. A community participatory project to restore a native grassland. *Nat. Area J.* **2005**, *25*, 137–146.
19. Reinartz, J.A. Controlling glossy buckthorn (*Rhamnus frangula* L.) with winter herbicide treatments of cut stumps. *Nat. Area J.* **1997**, *17*, 38–41.
20. Post, T.W.; McCloskey, E.; Klick, K.F. Two-year study of fire effects on *Rhamnus frangula* L. *Nat. Area J.* **1989**, *9*, 175–176.
21. Nagel, L.M.; Corace, R.G.; Storer, A.J. An experimental approach to testing the efficacy of management treatments for glossy buckthorn at Seney National Wildlife Refuge, upper Michigan. *Ecol. Restor.* **2008**, *26*, 136–142. [CrossRef]

22. Kettenring, K.M.; Adams, C.R. Lessons learned from invasive plant control experiments: A systematic review and meta-analysis. *J. Appl. Ecol.* **2011**, *48*, 970–979. [CrossRef]
23. Wenning, B. Glossy Buckthorn: An Exotic Invasive Plant Fact Sheet. Ecological Landscape Alliance. 2011. Available online: http://www.ecolandscaping.org/11/invasive-plants/glossy-buckthorn-an-exotic-invasive-plant-fact-sheet/ (accessed on 28 December 2015).
24. Forest Invasive Plants Resource Center. Common Buckthorn (*Rhamnus cathartica*), Glossy Buckthorn (*Frangula alnus*). Northeastern Area State and Private Forestry. USDA Forest Service. 2015. Available online: http://na.fs.fed.us/spfo/invasiveplants/factsheets/pdf/common-and-glossy-buckthorn.pdf (accessed on 13 October 2015).
25. Ward, J.; Williams, S. Controlling an invasive shrub, Japanese barberry (*Berberis thunbergii* DC), using directed heating with propane torches. *Nat. Area J.* **2011**, *31*, 156–162. [CrossRef]
26. Ward, J.S.; Williams, S.C.; Worthley, T.E. Comparing effectiveness and impacts of Japanese barberry (*Berberis thunbergii*) control treatments and herbivory on plant communities. *Invasive Plant Sci. Manag.* **2013**, *6*, 459–469. [CrossRef]
27. Godwin, H. *Frangula alnus* Miller (*Rhamnus frangula* L.). *J. Ecol.* **1943**, *31*, 77–92. [CrossRef]
28. Kays, J.S.; Canham, C.D. Effects of time and frequency of cutting on hardwood root reserves and sprout growth. *Forest Sci.* **1991**, *37*, 524–539.
29. Zuidema, A. Cutting frequency as a measure to control invasive glossy buckthorn (*Frangula alnus* P. Mill.) and facilitate eastern white pine (*Pinus strobus* L.) regeneration. M.S. Thesis, University of New Hampshire, Durham, NH, USA, 2014; p. 51.
30. National Oceanic and Atmospheric Administration. National Climatic Data Center. 2013. Available online: http://www.ncdc.noaa.gov/ (accessed on 21 February 2013).
31. Natural Resources Conservation Service. Web Soil Survey. 2015. Available online: http://websoilsurvey.sc.egov.usda.gov/App/WebSoilSurvey.aspx (accessed on 28 December 2015).
32. Dovciak, M.; Reich, P.B.; Frelich, L.E. Seed rain, safe sites, competing vegetation, and soil resources spatially structure white pine regeneration and recruitment. *Can. J. For. Res.* **2003**, *33*, 1892–1904. [CrossRef]
33. Pitt, D.G.; Morneault, A.; Parker, W.C.; Lanteigne, L.; Hoepting, M.K.; Stinson, A. Influence of herbaceous and woody competition on white pine regeneration in a uniform shelterwood. *For. Chron.* **2011**, *87*, 653–668. [CrossRef]
34. Sitzia, T.; Campagnaro, T.; Kowarik, I.; Trentanovi, G. Using forest management to control invasive alien species: Helping implement the new European regulation on invasive alien species. *Biol. Invasions* **2016**, *18*, 1–7. [CrossRef]

5

Effects of Repeated Growing Season Prescribed Fire on the Structure and Composition of Pine–Hardwood Forests in the Southeastern Piedmont, USA

Matthew J. Reilly [1,2,*], Kenneth Outcalt [2], Joseph J. O'Brien [2] and Dale Wade [2]

1 Department of Forest Ecosystems and Society, College of Forestry, Oregon State University, Corvallis, OR 97330, USA

2 USDA Forest Service, Southern Research Station, Athens, GA 30602, USA; koutcalt@fs.fed.us (K.O.); jjobrien@fs.fed.us (J.J.O'B.); pyrowade@gmail.com (D.W.)

* Correspondence: matthew.reilly@oregonstate.edu

Abstract: We examined the effects of repeated growing season prescribed fire on the structure and composition of mixed pine–hardwood forests in the southeastern Piedmont region, Georgia, USA. Plots were burned two to four times over an eight-year period with low intensity surface fires during one of four six-week long periods from early April to mid-September. Density of saplings (0.25–11.6 cm diameter at breast height) was significantly reduced after one or two fires during the first four-year period. Sapling density declined with additional burning over the next four years, but density of mesic hardwoods including sweetgum (*Liquidambar styraciflua*) and red maple (*Acer rubrum*) remained relatively high (~865 stems ha^{-1}). Repeated burning had little effect on density or basal area of trees ($\geq$11.7 cm dbh) and changes in overstory structure were limited to small increases in the quadratic mean diameter of all trees and pines. We found little evidence to suggest differential effects on structure or composition due to timing of burn within the growing season. Although repeated growing season burning alters midstory structure and composition, burning alone is unlikely to result in immediate shifts in overstory composition or structure in mixed pine–hardwood forests of the southeastern Piedmont region.

Keywords: prescribed fire; season of burn; southeastern Piedmont; growing season burn; pine–hardwood forests

---

1. Introduction

Prescribed fire is an important tool for management and restoration of forests around the world [1–7]. The objectives of prescribed fire are diverse and vary among forest ecosystems, but often include manipulation of stand structure to decrease hazardous fuels [7], increase diversity and productivity in the understory [8], and promote open habitat conditions [4]. Responses to prescribed fire also vary among forested ecosystems depending on productivity, species composition, and topography [9]. Accordingly, region specific information is required to provide guidelines and inform managers on the potential outcomes of prescribed fire as a landscape process [10].

The effects of prescribed fire usually include reductions in basal area and density, but can vary among species with different life history traits (i.e., re-sprouters vs. obligate seeders), as well as among seasons due to differences in phenology and fire intensity [11]. In most cases, growing season burns are considered more effective at reducing woody vegetation than dormant season burns [11–14]. Less is known about how the effects of prescribed fire differ due to variation in timing of application within the growing season, and there is little consensus among studies from different ecosystems [13–18]. Furthermore, there is evidence supporting variable response among species within a single ecosystem [14,18]. There is some evidence that hardwoods are more sensitive to early growing season fires, potentially because root carbohydrate stores may be reduced following leaf production [13,14]. Pines (*Pinus* spp.) are sometimes thought to be more susceptible to late growing season fire, but results are mixed [14,19]. Thus, few generalizations are available to support management across a wide range of forest ecosystems.

In the southeastern United States, more land is treated with prescribed fire than the rest of the U.S. combined [4]. Despite a long history of research on prescribed fire in the Coastal Plain and Southern Appalachian Mountain regions [19–21], there is little recent work in the Piedmont region where prescribed fire is an important management practice on federal forest lands. The objectives of prescribed fire on federal lands in the Piedmont region generally include thinning shrubs and hardwood saplings [5], increasing diversity in the understory [6], and promoting habitat for the endangered red cockaded woodpecker (*Picoides borealis* Vieillot). Red cockaded woodpeckers require frequent fire to maintain open stands [22–24] and will abandon colonies if midstory density gets too high [25].

Most information on the effects of fire in the Piedmont region dates back to the 1950s and 1960s when prescribed fire was used as a silvicultural tool to reduce hardwood competition in loblolly pine (*Pinus taeda* L.) stands [26–29]. Many of these early studies showed that single or a few repeated burns result in a temporary reduction of hardwoods, but some suggest that the net result is typically an increase in hardwoods a few years following a burn [30,31]. However, long-term studies on repeated fire in the Piedmont region are limited [32]. Thus, managers must draw from studies in the Coastal Plain region where variables such as fuels, burning conditions, and species composition often differ drastically.

We compared effects of repeated growing season prescribed fires on the structure and composition of mixed pine–hardwood forests in the southeastern Piedmont, Georgia, USA. Twenty-four plots were burned two to four times with low intensity surface fires over an eight-year period during one of four six-week long periods from early April to mid-September. We address the following questions:

(1) How do the density and composition of saplings in the midstory change following repeated growing season burns?
(2) How do the structure (i.e., basal area, density, and quadratic mean diameter) and composition of overstory trees change following repeated growing season burns?
(3) How do the effects of repeated growing season burns on overstory and midstory structure and composition differ when fires are applied at different times within the growing season?

2. Materials and Methods

2.1. Study Site

The 13,355 ha Piedmont National Wildlife Refuge (PNWR) is located in central Georgia near the Ocmulgee River in Jones County, approximately 40 kilometers north of Macon (Figure 1). Elevations range from a low of 100 m in creek bottoms to a high of 205 m on upper ridges. Most of the refuge is on severely eroded former agricultural land that was abandoned in the early 1900s. The average annual rainfall is approximately 115 cm and occurs fairly evenly throughout the year, but short (several weeks) dry spells are common in the summer, with more extended droughts occurring every decade or so [33]. The growing season extends from April to October. Temperature maximums average about 15 °C during the winter to above 32 °C in the summer. Soils are primarily Davidson clay loams [34].

There is little information available on historical fire regimes in the Piedmont region. Cowell [35] used historical land-survey data from the early 19th century and suggested a historical fire return interval from 20–50 years, but Guyette et al. [36] predict a much lower frequency of 2–4 years based on climate and physical chemistry of fuels. The PNWR instituted a 4-year winter burn schedule in the 1960s to improve wildlife habitat on upland pine sites. Within the past two decades, the refuge started to burn many sites on a 3-year rotation, primarily for red-cockaded woodpecker habitat improvement.

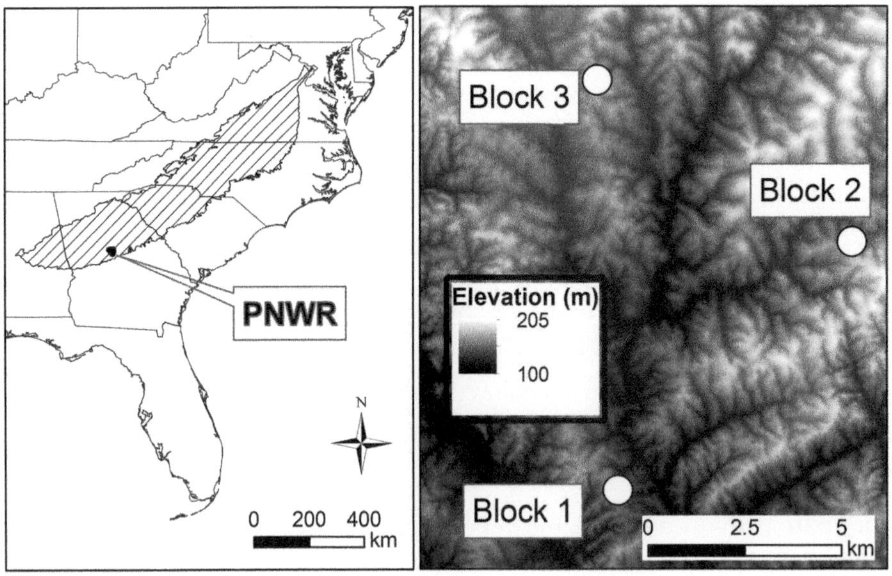

Figure 1. Location of the study site and three blocks within the Piedmont National Wildlife Refuge near Macon, Georgia, USA.

Similar to most of the southeastern Piedmont region, upland forests of the PNWR are dominated by second and third growth mixed pine–hardwood forests that established after agricultural abandonment. The pine component is mostly loblolly pine with some shortleaf pine (*Pinus echinata* Miller). The hardwood component is diverse and consists of two major groups: mesic hardwoods and upland hardwoods. Mesic hardwood species include sweetgum (*Liquidambar styraciflua* L.), red maple (*Acer rubrum* L.), winged elm (*Ulmus alata* Michaux), and tulip poplar (*Liriodendron tulipifera* L.). Mesic hardwood species in the midstory include Florida maple (*Acer barbatum* Michaux), dogwood (*Cornus florida* L.), and persimmon (*Diospyros virginiana* L.). Upland hardwood species include oaks and hickories such as white oak (*Quercus alba* L.), southern red oak (*Quercus falcata* Michaux), post oak (*Quercus stellata* Wang), mockernut hickory (*Carya tomentosa* Nuttall), and pignut hickory (*Carya glabra* Miller). Nomenclature follows Radford et al. [37].

2.2. Study Design

The study consisted of three sites which served as blocking units, each with a series of eight plots approximately one hectare in size. Individual blocks varied in overstory structure and composition as well as disturbance history. Basal area in all blocks was composed primarily of loblolly pine but each differed in the amount of oak-hickory and mesic hardwoods (Figure S1). Block 1 had not experienced fire since at least prior to 1980. Blocks 2 and 3 were burned once in the early 1980s and once in the early 1990s, with the exception of two plots.

Two plots within each block were randomly assigned treatments and scheduled to be burned biennially for eight years within one of four six-week windows between early April and the middle of September (Table 1). The windows roughly correspond to late spring (1 April to 12 May), early summer (13 May to 25 June), mid-summer (26 June to 6 August), and late summer (8 August to 18 September). Managers were unable to burn according to schedule due to drought and four burns were postponed from 2000 to 2001 and two burns occurred outside their assigned window. These deviations resulted in a few systematic biases. All plots assigned to burn in late spring were burned four times, while those assigned to late summer were burned only three times (and one plot only twice). In addition, both burns that occurred outside of their given range of dates were supposed to be early summer burns.

All fires were lit as strip-head fires that burned as low intensity surface fires. Burning conditions varied but relative humidity was generally 45%–80% with temperatures of 21–32 °C. Fires generally top-killed shrubs and consumed >90% of the vegetation in the understory. Flame lengths generally averaged below 0.5 m.

Table 1. Burn windows and dates of growing season prescribed fires in the Piedmont National Wildlife Refuge near Macon, Georgia, USA. Dates are represented as Month/Day.

Block		Late Spring (4/1 to 5/12)	Early Summer (5/13 to 6/25)	Mid Summer (6/26 to 8/6)	Late Summer (8/7 to 9/18)
Block 1	1998	4/2, 4/27	5/13, 6/9	7/29, 7/29	8/24, 9/9
	2000	4/5, 4/26	7/26[1]	7/26	
	2001		6/7	7/12	9/18, 9/18
	2002	4/16, 5/7	5/15, 6/5	8/1	
	2003			6/26	9/10
	2004	4/15, 4/29	5/24	8/4	
Block 2	1998	4/2, 4/27	5/13, 6/8	7/21, 7/21	8/11, 9/8
	2000	4/5, 4/26	7/25[1]	7/25	
	2001		6/6	7/12	9/18, 9/18
	2002	4/16, 5/7	5/15, 6/5	8/1	
	2003			6/26	9/10, 9/10
	2004	4/19, 4/29	5/24	8/4	
Block 3	1998	4/2, 4/25	5/13, 6/8	7/22, 8/4	8/24, 9/8
	2000	4/16, 5/8	5/15	7/23	
	2001		6/5	7/9	9/17, 9/17
	2002	4/16, 5/8	5/15, 6/4	8/1	
	2003			6/26	8/28, 8/28
	2004	4/19, 4/29	5/24	8/4	

[1] Burned out of assigned window.

2.3. Field Sampling

One circular 0.08 ha plot was located in the center of each treatment plot and all trees >11.7 cm dbh (diameter at breast height) were identified to species and measured at the beginning of the study in 1997 and again at the end in 2005. In addition, a 30.5 m transect was established on an upper slope in each plot. At the midway point, all saplings (woody stems 0.25–11.6 cm dbh) within a radius of 5.1 m (81.7 m^2) were identified to species and dbh was measured in 1997, 2001, and 2005.

2.4. Data Analysis

We used a one-way repeated measures ANOVA to look at the effects of repeated burning and among seasonal burn window treatments on aspects of forest structure in the overstory and midstory. We chose to use an alpha level of $\alpha = 0.05$ to control for the potential to inflate Type I error due to the large number of dependent variables we were interested in testing. We chose a simple repeated measures ANOVA over other uni- and multivariate approaches in order to maximize the power of our tests since we had only three replicates for window of burn [38]. In instances where the assumption of circularity in the covariance matrix was not met, probability levels for F-Tests were adjusted using the more conservative Geisser–Greenhouse method [39]. The Tukey–Kramer multiple comparison test was used for post-hoc comparisons.

Analysis of overstory tree data included changes in stand density, basal area, and quadratic mean diameter of all trees and species groups between years (1997–2005) and windows. Species groups included pine, mesic hardwoods, and oak-hickory (see study site description for species composition of each group). The second analysis included the density of all saplings and species groups between years (1997, 2001, and 2005) and burn windows.

3. Results

3.1. Saplings

The density of all saplings decreased significantly over time (Table 2, Figure 2a). Most of the sapling component was made up of mesic hardwoods which also decreased significantly over both time periods ($p < 0.00001$) (Figure 2a). Pines decreased significantly ($p = 0.021$) between 1997 and 2001 but there was no significant difference between 2001 and 2005. There were no significant differences in density between years for oak-hickory saplings.

Results also revealed compositional differences among blocks (Table 2, Figure 2b). These included all saplings: ($p = 0.036$), pine ($p = 0.0067$), oak hickory ($p = 0.005$), and mesic hardwoods ($p = 0.014$) (Figure 2b). Differences in pine saplings among blocks was highly significant ($p = 0.0067$). No significant differences were found among windows, but there was significant three-way interaction between time, block, and window for all saplings ($p = 0.021$) and pine ($p = 0.038$).

Table 2. Analysis of Variance for effects of repeated growing season burn effects on sapling density over time, block, and timing of burn window (Win) in mixed pine–hardwood forests of the Piedmont National Wildlife Refuge near Macon, Georgia, USA. *p*-values ≤0.05 are noted with an *. Greenhouse–Geisser adjusted *p*-values (GG) are given for those repeated measures data that did not meet circularity assumptions.

Sapling Density	Time (df = 2)	Block (df = 2)	Time*Block (df = 4)	Win (df = 3)	Time*Win (df = 6)	Block*Win (df = 6)	Time*Block*Win (df = 12)
All	F = 49.5 GG = 0.0000*	F = 4.44 p = 0.036*	F = 1.9 GG = 0.152	F = 1.82 p = 0.197	F = 1.04 GG = 0.423	F = 1.53 p = 0.49	F = 2.63 GG = 0.021*
Pine	F = 7.00 GG = 0.021*	F = 7.81 p = 0.0067*	F = 5.16 GG = 0.024*	F = 2.15 p = 0.147	F = 1.81 GG = 0.198	F = 2.68 p = 0.069	F = 2.32 GG = 0.038*
Oak Hickory	F = 3.1 GG = 0.1	F = 8.73 p = 0.005*	F = 1.75 GG = 0.2	F = 1.88 p = 0.19	F = 0.70 GG = 0.59	F = 1.77 p = 0.19	F = 0.95 GG = 0.5
Mesic Hardwood	F = 36.35 p = 0.0000*	F = 6.25 p = 0.014*	F = 2.67 p = 0.06	F = 2.32 p = 0.13	F = 2.02 p = 0.1	F = 1.56 p = 0.24	F = 1.62 p = 0.15

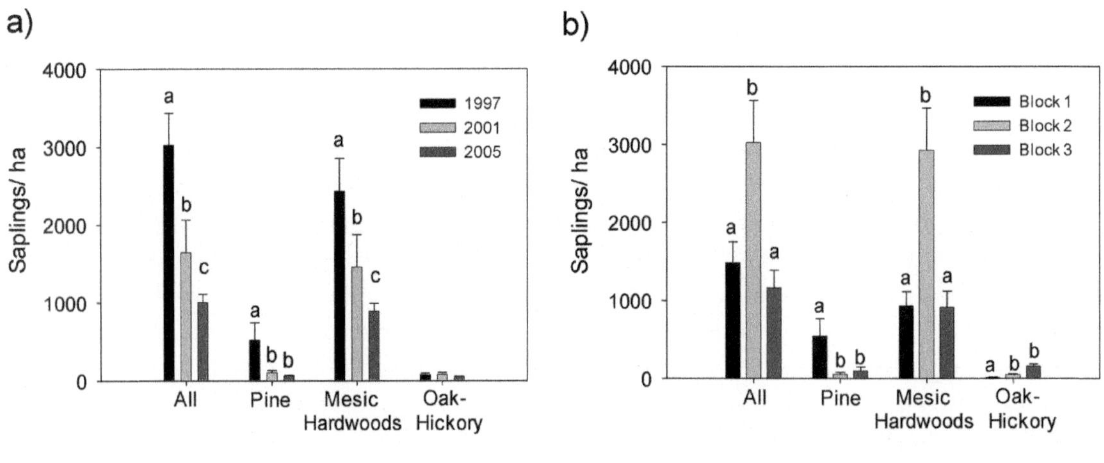

Figure 2. Effects of repeated growing season prescribed fire on density of saplings (0.25 to 11.6 cm dbh) (**a**) over time and (**b**) by block in mixed pine–hardwood forests of the Piedmont National Wildlife Refuge near Macon, Georgia, USA. Letters represent significant differences ($p \leq 0.05$) for post hoc Tukey's Tests on the main effect of time between 1997, 2001, and 2005.

3.2. Overstory Trees

There was little change over time in basal area of all trees or species groups between 1997 and 2005 (Figure 3, Table 3). Basal area of all trees increased only slightly from 24.6 m^2 ha^{-1} in 1997 to 25.1 m^2 ha^{-1} in 2005 (Figure 3a). Density of all trees decreased significantly ($p = 0.03$) from 440.9 to 395.7 trees ha^{-1} (Figure 3b). Density of pine trees also decreased significantly ($p = 0.012$) from 265.5 to 228 trees ha^{-1}. Quadratic mean diameter of all trees increased significantly ($p < 0.001$) from 27.2 to 28.9 cm (Figure 3c). Quadratic mean diameter of pines also increased significantly ($p < 0.001$) from 32.1 to 34.3 cm.

Differences in overstory forest structure and composition among blocks were significant ($p \leq 0.05$) (Table 3). Attributes of forest structure with significant differences included basal area ($p = 0.05$) and density ($p = 0.04$) of all trees, which were both greatest in Block 3 and lowest in Block 1 (Figure S1). Mesic hardwood abundance was also highest in Block 3 and lowest in Block 1, and significant differences among blocks included basal area ($p = 0.02$), quadratic mean diameter ($p = 0.019$), and density ($p = 0.03$) (Figure S1).

Table 3. Analysis of Variance table for effects of repeated growing season burn effects on overstory structure and composition over time, block, and timing of burn window (Win) in mixed pine–hardwood forests of the Piedmont National Wildlife Refuge near Macon, Georgia, USA. Quadratic mean diameter is abbreviated by QMD. *p*-values ≤0.05 are noted with an *. Greenhouse–Geisser adjusted *p*-values (GG) are given for those repeated measures data that did not meet circularity assumptions.

Structure/ Composition	Time (df = 1)	Block (df = 2)	Time*Block (df = 2)	Win (df = 3)	Time*Win (df = 3)	Block*Win (df = 6)	Time*Block*Win (df = 6)
Basal Area	F = 0.57 p = 0.46	F = 4.06 p = 0.05*	F = 0.1 p = 0.9	F = 1.07 p = 0.4	F = 0.77 p = 0.53	F = 0.71 p = 0.65	F = 0.45 p = 0.83
PineBasal Area	F = 0.32 p = 0.58	F = 1.16 p = 0.35	F = 0.24 p = 0.79	F = 1.00 p = 0.43	F = 0.27 p = 0.84	F = 0.21 p = 0.97	F = 0.56 p = 0.76
Oak-Hickory Basal Area	F = 0.28 p = 0.61	F = 0.93 p = 0.42	F = 1.05 p = 0.38	F = 1.43 p = 0.28	F = 0.63 p = 0.61	F = 0.98 p = 0.48	F = 2.13 p = 0.12
Mesic Hardwood Basal Area	F = 3.01 p = 0.36	F = 5.54 p = 0.02*	F = 0.54 p = 0.6	F = 1.81 p = 0.2	F = 3.44 p = 0.05*	F = 4.54 p = 0.013*	F = 0.42 p = 0.85
Density	F = 6.45 p = 0.03*	F = 4.47 p = 0.04*	F = 0.29 p = 0.76	F = 1.79 p = 0.20	F = 0.43 p = 0.73	F = 0.64 p = 0.70	F = 0.72 p = 0.64
Pine Density	F = 8.72 p = 0.012*	F = 0.96 p = 0.41	F = 1.12 p = 0.36	F = 1.36 p = 0.30	F = 0.81 p = 0.51	F = 0.36 p = 0.89	F = 1.02 p = 0.46
Oak-Hickory Density	F = 0.05 p = 0.82	F = 1.22 p = 0.33	F = 1.32 p = 0.30	F = 1.65 p = 0.23	F = 0.05 p = 0.98	F = 0.64 p = 0.69	F = 1.11 p = 0.41
Mesic Hardwood Density	F = 2.25 p = 0.16	F = 4.79 p = 0.03*	F = 0.06 p = 0.94	F = 2.84 p = 0.08	F = 2.46 p = 0.11	F = 3.47 p = 0.03*	F = 1.22 p = 0.36
QMD	F = 24.02 p < 0.001*	F = 0.70 p = 0.52	F = 0.92 p = 0.43	F = 0.41 p = 0.75	F = 0.51 p = 0.68	F = 0.40 p = 0.87	F = 0.55 p = 0.76
Pine QMD	F = 36.1 p < 0.001*	F = 0.50 p = 0.62	F = 1.72 p = 0.22	F = 0.45 p = 0.72	F = 0.37 p = 0.78	F = 0.36 p = 0.89	F = 0.47 p = 0.82
Oak-Hickory QMD	F = 0.06 p = 0.82	F = 3.13 p = 0.08	F = 0.04 p = 0.96	F = 1.56 p = 0.25	F = 1.39 p = 0.29	F = 0.78 p = 0.60	F = 3.81 p = 0.023*
Mesic Hardwood QMD	F = 3.13 p = 0.102	F = 5.57 p = 0.02*	F = 0.01 p = 0.99	F = 31.9 p < 0.001*	F = 0.33 p = 0.80	F = 27.71 p < 0.001*	F = 0.47 p = 0.82

There were no differences among burning windows for density or basal area of all trees or species groups (Table 3). Mesic hardwoods' quadratic mean diameter differed among windows ($p < 0.001$), but there was a significant block by window interaction ($p < 0.001$) (Table 3). Significant two-way interactions included time and window for mesic hardwood basal area ($p = 0.05$), block and window for mesic hardwood basal area ($p = 0.013$) and density ($p = 0.03$). Three-way interactions among time, block, and window for oak-hickory quadratic mean diameter were also significant ($p = 0.023$). These results suggest a large degree of variation in composition (particularly abundance of mesic hardwoods) among windows.

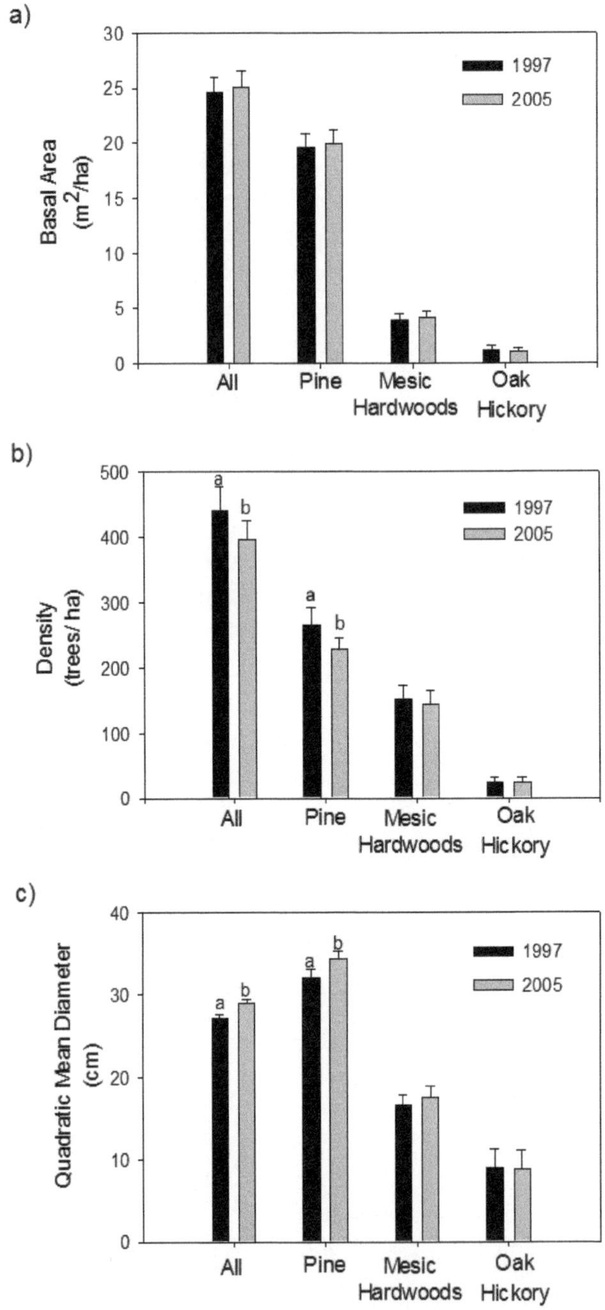

Figure 3. Effects of repeated growing season prescribed fire on (**a**) basal area, (**b**) density, and (**c**) quadratic mean diameter of overstory trees ($\geq$11.7 cm) in mixed pine–hardwood forests of the Piedmont National Wildlife Refuge near Macon, Georgia, USA. Letters represent significant differences ($p \leq 0.05$) for post hoc Tukey's Tests on the main effect of time between 1997 and 2005.

4. Discussion

Our results agree with early work in the southeastern Piedmont that the most consistent and readily observable effect of fire is an immediate reduction in density of saplings and smaller trees [26–29]. However, our results also demonstrate that repeated burning can be effective at maintaining and even furthering initial decreases in midstory density. Differential responses among species groups suggest minor compositional changes, specifically the loss of some smaller pine trees and saplings and mesic hardwood saplings in the midstory, but changes in overstory trees >11.7 cm dbh were limited. Overall, we found little evidence to suggest that repeated burning during different time periods within the growing season resulted in different structural or compositional changes.

4.1. Effects on Midstory Saplings

While other studies in the region have reported net increases in hardwood saplings after single or few burns [28,30,31], our results suggest that three or four burns may effectively reduce sapling density. These results are consistent with Dey and Hartman [40] who found that the probability of surviving a fire decreases with increasing number of burns for many of the species in our study. Differences between pines and hardwoods were apparent and related to the ability to re-sprout following top-kill. In Block 1 where density was initially high, pine saplings were reduced by almost four times between 1997 and 2001, followed by only a minor reduction at the final sampling period. Mesic hardwood saplings, the most abundant group, declined at a similar rate over both time periods but final densities were still relatively high (~800 stems/ha) as a result of sprouting. Although the least abundant, oak-hickory saplings were also maintained by re-sprouting and appear to be the most resilient as repeated burning did little to reduce sapling density except in Block 1. While continued burning at frequencies of 2 to 3 years would likely maintain lower densities of saplings, evidence from the coastal plain suggests that annual burning would be required to entirely eliminate sprouting hardwoods [20].

4.2. Effects on Overstory Trees

Repeated growing season fires had very little effect on the structure of overstory trees. Despite a slight increase in the quadratic mean diameter, there was very little change in density and basal area. Any losses of basal area to mortality were evidently offset by growth of surviving trees. Significant decreases in quadratic mean diameter of all trees and pines alone suggest some mortality of smaller pine trees, but it is unlikely that fire alone accounts for the observed changes. Past work has shown that low intensity fires such as those in this study rarely kill loblolly pines $\geq$11.7 cm dbh [41–43]. Instead, slight changes in density and quadratic mean diameter are most likely the result of density-dependent mortality of smaller, suppressed pines. Mortality of smaller pines may be related to self- thinning (e.g., Peet and Christensen [44]) that may have been accelerated by fire and drought during the study period. Assessing the relative contribution of burning and drought to observed dynamics is difficult, however, especially without unburned controls.

Mesic hardwoods and species of oak-hickory showed very little evidence of change compared to loblolly pine, and our results are consistent with other studies that found little effect of fire on hardwood species $\geq$11.7 cm dbh [26,45]. Once hardwood establish and reach this size, it appears unlikely that low intensity fire will be sufficient to affect composition of these species. Some species of mesic hardwoods may be less susceptible to suppression and thinning because they are more shade tolerant (e.g., red maple, Florida maple, and flowering dogwood), while oaks may be more tolerant of drought [46]. Although our results suggest that repeated burning may have little short-term effect on overstory structure and composition, lack of recruitment into smaller size classes from the decreased sapling pool may result in longer-term changes with continued application of prescribed fire.

4.3. Timing of Growing Season Burn

Past studies on the effects of seasonality and timing of burn have been limited due to confounding variables and problems in experimental design [18]. Likewise, this study has some experimental flaws primarily due to difficulties in following the original burn schedule. Despite these issues, we found little evidence to suggest that time of burn within the growing season affects forest structure or composition in the understory after three or four repeated fires have been applied. Our results provide little evidence to support differential effects of burn timing after eight years, but do not necessarily discount the potential importance of growing season burn window which may have longer term consequences or interact with reproduction of individual species.

4.4. Management Considerations

One of the foremost goals of burning on public land in the Piedmont of the southeastern US is the promotion of habitat for red-cockaded woodpeckers which are strongly associated with open park-like pine savannahs with little woody vegetation in the midstory. Woodpecker cluster stands in loblolly/shortleaf pine forests generally have basal areas between 9–14 m^2 ha^{-1} [22] with less than 3 m^2/ha of the total being hardwood [23]. Given the lack of overstory change observed in this study, our data suggest that repeated growing season burns alone are insufficient to establish suitable habitat for red-cockaded woodpeckers in an eight-year period. Basal area in our plots was still more than twice that recommended for woodpecker habitat. Controlling woody vegetation in the midstory is also of particular importance as birds will abandon cluster sites when midstory basal area reaches 5 m^2/ha [24,25]. Although burning has the potential to eliminate saplings and prevent recruitment into the midstory, mechanical thinning would be necessary to reduce basal area, particularly that of mesic hardwoods, to promote suitable habitat.

The lack of structural change after repeated fire suggests very low intensity of the growing season burns in the Piedmont region. A large contribution of mesic hardwood litter to live and dead fuels can decrease flammability of surface fuels [41] and limit fire intensity, hindering the ability of managers to achieve desired outcomes under an acceptable prescription. Even during drought conditions, relative humidity and fuel moisture in live vegetation may remain high during the growing season and more aggressive ignition techniques (i.e., headfires) may be required for greater success at reducing hardwood density. Other work on prescribed fire in the southeastern United States found that despite more drought-like conditions during the growing season, dormant season burns were higher in intensity due to lower amounts of live fuels and greater total fuel loads [47]. Thus, dormant season burns may also have greater potential to achieve desired management outcome.

5. Conclusions

Repeated growing season burns have the potential to alter forest structure in Piedmont forests by decreasing sapling density in the midstory, but low intensity fires such as the ones in our study result in little change in overstory structure and composition. A continued regime of frequent growing season burns may result in forest structure resembling Coastal Plain pine savannahs at longer time scales, but long-term compositional changes are unclear. Greater fire free intervals are necessary to allow for loblolly pine regeneration to attain a size capable of surviving fire but could allow for competing hardwoods to reach fire tolerant stature unless more fires can completely kill and eliminate sprouting individuals. Although we found no evidence that burning at different times throughout the growing season has differential effects on forests of the Piedmont region, we cannot discount the potential for differences in effects with more intense fires.

Supplementary Materials
Figure S1: Differences among blocks in the effects of repeated growing season prescribed fire on (a) basal area, (b) density, and (c) block quadratic mean diameter of overstory trees (≥11.7 cm dbh) in mixed pine–hardwood forests of the Piedmont National Wildlife Refuge near Macon, Georgia, USA.

Acknowledgments: We would like to acknowledge Thomas Phillips, Jimmy Rickard, David Combs, Mike Allen, Mark Yates, and Bill O'Connell for conducting vegetation surveys throughout the duration of this project. Carl Schmidt and John Mason of the Piedmont National Wildlife Refuge were responsible for application of burn treatments. Funding was provided by U.S. Fish and Wildlife Service/Boise Interagency Fire Center.

Author Contributions: D.W. conceived, designed, and performed the experiments; M.R. and J.O'B. analyzed the data; M.R. wrote the paper and K.O. and J.O' B. advised throughout.

Conflicts of Interest: The authors declare no conflict of interest.

References

1. Moore, M.M.; Wallace Covington, W.; Fule, P.Z. Reference conditions and ecological restoration: A southwestern ponderosa pine perspective. *Ecol. Appl.* **1999**, *9*, 1266–1277. [CrossRef]
2. Peterson, D.W.; Reich, P.B. Prescribed fire in an oak savanna: Fire frequency effects on stand structure and dynamics. *Ecol. Appl.* **2001**, *11*, 914–927. [CrossRef]
3. Sparks, J.C.; Masters, R.E.; Engle, D.M.; Bukenhofer, G.A. Season of burn influences fire behavior and fuel consumption in restored shortleaf pine-grassland communities. *Rest. Ecol.* **2002**, *10*, 714–722. [CrossRef]
4. Wade, D.D.; Brock, B.L.; Brose, P.H.; Grace, J.B.; Hoch, G.A.; Patterson, W.A. Fire in eastern ecosystems. In *Wildland Fire in Ecosystems: Effects of Fire on Flora*; GTR RMRS-42; Brown, J.K., Smith, J.K., Eds.; USDA forest Service, Rocky Mountain Research Station: Ogden, UT, USA, 2000.
5. McCaw, W.L. Managing forest fuels using prescribed fire-A perspective from southern Australia. *For. Ecol. Manag.* **2013**, *294*, 217–224. [CrossRef]
6. Govender, N.; Trollope, W.S.W.; Van Wilgen, B.W. The effect of fire season, fire frequency, rainfall and management on fire intensity in savannah vegetation in South Africa. *J. Appl. Ecol.* **2006**, *43*, 748–758. [CrossRef]
7. Fernandes, P.M.; Botehlo, H.S. A review of prescribed burning effectiveness in fire hazard reduction. *Int. J. Wildland Fire* **2003**, *12*, 117–128. [CrossRef]
8. Brockway, D.G.; Lewis, C.E. Long-term effects of dormant-season prescribed fire on plant community diversity, structure and productivity in a longleaf pine wiregrass ecosystem. *For. Ecol. Manag.* **1997**, *96*, 167–183. [CrossRef]
9. Elliott, K.J.; Hendrick, R.L.; Major, A.E.; Vose, J.M.; Swank, W.T. Vegetation dynamics after a prescribed fire in the southern Appalachians. *For. Ecol. Manag.* **1999**, *114*, 199–213. [CrossRef]
10. Ryan, K.C.; Knapp, E.E.; Varner, J.M. Prescribed fire in North American forests and woodlands: History, current practice, and challenges. *Front. Ecol. Environ.* **2013**, *11*, e15–e24. [CrossRef]
11. Knapp, E.E.; Estes, B.L.; Skinner, C.N. *Ecological Effects of Prescribed Fire Season; a Literature Review and Synthesis for Managers*; USDA Forest Service General Technical Report PSW-GTR-224; USDA Forest Service, Pacific Southwest Research Station: Albany, CA, USA, 2009.
12. Streng, D.R.; Glitzenstein, J.S.; Platt, W.J. Evaluating effects of season of burn in longleaf pine forests: A acritical literature and some results from an ongoing long-term study. In Proceedings of the Tall Timbers Fire Ecol. Conf. No. 18, Tallahassee, FL, USA, 30 May–2 June 1991; Tall Timbers Research Station: Tallahassee, FL, USA, 1993.
13. Glitzenstein, J.S.; Streng, D.R.; Platt, W.J. *Evaluating Effects of Season of Burn on Vegetation in Longleaf Pine Savannahs*; Final Report; Florida Game and Freshwater Fish Committee: Tallahassee, FL, USA, 1990.
14. Glitzenstein, J.S.; Platt, W.J.; Streng, D.R. Effects of fire regime and habitat on tree dynamics in north Florida longleaf pine savannas. *Ecol. Monogr.* **1995**, *65*, 441–476. [CrossRef]
15. De Groot, W.J.; Wein, R.W. Effects of fire severity and season of burn on *Betula glandulosa* growth dynamics. *Int. J. Wildland Fire* **2004**, *13*, 287–295. [CrossRef]
16. Kaufman, J.B.; Marin, R.E. Sprouting shrub response to different season and fuel consumption levels of prescribed fire in Sierra Nevada mixed conifer ecosystems. *For. Sci.* **1990**, *36*, 748–764.
17. Williams, R.J.; Cook, G.D.; Gill, A.M.; Moore, P.H.R. Fire regime, fire intensity and tree survival in a tropical savanna in northern Australia. Aust. *J. Ecol.* **1999**, *24*, 50–59.
18. Robbins, L.E.; Myers, R.L. *Seasonal Effects of Prescribed Burning in Florida*; Miscellaneous publication/Tall Timbers Research, Inc.: Tallahassee, FL, USA, 1992; Volume 8, pp. 1–97.
19. Stoddard, H.L. Use of controlled fire in southeastern upland game management. *J. For.* **1935**, *33*, 346–351.
20. Waldrop, T.A.; Van Lear, D.H.; Lloyd, F.T.; Harms, W.R. *Long-Term Studies of Prescribed Burning in Loblolly Pine Forests of the Southeastern Coastal Plain*; USDA Forest Service General Technical Report SE-45; USDA Forest Service, Southeastern Forest Experiment Station: Asheville, NC, USA, 1987.
21. Reilly, M.J.; Waldrop, T.A.; O'Brien, J.J. Fuels management in the southern appalachian mountains, hot continental division. In *Cumulative Watershed Effects of Fuel Management in the Eastern United States;* USDA Forest Service General USDA Forest Service General Technical Report SRS-161; Southern Research Station: Asheville, NC, USA, 2012.

22. Conner, R.N.; Snow, A.E.; O'Halloran, K.A. Red-cockaded woodpecker use of seedtree/shelterwood cuts in eastern Texas. *Wildlife Soc. Bull.* **1991**, *19*, 67–73.
23. Conner, R.N.; O'Halloran, K.A. Cavity-tree selection by red-cockaded woodpeckers as related to growth dynamics of southern pines. *Wilson. Bull.* **1987**, *99*, 398–412.
24. Conner, R.N.; Rudolph, D.C. *Red-Cockaded Woodpecker Colony Status and Trends on the Angelina, Davy Crockett, and Sabine National Forests*; USDA Forest Service Research Paper SO-250; Southern Forest Experimental Station: New Orleans, LA, USA, 1989.
25. Loeb, S.C.; Pepper, W.D.; Doyle, A.T. Habitat characteristics of active and abandoned Red-cockaded woodpecker colonies. *South J. Appl. For.* **1992**, *16*, 120–125.
26. Chaiken, L.E. *The Behavior and Control of Understory Hardwoods in Loblolly Pine Stands*; USDA Forest Service Research Tech. Note 72; Southern Forest Experimental Station: New Orleans, LA, USA, 1949.
27. Harrington, T.A.; Stevenson, G.K. Repeat burns reduce small stems in Texas Big Thicket. *J. For.* **1955**, *53*, 847.
28. Hodgkins, E.J. Effects of fire on undergrowth vegetation in upland southern pine forests. *Ecology* **1958**, *39*, 36–46. [CrossRef]
29. Brender, E.V.; Cooper, R.W. Prescribed burning in Georgia's Piedmont loblolly pine stands. *J. For.* **1968**, *66*, 31–36.
30. Oosting, H.J. The comparative effect of surface and crown fire on the composition of a loblolly pine community. *Ecology* **1944**, *25*, 61–69. [CrossRef]
31. Oosting, H.J.; Livingston, R.B. A resurvey of a loblolly pine community twenty-nine years after ground and crown fire. *Bull. Torrey Bot. Club* **1964**, *91*, 387–395. [CrossRef]
32. Wade, D.D.; Weise, D.; Shell, R. *Some Effects of Periodic Winter Fire on Communities on the Georgia Piedmont, Proceedings of the 5th Biennial Southern Silvicultural Research Conference, Memphis, TN, USA, 1–3 November 1988*; General Tech. Rep. SO-74; Dept. of Agriculture, Forest Service, Southern Forest Experiment Station: New Orleans, LA, USA, 1989.
33. Cook, E.R.; Meko, D.M.; Stahle, D.W.; Cleaveland, M.K. Drought reconstructions for the continental United States. *J. Clim.* **1999**, *12*, 1145–1162. [CrossRef]
34. USDA Soil Conservation Service and Forest Service. *Soil Survey of Baldwin, Jones, and Putnam Counties, Georgia*; USDA Soil Conservation Service and Forest Service: Washington, DC, USA, 1976.
35. Cowell, C.M. Presettlement Piedmont forests: Patterns in composition and disturbance in central Georgia. *Ann. Assoc. Am. Geogr.* **1995**, *85*, 65–83.
36. Guyette, R.P.; Stambaugh, M.C.; Dey, D.C.; Muzika, R.-M. Predicting fire frequency with chemistry and climate. *Ecosystems* **2012**, *15*, 322–335. [CrossRef]
37. Radford, A.E.; Ahles, H.E.; Bell, C.R. *Manual of the Vascular Flora of the Carolinas*; The University of North Carolina Press: Chapel Hill, NC, USA, 1968.
38. Scheiner, S.M. *MANOVA*: Multiple response variables and multispecies interactions. In *The Design and Analysis of Ecological Experiments*; Scheiner, S.M., Gurevitch, J., Eds.; Oxford University Press: New York, NY, USA, 2001; pp. 134–157.
39. Von Ende, C.N. Repeated-measures analysis: Growth and other time dependent measures. In *The Design and Analysis of Ecological Experiments*; Scheiner, S.M., Gurevitch, J., Eds.; Oxford University Press: New York, NY, USA, 2001; pp. 134–157.
40. Dey, D.C.; Hartman, G. Returning fire to Ozark Highland forest ecosystems: Effects on advance regeneration. *For. Ecol. Manag.* **2005**, *217*, 37–53. [CrossRef]
41. McNab, W.H. An overcrowded loblolly pine stand thinned with fire. *South. J. Appl. For.* **1977**, *1*, 24–26.
42. Waldrop, T.A.; Lloyd, F.T. Precommercial thinning a sapling-sized loblolly pine stand with prescribed fire. *South. J. Appl. For.* **1988**, *12*, 203–207.
43. Wade, D.D. Thinning young loblolly pine stands with fire. *Int. J. Wildland Fire* **1993**, *3*, 169–178. [CrossRef]
44. Peet, R.K.; Christensen, N.L. Competition and tree death. *Bioscience* **1987**, *37*, 586–595. [CrossRef]
45. Lotti, T. Eliminating understory hardwoods with summer prescribed fires in Coastal Plain loblolly pine stands. *J. For.* **1956**, *54*, 191–192.

46. Burns, R.M.; Honkala, B.H. *Silvics of North America: 2. Hardwoods*; Agricultural Handbook; USDA: Washington, DC, USA, 1990.
47. Kreye, J.K.; Varner, J.M.; Hiers, J.K.; Mola, J. Toward a mechanism for eastern North American forest mesophication: Differential litter drying across 17 species. *Ecol. Appl.* **2013**, *23*, 1976–1986. [CrossRef] [PubMed]

6

Resource Limitations Influence Growth and Vigor of Idaho Fescue, a Common Understory Species in Pacific Northwest Ponderosa Pine Forests

Craig A. Carr [1,*] and William C. Krueger [2]

[1] Department of Animal and Range Sciences, Montana State University, PO Box 172900, Bozeman, MT 59717, USA

[2] Department of Animal and Rangeland Sciences, Oregon State University, 112 Withycombe Hall, Corvallis, OR 97331, USA; william.c.krueger@oregonstate.edu

* Correspondence: craig.carr@montana.edu

Abstract: Alterations in under-canopy resource availability associated with elevated ponderosa pine (*Pinus ponderosa* Dougl.) abundance can negatively influence understory vegetation. Experimental evidence linking under-canopy resource availability and understory vegetation is scarce. Yet this information would be beneficial in developing management strategies to recover desired understory species. We tested the effects of varying nitrogen (N) and light availability on Idaho fescue (*Festuca idahoensis* Elmer), the dominant understory species in ponderosa pine/Idaho fescue plant associations in eastern Oregon. In a greenhouse experiment, two levels of N (50 $kg{\cdot}N{\cdot}ha^{-1}$ and 0 $kg{\cdot}N{\cdot}ha^{-1}$) and shade (80% shade and 0% shade) were applied in a split-plot design to individual potted plants grown in soil collected from high abundance pine stands. Plants grown in unshaded conditions produced greater root (p = 0.0027) and shoot (p = 0.0017) biomass and higher cover values (p = 0.0378) compared to those in the shaded treatments. The addition of N had little effect on plant growth (p = 0.1602, 0.5129, and 0.0853 for shoot biomass, root biomass, and cover, respectively), suggesting that soils in high-density ponderosa pine stands that lack understory vegetation were not N deficient and Idaho fescue plants grown in these soils were not N limited. Management activities that increase under-canopy light availability will promote the conditions necessary for Idaho fescue recovery. However, successful restoration may be constrained by a lack of residual fescue or the invasion of more competitive understory vegetation.

Keywords: ponderosa pine; Idaho fescue; overstory-understory interactions; nitrogen; light availability

---

1. Introduction

In ponderosa pine (*Pinus ponderosa* Dougl.) forests, understory herbaceous vegetation is important for maintaining soil stability, site hydrology, and nutrient dynamics, it provides important wildlife habitat and livestock forage, and it is a major component of the fire regime, natural variability, and aesthetic value of these ecosystems. Since European settlement, many ponderosa pine forests have undergone a substantial structural change, moving from open-canopy, low-density, older-aged forests to high-density, closed-canopy, younger forests, primarily in response to increases in the fire return interval [1,2]. Along with overstory alterations, reductions in herbaceous understory growth, biomass, and vigor have occurred, and in many cases, ponderosa pine stands have experienced losses of important understory species [3–8]. Management of ponderosa pine forests toward a structure that more closely resembles presettlement conditions is often prescribed, however, many restoration programs focus on the manipulation of overstory structure without consideration of the treatment effects on understory species. Understory dynamics should also be considered in ponderosa pine restoration activities [9].

Although existing research has provided insight into the mechanisms associated with understory vegetation response to altered pine abundance [6,10–13], the results have been inconsistent and have not addressed climax Pacific Northwest ponderosa pine stands, therefore a complete understanding of the management needs in these forests is lacking. Carr and Krueger [3] observed reductions in under-canopy light and nitrogen availability associated with increasing tree abundance in a case study of a ponderosa pine forest in the Pacific Northwest, however no experimental evidence linking under-canopy resource availability to understory vegetation dynamics was available. Thus, we initiated an experiment designed to corroborate observational data and evaluate if under-canopy light and nitrogen availability influence the growth and vigor of understory vegetation and to consider the implications for the management of Pacific Northwest ponderosa pine forests. We selected the perennial bunchgrass Idaho fescue (*Festuca idahoensis* Elmer) for this study as it is the dominant understory species in ponderosa pine/Idaho fescue plant associations in eastern Oregon [14] and has been shown to decline significantly under conditions of high ponderosa pine abundance [3].

2. Materials and Methods

A greenhouse experiment designed to test the effects of variable light and N availability on Idaho fescue growth rate, biomass, and vigor was conducted over 83 days spanning 11 April through 3 July 2007.

2.1. Field Protocol

Idaho fescue plants were collected in early April from a ponderosa pine stand in eastern Oregon (44°12′52″ N, 118°59′16″ W). The site was located at 1500 m elevation and situated in an area of gently sloping (~8% slope gradient) north-facing hillsides. The average daily high and low summer and winter temperatures, recorded between 1971 and 2000 at Seneca, Oregon (~8 km south of the study area) were 24 °C and 1.5 °C, and 4 °C and −9 °C, respectively [15]. Approximately 32 cm of precipitation is received annually, most occurring in the winter months and primarily in the form of snow [15]. Soils of the area were composed of an ash mantle overlaying serpentinite gravel and cobble and were classified as fine-loamy, mixed, superactive, and frigid Vitrandic Haploxerolls. The site exhibited an open canopy structure that was relatively free of trees. The adjacent forest was composed primarily of small-diameter ponderosa pine trees (15–20 cm diameter at 1.4 m height) growing at a density of approximately 260 trees·ha^{-1}. Idaho fescue plants (2–3 cm basal diameter) were removed to a depth of 10 cm using a 5.08 cm diameter soil core centered on the plant and then placed in a container of native soil for transportation to the greenhouse where they were replanted into 3.79 L pots. Prior to replanting, the foliage of each plant was clipped at a height of 5 cm and the soil below 5 cm depth was removed.

Soil from a dense ponderosa pine stand (2800 trees·ha^{-1}) neighboring the location of the Idaho fescue harvest was used for potting media. This soil was selected because it represented the soil conditions of high pine density stands that lack the characteristic perennial bunchgrass dominated understory. This soil provided the baseline soil conditions from which increasing plant resource availability would provide insight into potential nutrient limitations for understory vegetation growth in severely degraded under-canopy environments. The soil was harvested to a depth of 10 cm (mineral soil) and prior to potting the soil was uniformly mixed and passed through a 7 mm screen to remove rocks, litter, and root fragments.

2.2. Greenhouse Protocol and Experimental Design

Potted Idaho fescue plants were grown in a greenhouse located on the Oregon State University campus at Corvallis, Oregon (44°34′12″ N, 123°16′48″ W). The average daytime (06:00–20:00 h) temperature of the greenhouse throughout the study was 23.7 °C (SD = 2.19).

Two levels of shade and N were applied in a split plot design. Whole plots (80% shade or 0% shade) were randomly assigned to greenhouse bench locations (replications) and subplots (50 $kg{\cdot}N{\cdot}ha^{-1}$ or 0 $kg{\cdot}N{\cdot}ha^{-1}$) were randomly assigned to two pots randomly allocated within each whole plot. Five replications of each treatment combination were performed.

Shade treatments were applied using tents made of 80% shade cloth draped over a 76 cm $\times$ 76 cm $\times$ 76 cm polyvinyl chloride (PVC) pipe frame. Eighty percent shade represented the average canopy closure (79.14%) observed in the field location for high density pine stands that lack characteristic perennial bunchgrass dominated vegetation [3].

Urea (46-0-0) was applied at 50 $kg{\cdot}N{\cdot}ha^{-1}$ to each pot selected for N treatment. Urea pellets were covered with soil and watered-in to prevent N loss through ammonia (NH_3) volatilization. Fifty $kg{\cdot}N{\cdot}ha^{-1}$ was deemed adequate to meet the needs of eastern Oregon Idaho fescue. A plant tissue N content of 2%–5% (dry weight) is required for optimal plant growth [16] and the average dry weight biomass of Idaho fescue in ponderosa pine/Idaho fescue plant associations is 365 $kg{\cdot}ha^{-1}$ [14], therefore, approximately 7.3–18.25 $kg{\cdot}ha^{-1}$ was needed to support typical growth of Idaho fescue in the region of our field sites. We used 50 $kg{\cdot}N{\cdot}ha^{-1}$ to ensure that N supply was not limiting and to allow for some N loss through leaching. All pots were kept moist throughout the experiment with regular watering.

Pretreatment Idaho fescue size, indexed by the product of the length of the longest axis through the base of the plant and the axis perpendicular to the longest axis, was measured for use as a covariate in the analyses. Measurements of plant height and foliar cover were performed weekly and biweekly, respectively. Plant height values used in these analyses were averages of the shortest, tallest, and two intermediate length leaves. Foliar cover was evaluated using downward looking digital photographs of each plant. A digital grid of 3400 uniformly spaced dots was overlaid on each digital photo in ArcMap version 9.1 [17]. The number of points touching the leaf material was tallied for each plant and used as an index of foliar cover. Images were acquired using a Nikon Coolpix 950 digital camera (Nikon Corporation, Tokyo, Japan) mounted 1 m above the ground on a PVC frame modified from Louhaichi et al. [18]. This set up along with the uniform focal length (7.2 mm) and resolution (1200 $\times$ 1600 pixels) for each photo provided standard image dimensions of 89 cm $\times$ 67 cm with a pixel size of 0.56 mm. At the end of the experiment, shoot and root biomass were measured. Plants were clipped at ground level and all leaf material weighed after drying for 48 h at 60 °C. Root material was collected by passing the potting media through a 7 mm screen and then washing the remaining material through a 0.59 mm (#30) sieve. The residual root material was dried for 48 h at 60 °C and then weighed. Three shoot biomass samples measured below the resolution of the scale were used to weigh the samples (0.1 g). To provide a value for trace biomass, 0.09 g was used in the analyses.

Analyses of shoot biomass, root biomass, final shoot length, and final cover were performed using PROC MIXED [19]. Random effects for replication and whole plot were included in the model along with the fixed effects of the shade and N treatments and their interaction. Root and shoot biomass values were log transformed prior to analysis to better meet the assumptions associated with ANOVA analyses.

The responses of foliar cover and leaf length over time were evaluated using PROC MIXED [19]. The models included fixed effects for shade, N, date, and their interactions, and random effects for replication, whole plot, and subplot and covariance structures selected to best model the within plant variability associated with the repeated measures nature of the experiment [20]. Because of a consistent lack of significance of the three-way interaction among shade, N, and date, and two-way interaction between shade and N, models for both responses were reduced to the main effects of shade, N, and date, and the interactions between shade and date and N and date. To meet the assumptions of equal variance both the cover and leaf length data were log transformed prior to analyses.

3. Results

The only statistically significant effect in the analyses of shoot biomass, root biomass, shoot length, and foliar cover was associated with the shade treatment (Table 1). Plants growing under shade had lower shoot and root biomass and foliar cover and exhibited longer shoot lengths compared to those plants growing in the un-shaded conditions (Figure 1). Although N addition appeared to show a positive trend in all responses (Figure 1), the effect was not statistically significant.

Table 1. *F* statistics and *p*-values from Type III tests of the fixed effects of shade, nitrogen, and shade and nitrogen interaction from mixed effects models with response variables of Idaho fescue shoot and root weight, final shoot length, and final plant cover.

	Shade		Nitrogen		Shade * Nitrogen	
	F	*p*-Value	*F*	*p*-Value	*F*	*p*-Value
Shoot Weight	56.61	0.0017	2.40	0.1602	1.10	0.3244
Root Weight	433.46	0.0027	0.47	0.5129	0.10	0.7567
Final Length	50.19	0.0021	2.57	0.1478	0.03	0.8596
Final Cover	9.35	0.0378	3.85	0.0853	0.02	0.8932

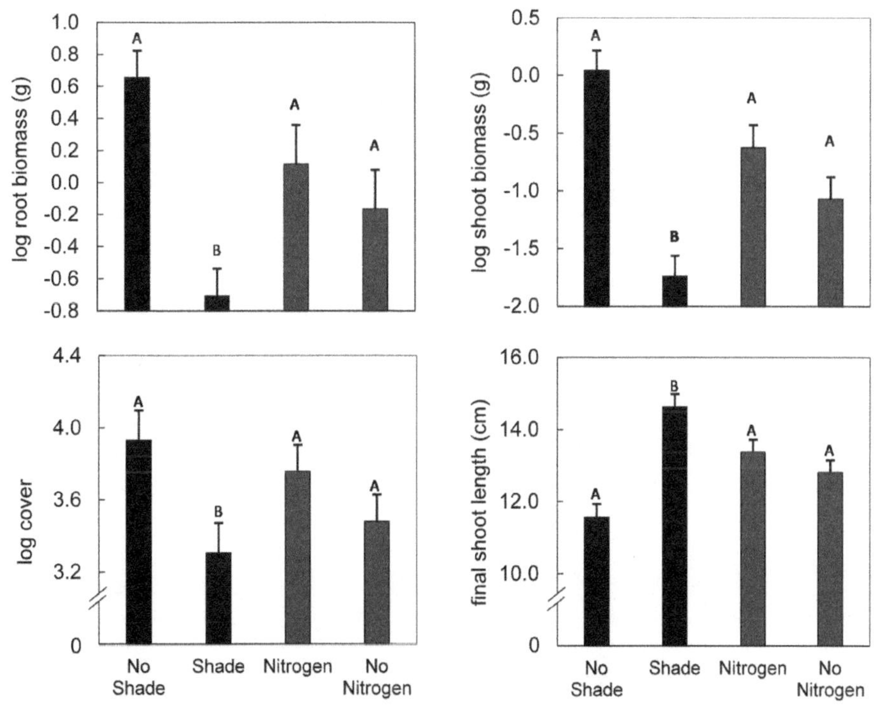

Figure 1. The effects of shade and nitrogen treatments on Idaho fescue shoot biomass, root biomass, shoot length, and cover. Columns represent group least squares means while error bars. Differing letters represent differences within treatment groups at the $p < 0.05$ level

When including time as a factor in the analyses of shoot length, the main effect of shade remained significant while differences in shoot length over time in response to N addition were also apparent (Table 2). No interaction between shade and N was evident, however significant interactions between shade and time and N and time were evident (Table 2). The change in shoot length over time tended to follow a curvilinear pattern showing rapid elongation in early growth and stabilizing as maximum shoot length was approached in the later weeks of the experiment (Figure 2). Differences in shoot length among shade treatment groups were not evident in the first 2 weeks of growth, however the subsequent time periods showed successive increases in shoot length in the shaded treatment (Figure 2). The addition of N stimulated an increase in the early rate of leaf elongation as significant differences among N treatment groups occurred in weeks 2 through 5. However, by the end of the study, plants in the unfertilized treatment had shoot lengths that equaled that of the fertilized plants (Figure 2).

Table 2. *F* statistics and P-values from Type III tests of the fixed effects of shade, nitrogen, time, and the interaction effects of shade and time and nitrogen and time from mixed effects models with response variables of Idaho fescue shoot length and foliar cover.

	Shade		Nitrogen		Time		Shade *Time		Nitrogen * Time	
	F	*p*-Value	*F*	*p*-Value	*F*	*p*-Value	*F*	*p*-Value	*F*	*p*-Value
Length	29.19	0.0001	7.50	0.0169	295.16	<0.0001	10.36	<0.0001	3.50	0.0006
Cover	3.57	0.1309	5.61	0.0422	183.57	<0.0001	5.63	0.0009	4.13	0.0059

Including time in the analysis of foliar cover indicated significant effects of the interactions between N and time and shade and time. The main effect of N was weak while the main effect of shade was nonsignificant (Table 2). Plant cover change over time generally followed a curvilinear pattern with rapid increase in foliar cover through the early portion of the experiment followed by a much reduced rate of increase toward the end of the trial (Figure 2). Differences in Idaho fescue cover among shade treatment groups were not evident in the first 6 weeks of the experiment, however the subsequent two measurements, in week 7 and week 9, indicated successive increases in plant cover and identified unshaded plants as eventually accumulating greater plant cover compared to shaded individuals (Figure 2). The addition of N stimulated an increase in foliar cover earlier in the experiment as significant differences among N treatment groups occurred in weeks 3 and 5 (Figure 2). However, by the end of the experiment, plants in the un-fertilized treatment had foliar cover values that rivaled that of the fertilized plants (Figure 2).

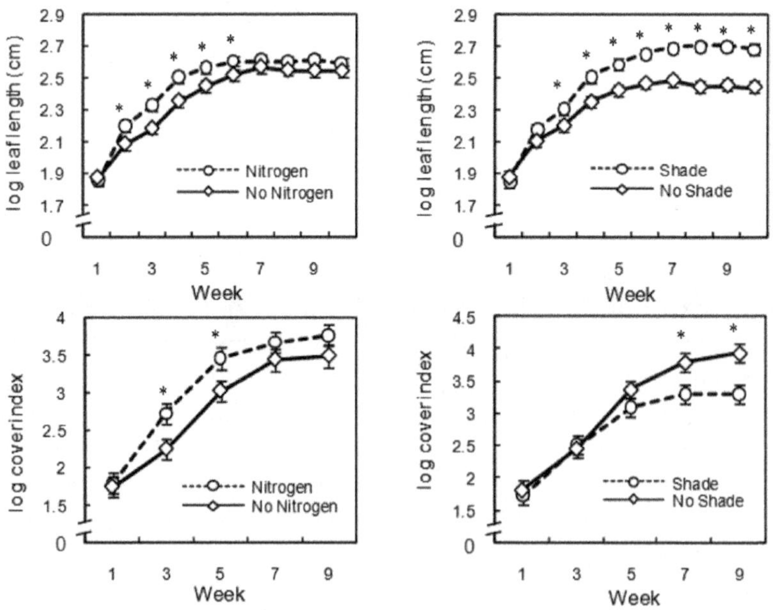

Figure 2. The effects of shade and nitrogen treatments on Idaho fescue leaf length and foliar cover. Dots represent the nitrogen or shade treatment group least squares means at each time interval while error bars are ±1 standard error. Length was measured weekly throughout the experiment while cover was measured bi-weekly. Although not continuous data, lines between data points display the interaction between treatments and time. Significant differences ($p < 0.05$) within time periods are identified by *.

4. Discussion

It was apparent from these analyses that Idaho fescue plants grew larger under high light intensity environments. Plants in the unshaded treatments exhibited higher shoot and root biomass and larger cover indices than those in the shaded environments. Plants commonly exhibit greater biomass and growth when developing under high irradiance [21,22] and others have observed decreases in biomass of ponderosa pine forest understory vegetation when grown under shaded conditions [4,10,23]. Approximately 40% of plant dry matter comes from carbon fixed through photosynthesis [24], therefore changes in plant biomass may be attributed to changes in the photosynthetic activity of the plant. Although numerous factors can affect photosynthesis including light, temperature, water, CO_2 availability, and plant health and age [25], it was evident that light availability influenced the photosynthetic ability of the Idaho fescue plants used in this experiment. The light saturation point for a plant is reached when carbon fixation and photosynthetic activity is maximized and further increases in light intensity do not generate associated increases in photosynthetic activity. Coyne et al. [25] indicated the light saturation point for C_3 species, such as Idaho fescue, is around 50% of full sunlight. Our 80% shade treatment corresponded to a photosynthetically active radiation intensity approximately 22% of that experienced in the un-shaded treatments [26] and therefore substantially lower than the light saturation point.

The increase in leaf length observed in the shaded treatments is a common response to low irradiance environments and is a leaf-level adaptation that increases leaf area [21,22,27]. Leaf elongation is one of several plant morphological and physiological adaptations associated with conservative resource use that promotes efficient light energy capture while reducing respiratory losses [22,28,29]. These changes enable plants to maintain a positive carbon balance under conditions of limited carbon input brought on by the diminished light energy availability [28].

The shade effect on leaf length was consistent over time with the exception of the first two measurement periods, although divergence in leaf length was evident by week 2. Plants in the shaded treatment exhibited longer leaf lengths over weeks 3–10. The lack of a significant shade effect over the first 2 weeks was likely a reflection of the period of plant growth initiation.

There was also a lag period prior to observing a significant difference in foliar cover between plants in the two shade treatment groups. It appears as though initial rates of increase in foliar cover were similar, however at around week 5, the plants in the shaded treatment substantially reduced the rate of increase in cover while those in the un-shaded group continued with a strong growth rate. The decline in the growth rate of shaded plants was likely related to the inability of these plants to fix carbon at a rate required to support further growth [24].

Longer leaves were evident in plants subjected to the N addition treatment during weeks 2–6, indicating that leaf growth rate was higher under the N treatment during this time period. As this was not a persistent effect and final leaf length did not differ among N treatment groups, it is likely that this temporal effect was related to increased mineral N supply to the plants in the fertilized treatments. Mechanisms associated with root-nutrient contact including mass transport, diffusion, and root interception are enhanced under conditions of higher soil nutrient concentrations [30]. Therefore, plants in the fertilized treatment were able to acquire the N needed to meet their growth needs at a faster rate than those in the un-fertilized treatments simply as a function of increased N availability associated with higher soil concentrations.

Nitrogen addition also appeared to stimulate a short-term increase in the growth rate, as indexed by cover, of Idaho fescue plants. Plants in the N addition treatment expressed larger foliar cover values than those in the un-fertilized group during weeks 3, 5, and 7. Cover values from the two groups were statistically similar in earlier (i.e., week 1) and later (week 9) time periods. This pattern was similar to the growth rate indexed by leaf elongation and was also likely in response to increased availability of mineral N at the root-soil interface associated with elevated soil N concentrations in the fertilized treatment. The unfertilized plants were eventually able to accumulate enough N to meet their growth requirements as foliage cover values for both N treatment groups were similar by the end of the experiment.

The general lack of a lasting N effect on Idaho fescue growth indicated that soils from dense ponderosa pine forests were not N deficient and Idaho fescue plants grown in these soils were not N limited. Although others have reported decreases in net N mineralization associated with increasing ponderosa pine abundance [23,31], the reduction may not have a substantial effect on the growth of desired understory species. In our study, soils were sieved prior to use to remove all litter and coarse organic material which likely reduced the potential for N immobilization and this may explain why N mineralization rates were high enough to support vigorous plant growth, even in un-fertilized treatments. Kaye et al. [32] reported similar findings to ours where increased net N mineralization associated with restoration treatments in formerly dense ponderosa pine stands did not correspond to increased total plant biomass or total plant N uptake.

Similar relationships between overstory and understory herbaceous species have been explored in other forest types (see [33]). Light availability is a primary limiting resource in understory environments in eastern deciduous forests [34]. Understory plants that persist under low light conditions incorporate various physiological and morphological adaptations, including phenology, leaf morphology, growth form, and photosynthetic pathway [34,35]. A lack of adaptations may explain why understory plants that persist under more open canopies, are substantially diminished as canopy closure increases (e.g., [3]). Moreover, in contrast to our single species experiment, elevated nitrogen availability can influence understory vegetation properties. Gilliam [36], in a review of the literature on understory response to increased nitrogen deposition, indicated that increased nitrogen availability can influence understory vegetation through reductions in biodiversity, altered competitive interactions among species, and enhanced exotic plant invasion.

Discretion in directly applying these results to field situations is recommended as other factors including pine needle litter accumulation and reduced soil temperatures, along with more natural shade characteristics under dense ponderosa pine stands, may alternatively influence Idaho fescue growth patterns from what was observed in this greenhouse experiment. In addition, our single species experiment does not account for the suite of vegetation that make up the understory of these ponderosa pine forests, however Idaho fescue is the dominant species providing over 40% of the understory cover in this plant association [14] and represents a general response to changing undercanopy resources in this community.

5. Conclusions

From these analyses, it was apparent that Idaho fescue plants grown in shaded conditions were less vigorous; exhibiting lower root biomass, shoot biomass, and foliar cover values. This corroborates observational data which indicated that understory vegetation abundance was directly related to soil nitrogen and under-canopy light availability [3]. However, our experiment indicated that only light limited Idaho fescue growth and vigor. Extending our findings more broadly to understory vegetation, we speculate that management activities that focus on reducing tree density should increase undercanopy light and may improve conditions for understory vegetation. However, the recovery of understory species may not occur simply through stand thinning treatments as perennial herbaceous species tend to exhibit poor seed banking abilities [37], and in the absence of sufficient residual understory vegetation, recovery will require off-site sources [38]. Moreover, the potential exists for shifts in understory vegetation to more competitive species [3,39], including exotic annual grasses, which may further limit desired vegetation recovery and exacerbate the degradation in ecosystem resilience.

Acknowledgments: Research was funded by the Oregon Agricultural Experiment Station, the United States Department of Agriculture, Natural Resources Conservation Service, and the Montana Agricultural Experiment Station. We appreciate the efforts of the anonymous reviewers whose comments improved the quality of the manuscript.

Author Contributions: C.A.C. and W.C.K. were involved in conceiving the experiment, developing appropriate experimental design and analyses, and in writing the manuscript. C.A.C. was responsible for data collection and analyses.

Conflicts of Interest: The authors declare no conflict of interest.

References

1. Covington, W.; Moore, M. Southwestern ponderosa forest structure: Changes since Euro-American settlement. *Forestry* **1994**, *92*, 39–47.
2. Johnson, C.G. *Forest Health in the Blue Mountains: A Plant Ecologist's Perspective on Ecosystem Processes and Biological Diversity*; Pacific Northwest Research Station PNW-GRR-339; USDA Forest Service: Portland, OR, USA, 1994; p. 23.
3. Carr, C.A.; Krueger, W.C. Understory vegetation and ponderosa pine abundance in eastern Oregon. *Rangel. Ecol. Manag.* **2011**, *64*, 533–542. [CrossRef]
4. Clary, W.P.; Kruse, W.H.; Larson, F.R. Cattle grazing and wood production with different basal areas of ponderosa pine. *J. Range Manag.* **1975**, 434–437. [CrossRef]
5. Moore, M.M.; Covington, W.W.; Fule, P.Z. Reference conditions and ecological restoration: A southwestern ponderosa pine perspective. *Ecol. Appl.* **1999**, *9*, 1266–1277. [CrossRef]
6. Naumburg, E.; DeWald, L.E. Relationships between *Pinus ponderosa* forest structure, light characteristics, and understory graminoid species presence and abundance. *Forest Ecol. Manag.* **1999**, *124*, 205–215. [CrossRef]
7. Uresk, D.W.; Severson, K.E. Response of understory species to changes in ponderosa pine stocking levels in the black hills. *Great Basin Nat.* **1998**, *58*, 312–327.
8. Wienk, C.L.; Sieg, C.H.; McPherson, G.R. Evaluating the role of cutting treatments, fire and soil seed banks in an experimental framework in ponderosa pine forests of the black hills, South Dakota. *Forest Ecol. Manag.* **2004**, *192*, 375–393. [CrossRef]
9. Moore, M.M.; Casey, C.A.; Bakker, J.D.; Springer, J.D.; Fule, P.Z.; Covington, W.W.; Laughlin, D.C. Herbaceous vegetation responses (1992–2004) to restoration treatments in a ponderosa pine forest. *Rangel. Ecol. Manag.* **2006**, *59*, 135–144. [CrossRef]
10. Naumburg, E.; Dewald, L.E.; Kolb, T.E. Shade responses of five grasses native to southwestern US *Pinus ponderosa* forests. *Can. J. Bot.* **2001**, *79*, 1001–1009.
11. Riegel, G.M.; Miller, R.F.; Krueger, W.C. Understory vegetation response to increasing water and nitrogen levels in a pinus-ponderosa forest in northeastern Oregon. *Northwest Sci.* **1991**, *65*, 10–15.
12. Riegel, G.M.; Miller, R.F.; Krueger, W.C. Competition for resources between understory vegetation and overstory pinus-ponderosa in northeastern Oregon. *Ecol. Appl.* **1992**, *2*, 71–85. [CrossRef] [PubMed]
13. Riegel, G.M.; Miller, R.F.; Krueger, W.C. The effects of aboveground and belowground competition on understory species composition in a pinus-ponderosa forest. *Forest Sci.* **1995**, *41*, 864–889.
14. Johnson, C.; Clausnitzer, R. *Plant Associations of the Blue and Ochoco Mountains*; USDA Forest Service R6-ERW-TP-036–92; USDA Forest Service: Portland, OR, USA, 1992.
15. Western Regional Climate Center(WRCC). Climatalogical Summary for Seneca, Oregon. Available online: http://www.wrcc.dri.edu/cgi-bin/cliMONtpre.pl?or7675 (accessed on 10 May 2010).
16. Marschner, H. *Mineral Nutrition of Higher Plants*, 2nd ed.; Academic Press Ltd.: San Diego, CA, USA, 1995.
17. Environmental Systems Research Institute. *Arcmap 9.1*; Environmental Systems Research Institute, Inc.: Redlands, CA, USA, 2005.
18. Louhaichi, M.; Johnson, D.E.; Borman, M.M. *Manual for Assessing Depredation on Grain and Seed Yields, Ver 1.1*; Department of Rangeland Resources, Oregon State University: Corvallis, OR, USA, 2003.
19. SAS Institute. *Sas for Windows Version 9.1*; SAS Institute, Inc.: Cary, NC, USA, 2003.
20. Littell, R.C.; Stroup, W.W.; Milliken, G.A.; Wolfinger, R.D.; Schabenberger, O. *SAS for Mixed Models*; SAS Institute: Cary, NC, USA, 2006.
21. Allard, G.; Nelson, C.J.; Pallardy, S.G. Shade effects on growth of tall fescue: 1. Leaf anatomy and dry-matter partitioning. *Crop Sci.* **1991**, *31*, 163–167. [CrossRef]
22. Boardman, N. Comparative photosynthesis of sun and shade plants. *Ann. Rev. Plant Physiol.* **1977**, *28*, 355–377. [CrossRef]
23. Moir, W.H. Influence of ponderosa pine on herbaceous vegetation. *Ecology* **1966**, *47*, 1045–1048. [CrossRef]

24. Lambers, H.; Chapin, F., III; Pons, L. *Plant Physiological Ecology*; Springer: New York, NY, USA, 1998; p. 540.
25. Coyne, P.; Trlica, M.; Owensby, C. Carbon and nitrogen dynamics in range plants. In *Wildland Plants: Physiological Ecology and Developmental Morphology*; Society for Range Management: Denver, CO, USA, 1995; pp. 59–167.
26. Carr, C.A. An Evaluation of Understory Vegetation Dynamics, Ecosystem Resilience and State and Transition Ecological Theory in an Eastern Oregon Ponderosa Pine Forest. Ph.D Dissertation, PhOregon State University, Corvallis, OR, USA, 2007.
27. Murchie, E.H.; Horton, P. Acclimation of photosynthesis to irradiance and spectral quality in British plant species: Chlorophyll content, photosynthetic capacity and habitat preference. *Plant Cell Environ.* **1997**, *20*, 438–448. [CrossRef]
28. Bjorkman, O. Some viewpoints on photosynthetic response and adaptation to environmental stress. In *Photosynthesis*; Briggs, W.R., Ed.; Alan R. Liss: New York, NY, USA, 1989.
29. Smith, H. Light quality, photoperception, and plant strategy. *Ann. Rev. Plant Phsyiol.* **1982**, *33*, 481–518. [CrossRef]
30. Havlin, J.L.; Tisdale, S.L.; Nelson, W.L.; Beaton, J.D. *Soil Fertility and Fertilizers*, 6th ed.; Prentice-Hall Inc.: Upper Saddle River, NJ, USA, 2007; p. 499.
31. Kaye, J.P.; Hart, S.C. Ecological restoration alters nitrogen transformations in a ponderosa pine bunchgrass ecosystem. *Ecol. Appl.* **1998**, *8*, 1052–1060. [CrossRef]
32. Kaye, J.P.; Hart, S.C.; Fule, P.Z.; Covington, W.W.; Moore, M.M.; Kaye, M.W. Initial carbon, nitrogen, and phosphorus fluxes following ponderosa pine restoration treatments. *Ecol. Appl.* **2005**, *15*, 1581–1593. [CrossRef]
33. Gilliam, F.S.; Roberts, M.R. *The Herbaceous Layer in Forests of Eastern North America*; Oxford Universtiy Press: New York, NY, USA, 2003.
34. Nuefeld, H.S.; Young, D.R. Ecophysiology of the herbaceous layer in temperate deciduous forests. In *The Herbaceous Layer in Forests of Eastern North America*; Gilliam, F.S., Roberts, M.R., Eds.; Oxford Universtiy Press: New York, NY, USA, 2003; pp. 38–90.
35. Pearcy, R.W. Photosynthetic gas exchange responses of Australian tropical forests trees in canopy, gap, and understory micro-environments. *Funct. Ecol.* **1987**, *1*, 169–178. [CrossRef]
36. Gilliam, F.S. Response of the herbaceous layer of forest ecosystems to excess nitrogen deposition. *J. Ecol.* **2006**, *94*, 1176–1191. [CrossRef]
37. Rees, M. Delayed germination of seeds—A look at the effects of adult longevity, the timing of reproduction, and population age/stage structure. *Am. Nat.* **1994**, *144*, 43–64. [CrossRef]
38. Korb, J.E.; Springer, J.D.; Powers, S.R.; Moore, M.M. Soil seed banks in *Pinus ponderosa* forests in Arizona: Clues to site history and restoration potential. *Appl. Veg. Sci.* **2005**, *8*, 103–112. [CrossRef]
39. Page, H.N.; Bork, E.W. Effect of planting season, bunchgrass species, and neighbor control on the success of transplants for grassland restoration. *Restor. Ecol.* **2005**, *13*, 651–658. [CrossRef]

7

Assessing Precision in Conventional Field Measurements of Individual Tree Attributes

Ville Luoma [1,4,*], Ninni Saarinen [1,4], Michael A. Wulder [2], Joanne C. White [2], Mikko Vastaranta [1,4], Markus Holopainen [1,4] and Juha Hyyppä [3,4]

[1] Department of Forest Sciences, University of Helsinki, P.O.Box 27 (Latokartanonkaari 7), 00014 Helsinki, Finland; ninni.saarinen@helsinki.fi (N.S.); mikko.vastaranta@helsinki.fi (M.V.); markus.holopainen@helsinki.fi (M.H.)

[2] Canadian Forest Service (Pacific Forestry Centre), Natural Resources Canada, 506 West Burnside Road, Victoria, BC V8Z 1M5, Canada; mike.wulder@canada.ca (M.A.W.); joanne.white@canada.ca (J.C.W.)

[3] Department of Remote Sensing and Photogrammetry, Finnish Geospatial Research Institute FGI, National Land Survey, Geodeetinrinne 2, 04310 Masala, Finland; juha.hyyppa@nls.fi

[4] Centre of Excellence in Laser Scanning Research, Finnish Geospatial Research Institute FGI, National Land Survey, 04310 Masala, Finland

* Correspondence: ville.luoma@helsinki.fi

Abstract: Forest resource information has a hierarchical structure: individual tree attributes are summed at the plot level and then in turn, plot-level estimates are used to derive stand or large-area estimates of forest resources. Due to this hierarchy, it is imperative that individual tree attributes are measured with accuracy and precision. With the widespread use of different measurement tools, it is also important to understand the expected degree of precision associated with these measurements. The most prevalent tree attributes measured in the field are tree species, stem diameter-at-breast-height (dbh), and tree height. For dbh and height, the most commonly used measuring devices are calipers and clinometers, respectively. The aim of our study was to characterize the precision of individual tree dbh and height measurements in boreal forest conditions when using calipers and clinometers. The data consisted of 319 sample trees at a study area in Evo, southern Finland. The sample trees were measured independently by four trained mensurationists. The standard deviation in tree dbh and height measurements was 0.3 cm (1.5%) and 0.5 m (2.9%), respectively. Precision was also assessed by tree species and tree size classes; however, there were no statistically significant differences between the mensurationists for dbh or height measurements. Our study offers insights into the expected precision of tree dbh and height as measured with the most commonly used devices. These results are important when using sample plot data in forest inventory applications, especially now, at a time when new tree attribute measurement techniques based on remote sensing are being developed and compared to the conventional caliper and clinometer measurements.

Keywords: precision; accuracy; diameter-at-breast-height; dbh; tree height; forest mensuration; clinometers; calipers; hypsometers; forest inventory

---

1. Introduction

Tree species, diameter-at-breast-height (dbh), and height are the most common individual tree attributes measured or determined in the field [1,2]. Dbh is the most fundamental of tree measurements and is defined as the outside-bark-stem diameter of a tree at a point on the stem that is 1.3 m above the ground from the base of the tree [1,3]. In forestry, the word "diameter" implies that the trees would have circular cross-sections. Yet, several studies have proved that exact circularity is rare and the form of a cross-section is more like a closed convex [4–7]. The most common methods for the field measurements of dbh are the use of calipers or diameter tapes. Diameter tapes are more commonly employed for permanent sample plots because they are perceived as being more consistent for repeated measures [8]. However, calipers are often used and preferred for dbh measurements in temporary plots or when measuring a large number of trees, as they are quick and efficient to use [9]. With calipers, the dbh is measured by placing the two arms of the caliper on perpendicular sides of the tree at the height of 1.3 m. One arm is fixed at the origin of the measurement scale and the other one can be moved along the scale beam of the caliper. When the two arms are pressed together against the tree trunk, a 90° angle must be formed with the scale beam. Thereafter, the diameter can be read from the scale. Although measurement of dbh using calipers is straightforward, it includes many sources of error that can cause variation in measurements. For example, when calipers are not oriented perpendicularly to the vertical axis of the stem, or when branches and other anomalies such as gnarls, wounds or branch bumps are located exactly at height of 1.3 m. Site conditions must also be considered, such as when there are steep slopes. In this context, dbh should be measured halfway between the measure at 1.3 m height on the slope's upper and lower sides. In addition, there may be variation in how personnel use the calipers for the measurements. Conventionally, a minimum dbh threshold for measurement is specified and only those trees in a sample plot with a dbh greater than this minimum threshold are measured. The trees with a diameter less than the specified threshold often will have little effect on the derived forest attribute estimates, such as total volume or basal area, and the measurement of these small trees can be extremely time consuming and can substantially add to the plot costs [10].

Tree height is defined as the distance between the base and the top of a standing tree [1]. Clinometers are instruments used to measure individual tree height [1]. These instruments are based on simple trigonometric relationships between the known planimetric distance from the instrument to the tree and the angles from the instrument to the base and top of the tree, which must be clearly visible when performing measurements [11]. A vertex clinometer is one of the most popular instruments for measuring tree height because it is efficient, easy to calibrate, and automatically measures the distance to the tree [1]. Height measurements in sample plots are time consuming, so it is typical to measure a sub-sample of tree heights within the sample plot; e.g., [12–15], rather than all trees in the plot [16–18]. If only a sub-sample of tree heights is measured, the sample must be representative of the dbh frequency distribution within the plot. The heights of the unmeasured trees can then be estimated through species-specific dbh-height regression models developed from the measured sub-sample e.g., [19–21].

Attributes of individual trees, usually collected from sample plots, are used in many forest inventory and modeling applications. A forest inventory could—in principle—be based on measuring every tree in a given area, but this would be time consuming, expensive, and unnecessarily impractical for most applications in forestry [22]. Thus, the acquisition of forest resource information is often more efficiently based on sampling, measuring, and modeling [1]. The most common sample units are a single tree and a plot, while the most commonly measured tree attribute is stem dbh. Some examples of the most frequently used models in forest mensuration are models for tree height or stem volume; see e.g., [19–21,23]. To obtain estimates over large areas, individual tree measures are summed at the sample-plot level and then sample plots are used in deriving stand-, landscape-, regional- and even global-scale estimates of forest resources [22]. Due to this hierarchy and the capacity to propagate and exaggerate small errors through multipliers, it is highly important to measure individual tree attributes accurately and precisely. In addition, when new measurement techniques, such as terrestrial or mobile laser scanning [24–27], close-range photogrammetry [28,29] and mobile phone applications [30,31], for tree attributes are developed, their measurements are often compared to conventional caliper and clinometer measurements, which are considered as a baseline.

There have been several studies on the precision of tree dbh and height measurements. Different measurement techniques and tools have also been compared to each other. The precision and variation between several re-measurements of dbh and height have been reported for example in [32–38]. The results of these studies have been reported using various means (e.g., mean, absolute and relative standard deviation, and bias) and straightforward comparisons of the resulting precision are not possible in all cases. In addition, the amount of measured trees and the number of repeat measurements varied widely amongst the studies, and these studies all used diameter tapes for dbh measurements instead of calipers. When collecting reference data for research and in operational use, the calipers are often preferred instead of diameter tape and thus characterizing the precision of caliper measurements is useful information for benchmarking emerging alternative measurement tools.

In non-destructive measurements, there are no unambiguous instructions for determining the reference value for dbh or tree height and the problem has been approached differently in previous studies. For example, in [37], the author's own measurements of dbh and height were used as the true value in the study and were assumed to be error free when compared to measurements conducted by others, whereas Hyppönen and Roiko-Jokela [33] admitted that measurement bias could not be calculated since it was not possible to obtain the exact dbh. Furthermore, Kitahara et al. [39] have noted that in tree height measurements, educated and experienced mensurationists provide more precise results than beginners with only basic knowledge.

In addition to studies where the measurements are repeated using the same instruments, there have been studies that have compared the measurements of the same trees using several devices, such as a caliper and diameter tape, or a laser height finder and a clinometer, but no major differences in precision have been reported; e.g., [4,40–42]. Guillemette and Lambert [43] found no significant difference in dbh measured with calipers and diameter tape and recommended against the mixing of models and databases developed using either instrument. They also stated that evaluation of the most precise method is extremely difficult without destructive sampling.

In forest mensuration, the vast majority of dbh and height measurements are carried out non-destructively with standing trees, which was the case in this study as well. In this context, and considering that the previous studies, especially in Finland, are dated and consequently were not carried out with modern versions of measurement devices, there is a need for a current investigation of the precision of individual tree dbh and height measurements in boreal forest conditions when caliper and clinometer measurements are used. The objective of this study was to determine the degree of precision for tree dbh and height measurements when four experienced mensurationists complete the same measurements on the same sample trees in conditions that are typical when collecting operational forest resource information in boreal Finland. In addition, we explore whether the use of dbh cross-measurements, where dbh is the mean of two dbh measurements from perpendicular directions (hereafter, dbh_{obs}), would have an effect on the precision of the measurements.

2. Materials and Methods

2.1. Study Area

The study area is located in Evo, southern Finland (61.19° N, 25.11° E) approximately 100 km north of Helsinki. The area consists of ~2000 ha of forest land with a range of stand conditions from intensively managed to natural southern boreal forests. The average stand size is a little less than 1 ha and the stands are mainly even-aged and single layer. The elevation in the area varies from 125 m to 185 m above the sea level. The dominant tree species in the area are Scots Pine (*Pinus sylvestris*, L.) and Norway spruce (*Picea Abies* (L.) H. Karst.), covering 44.7% and 33.5% of the total stem volume, respectively. From the total stem volume, 21.8% are deciduous species. The site type varies from groves to barren heaths.

2.2. Workflow for Sampling of the Trees to Be Measured

Field measurements were carried out in the study area during the summer of 2014. In total, 120 field sample plots (32 m × 32 m) were mapped and measured. The aim of the field measurements was to provide reference data for many remote sensing focused research activities slated for implementation. The locations of the field sample plots were selected based on canopy density and height information derived from airborne laser scanning (ALS) data. Finally, field crews measured individual tree attributes for all the trees inside the selected 120 field sample plots ($n = 9435$). For this specific study, we selected a sub-sample of trees out of these 120 sample plots. Sample trees ($n = 319$) were selected based on tree attribute measurements to retain the maximal variation in dbh (range between 5.6 cm and 46.6 cm) and height (range between 5.0 m and 33.1 m). These steps are described in detail in the following subsections.

2.2.1. Sample Plot Measurements

The sampling of the field sample plots was based on forest height- and canopy-density metrics derived from ALS-data; e.g., [44]. First, a systematic grid (32 m × 32 m) was placed over the study area and ALS metrics describing forest height and density were calculated for each grid cell from a canopy height model (CHM). ALS metrics included mean height of vegetation and proportional vegetation cover at the height of 2 m. Then, 120 field sample plots were selected to represent the entire height–density variation within the study area (Figure 1) excluding sapling stands and very sparse seed-tree locations. The sample plots were square, with dimensions of 32 m × 32 m and an area of 1024 m^2.

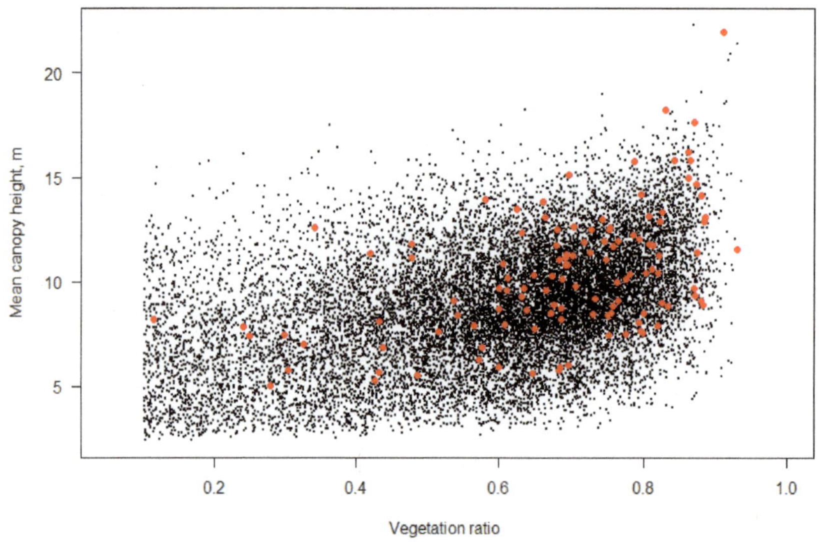

Figure 1. Variation in forest density (Vegetation ratio) and height (Mean canopy height) in the Evo study area. Black dots represent all possible field plots and the red dots describe the 120 field sample plots that were selected.

A preliminary stem map from the plot was used as guidance for the field measurements of tree attributes. The methods for producing the stem maps, based on terrestrial laser scanning (TLS) measurements, are described in more detail in [45,46]. The following variables were measured for all trees with a dbh larger than 5 cm within the field sample plots: tree species, dbh_{obs}, and height. A Haglöf Mantax Blue calipers and a Haglöf Vertex IV Ultrasonic (Haglöf Sweden AB, Långsele, Sweden) clinometer were used for measuring dbh_{obs} and height, respectively. The identity of the person doing the measurements was also recorded. The values for dbh_{obs} were obtained in millimetre scale using cross measurements with calipers. The mensurationist determined the height of 1.3 m above the base of the tree with a standardized 1.3 m long measure and then measured dbh from two directions perpendicular to each other on two sides of the tree (i.e., cross measurement) without any predetermined directions. Before tree height measurements, the clinometer was calibrated to current weather conditions using a measuring tape to measure the distance between the transponder and the clinometer. The clinometer was always calibrated when the mensurationist moved to a new plot or if the weather conditions changed.

First, the transponder was fastened at the stem of the tree that was the subject of the measurement at the height of 1.3 m. Then, the mensurationist selected the position of the measurement with distance to the tree being approximately the same as the tree height for optimal view of the whole tree. The mensurationist had to also pay attention to the visibility of the tree top when selecting the measurement position in a dense forest. After finding a suitable position, the mensurationist aimed the device first at the transponder and then to the tree top. The tree height was given by the device as a result of these distance and angle measurements. The average dbh_{obs} and height for all the trees on the sample plots was 19.1 cm (range between 5.0 cm and 73.1 cm) and 17.6 m (range between 1.4 m and 37.5 m), respectively.

2.2.2. Sample Trees

After measuring dbh_{obs} and height once for all the 9435 trees on 120 field plots, a sub-sample of the 319 trees was selected for re-measurement of the dbh_{obs} and height and to evaluate variability in the measurements. The sampling of the trees was based on tree species and dbh_{obs}. These sample trees included 156 Scots pine, 81 Norway spruce and 82 birch (*Betula pendula* and *B. pubescens*). The sample trees were measured independently by three other mensurationists to obtain four independent dbh_{obs} and height measures for each sample tree. Again, dbh_{obs} was based on cross measurements without any predefined measurement direction. All the mensurationists used measurement devices of same brand and model (i.e., Haglöf Mantax Blue calipers and Haglöf Vertex IV Ultrasonic clinometer) throughout the measurements. The measurement accuracy of the devices was tested with reference measurements and no differences were observed between the measurement devices used. The dbh_{obs} and height of the sample trees varied between 5.6 cm and 46.6 cm and from 5.0 m to 33.1 m, respectively (Table 1). These trees were located on 12 different field sample plots.

Table 1. The mean and standard deviation (std) of dbh_{obs} (diameter-at-breast-height) and height of the sample trees as well as the variation range of dbh and height among the sample trees.

Tree Species	*n*	mean dbh_{obs} (cm)	std dbh_{obs} (cm)	min dbh_{obs} (cm)	max dbh_{obs} (cm)	mean height (m)	std height (m)	min height (m)	max height (m)
Scots pine	156	20.1	6.9	6.2	46.6	17.7	4.3	5.0	32.1
Norway spruce	81	24.8	7.8	5.6	44.5	21.7	5.7	5.2	33.1
Birch	82	17.0	7.1	6.2	42.7	18.0	4.4	6.9	33.1
All sample trees	319	20.5	7.7	5.6	46.6	18.8	5.0	5.0	33.1

n = number of trees.

2.3. Evaluation of the Variance in the Field Measurements

Precision of dbh_{obs} and height measurements were analyzed by tree species, and dbh- and height-based tree size classes. The sample trees were divided into five classes containing comparably sized trees for analyzing the precision for dbh_{obs} and height measurements. The aim of this analysis was to investigate whether the variance in measurements changed between differently sized trees. For the dbh, the classes were 5.0–12.9 cm, 13.0–16.9 cm, 17.0–20.9 cm, 21.0–24.9 cm and over 25.0 cm. For the height, the classes were 5.0–14.9 m, 15.0–17.4 m, 17.5–19.9 m, 20.0–22.4 m and over 22.5 m. Classification of a tree into a size class was determined based on mean values of the four measurements. We tested whether the differences between the four measures were statistically significant on 95% level of significance using an analysis of variance (ANOVA). An ANOVA was performed for both dbh_{obs} and height measurements.

Precision was evaluated using absolute and relative standard deviations. The standard deviation s_{n-1} of four dbh_{obs} and height measurements for every sample tree was calculated using Equation (1).

$$s_{n-1} = \sqrt{\frac{\sum_{i=1}^{n}(x_i - \overline{x})^2}{n-1}} \quad (1)$$

where s_{n-1} is the unbiased estimator of standard deviation, n is the number of the measurements, x_i the measured value and $\overline{x}$ the mean of the measurements. The relative standard deviation $s_{\%}$ was calculated using Equation (2).

$$s_{\%} = \frac{s_{n-1}}{\overline{x}} \quad (2)$$

where s_{n-1} is the standard deviation and $\overline{x}$ the mean of the measurements.

The means of the standard deviations for both dbh_{obs} and height measurements of four separate measurements were calculated for the entire sample, as well as by tree species and for the dbh- and height-based size classes. We also investigated the size of the variation between the minimum and maximum value of the four dbh_{obs} measurements from each tree. The same investigation was done for the four height measurements. In addition, and to satisfy one of our stated objectives of this study, we investigated how much the standard deviation of dbh measurements varied if only one dbh measurement would have been taken by each mensurationist from all of the trees instead of the dbh_{obs} (i.e., cross measurements). For this part of the analysis, only the first of the two recorded dbh measurements from each mensurationist was used as the measured dbh value of the tree to simulate a scenario where each mensurationist recorded only one dbh value for each tree.

3. Results

Based on our analysis, there were no statistically significant differences between the measurements from the four members of the field crew for either dbh_{obs} or height ($\alpha = 0.05$). For dbh_{obs} measurements, it was assumed that the means of the dbh_{obs} measurements by all mensurationists (dbh_{obsm}) are equal to each other (Equation (3)).

$$\overline{dbh_{obs1}} = \overline{dbh_{obs2}} = \overline{dbh_{obs3}} = \overline{dbh_{obs4}} \quad (3)$$

The assumption of equality was supported by the results of the variance analysis ($p = 0.911$; Table 2). When performing the same analysis separately for tree species with the same assumptions of equal dbh_{obs}-means, the p-values were 0.952, 0.975 and 0.993 for pine, spruce, and birch, respectively.

Table 2. The results of ANOVA analysis for dbh_{obs} measurements for all sample trees as well as for different tree species with degrees of freedom (Df), sum of squares (Sum Sq), mean squared error (Mean Sq), F-value and p-value.

		dbh_{obs}			
Species	*Df*	Sum Sq	Mean Sq	*F*-value	*p*-value
All sample trees	3	3194	1064.8	0.179	0.911
Scots pine	3	1644	548	0.114	0.952
Norway spruce	3	1320	440.1	0.072	0.975
Birch	3	459	153.1	0.031	0.993

In measurements for tree height, it was assumed that the means of the height measurements by each mensurationist ($height_m$) were equal to each other:

$$\overline{height_1} = \overline{height_2} = \overline{height_3} = \overline{height_4} \quad (4)$$

The assumption of equality was supported by the variance analysis with a p-value of 0.140 (Table 3). The p-values for tree species with the same assumption were 0.342, 0.686 and 0.647 for pine, spruce, and birch, respectively.

Table 3. The results of ANOVA analysis for tree height measurements for all sample trees as well as for different tree species with degrees of freedom (*Df*), sum of squares (Sum Sq), mean squared error (Mean Sq), *F*-value and *p*-value.

		Height			
Species	*Df*	Sum Sq	Mean Sq	*F*-value	*p*-value
All sample trees	3	13,819	4606.3	1.832	0.140
Scots pine	3	6262	2087.2	1.117	0.342
Norway spruce	3	4844	1614.6	0.495	0.686
Birch	3	3280	1093.3	0.553	0.647

The largest difference between measured dbh_{obs} (i.e., 0.55 cm) was observed for spruce, whereas the largest difference between measured height values was observed for birch (0.82 m), but the differences were not significant (Table 4).

Table 4. Mean values of dbh_{obs} (diameter-at-breast-height, cross measurement) and height (h) measurements by different mensurationists.

		dbh_{obs} (cm)		
	Mensurationist 1	Mensurationist 2	Mensurationist 3	Mensurationist 4
Scots pine	19.95	20.03	20.09	20.38
Norway spruce	24.52	24.68	24.81	25.07
Birch	16.82	16.97	17.05	17.14
All sample trees	20.31	20.42	20.51	20.74
		h (dm)		
	Mensurationist 1	Mensurationist 2	Mensurationist 3	Mensurationist 4
Scots pine	17.91	18.02	17.83	17.21
Norway spruce	21.77	22.12	21.69	21.05
Birch	18.29	18.16	18.09	17.47
All sample trees	18.99	19.10	18.88	18.25

The mean dbh_{obs} and the mean height of all the sample trees were 20.5 cm and 18.8 m, respectively. The means of standard deviations for all dbh_{obs} and height measurements were 0.3 cm (1.5%) and 0.5 m (2.9%), respectively (Table 5). Analyzed tree-by-tree, the standard deviation in dbh_{obs} measurements varied from 0 cm to 1.0 cm and in height measurements from 0.1 m to 1.9 m (Figure 2). The relative standard deviation in the dbh_{obs} and height measurements was smallest for Norway spruce although the absolute standard deviation was smallest for birch in dbh_{obs} and Scots pine in height (Table 5).

Table 5. The species specific means of the standard deviations (std) calculated from the four independent dbh_{obs} (diameter-at-breast-height) and h (height) observations from each of the sample trees.

Species	*n*	**Std dbh_{obs} (cm)**	**Std dbh_{obs} (%)**	**Std h (m)**	**Std h (%)**
Scots pine	156	0.3	1.6	0.5	2.7
Norway spruce	81	0.3	1.3	0.6	2.6
Birch	82	0.3	1.5	0.7	3.6
All sample trees	319	0.3	1.5	0.5	2.9

The difference between the lowest and highest dbh_{obs} values from the same tree among the four measures varied in range from 0 cm to 2.1 cm (Figure 3). For 80.6% of the 319 trees measured, the range in variation remained less than 1.0 cm and for 94.0% less than 1.5 cm. For height, the difference between minimum and maximum values of the four individual measurements from the same tree was from 0.1 m to 4.2 m, respectively (Figure 3). For 73.3% of the trees, the largest difference within the height observations from the same tree was less than 1.5 m.

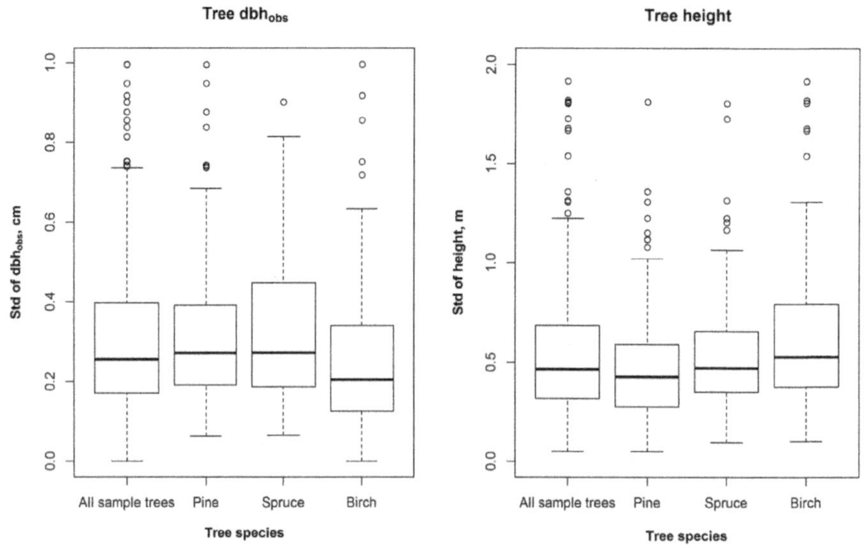

Figure 2. Boxplots demonstrating the variation range of the standard deviation (std) of tree dbh_{obs} (mean of cross measurements) and height measurements from the sample trees in cm and m, respectively. The standard deviations are presented for all the sample trees as well as separately for the tree species. In the figure, the bottom and the top of the box represent the first and third quartiles and the band inside the box is the median. The ends of the whiskers are within 1.5 interquartile of the lower and upper quartiles. The circles represent outliers.

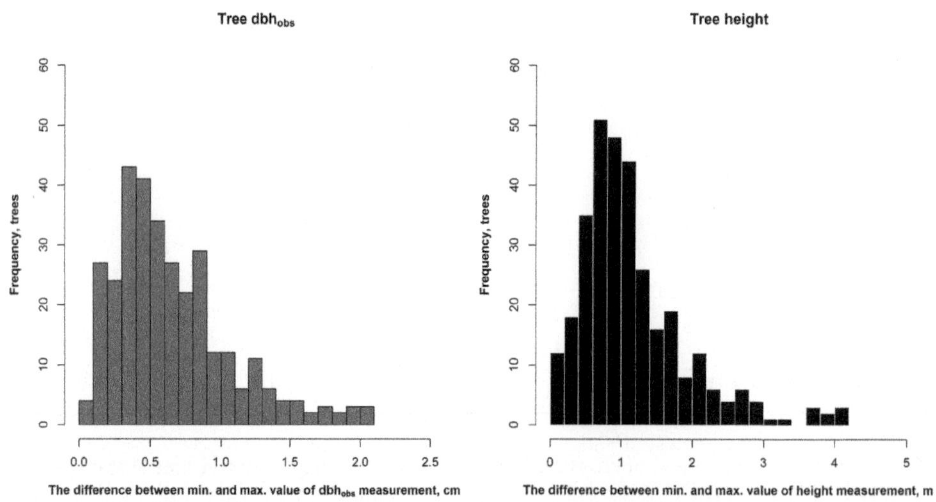

Figure 3. Histograms describing the frequency of largest difference among the sample tree measurements. On left: within the dbh_{obs} (mean of cross measurements) measurements from the same tree. On right: within the tree height measurements from the same tree.

When only the first dbh measurement from each mensurationist was used for the analysis instead of the dbh_{obs}, the standard deviation of the measurements was 0.5 cm (2.2%) (Figure 4). Likewise, the standard deviation was 0.5 cm for all tree species. The difference between the smallest and largest measured dbh value from the same tree varied from 0 cm to 6.6 cm; for 93.4% of the trees, the range of dbh measurements remained less than 2.0 cm.

The standard deviation of dbh_{obs} measurements varied between 0.2 cm and 0.4 cm (from 1.4% to 1.9%) in the five dbh-classes and between 0.2 cm and 0.4 cm (from 1.3% to 1.9%) within the five height classes (Tables 6 and 7). The absolute standard deviation values increased as the dbh increased. The reversed trend was detected with relative standard deviations (Table 6).

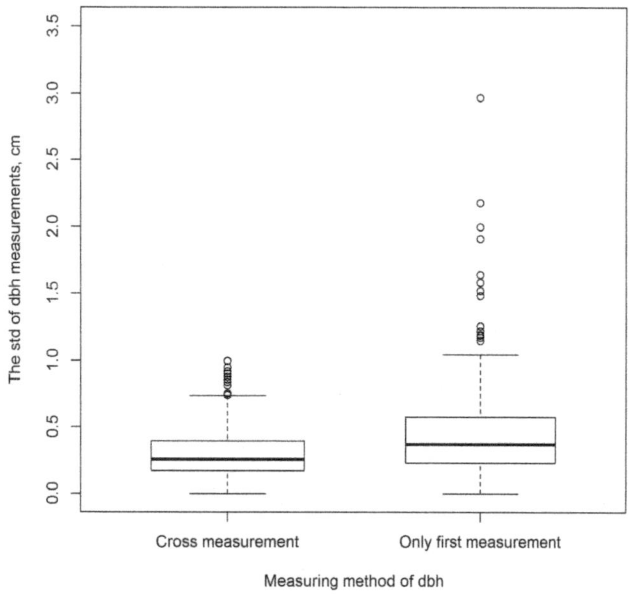

Figure 4. Boxplot describing the standard deviations (std) of sample tree dbh (diameter-at-breast-height) measurements with two different methods. On left, the standard deviation is a result of dbh values acquired using cross-measurements (i.e., dbh_{obs}), whereas on right, only one dbh measurement per mensurationist was used.

Table 6. The mean dbh_{obs} (mean of cross measurements) and h (height) of the five tree size classes that are based on the dbh, as well as the class-wise means of standard deviations (std_{obs}) calculated from the four observations of dbh_{obs} and h.

Dbh Classes							
		dbh_{obs}			h		
dbh classes (cm)	*n*	Mean (cm)	Std_{obs} (cm)	Std_{obs} (%)	Mean (m)	Std_{obs} (m)	Std_{obs} (%)
5.0–12.9	57	10.3	0.2	1.9	12.5	0.4	2.9
13.0–16.9	61	15.1	0.2	1.5	16.4	0.5	3.0
17.0–20.9	56	19.1	0.3	1.6	18.2	0.5	2.6
21.0–24.9	62	23.3	0.3	1.4	20.7	0.6	3.1
25.0+	83	30.3	0.4	1.4	23.9	0.7	2.8
All trees	319	20.5	0.3	1.5	18.8	0.5	2.9

Table 7. The mean dbh_{obs} (mean of cross measurements) and h (height) of five tree size classes that are based on tree height, as well as the class-wise means of standard deviations (std_{obs}) calculated from the four measurements of dbh_{obs} and h.

Height Classes							
		dbh_{obs}			h		
Height classes (m)	*n*	Mean (cm)	Std_{obs} (cm)	Std_{obs} (%)	Mean (m)	Std_{obs} (m)	Std_{obs} (%)
5.0–14.9	63	11.6	0.2	1.9	12.0	0.4	3.0
15.0–17.4	63	17.2	0.3	1.5	16.4	0.4	2.7
17.5–19.9	64	20.1	0.3	1.5	18.2	0.5	2.6
20.0–22.4	44	22.2	0.3	1.3	20.8	0.7	3.4
22.5+	85	29.0	0.4	1.4	25.0	0.7	2.9
All trees	319	20.5	0.3	1.5	18.8	0.5	2.9

The standard deviation of the height measurements varied between 0.4 m and 0.7 m (from 2.6% to 3.1%) within the five dbh classes and from 0.4 m to 0.7 m (from 2.6% to 3.4%) between the five height classes (Tables 6 and 7). The standard deviations of dbh_{obs} and height measurements by tree size are plotted in Figure 5.

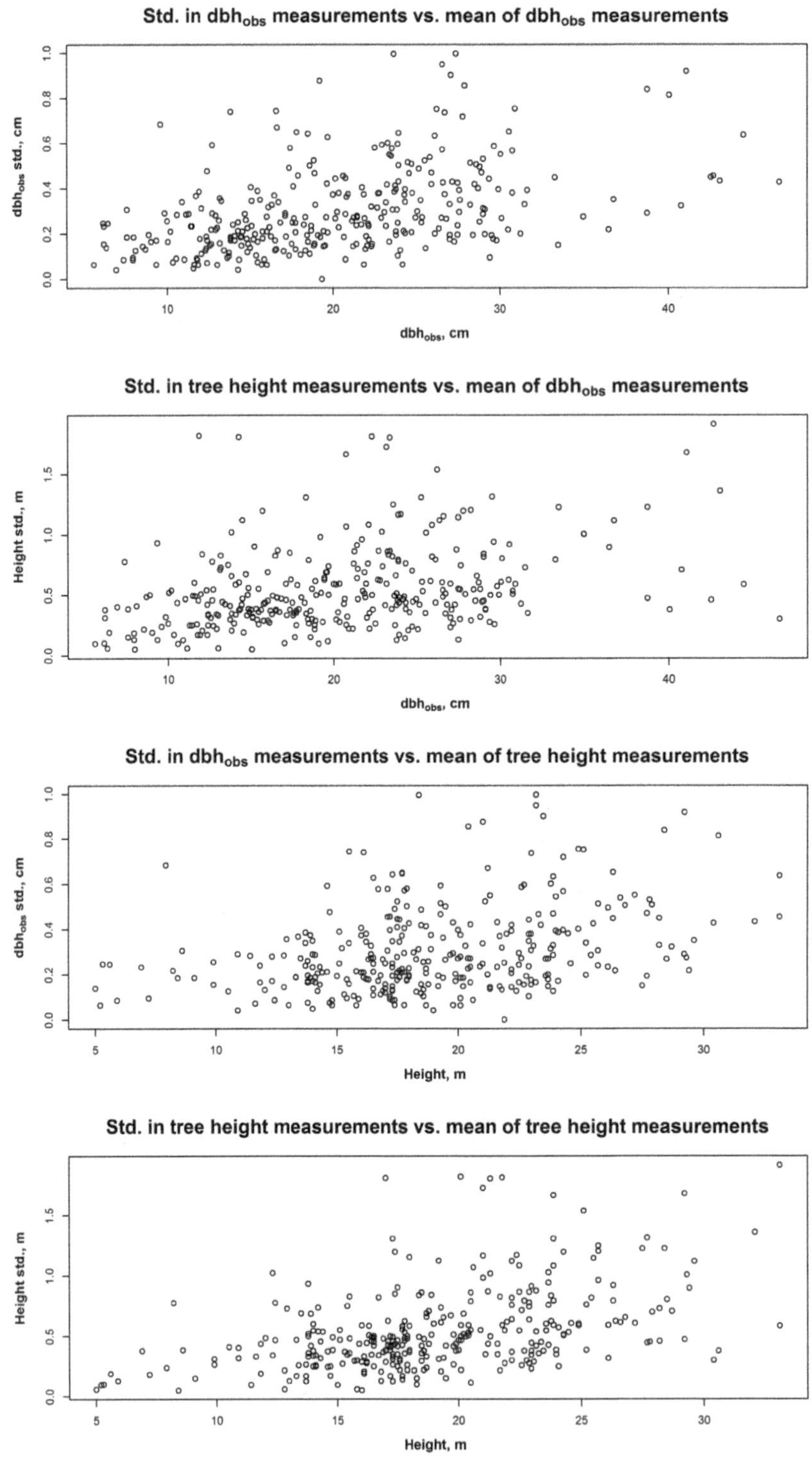

Figure 5. Standard deviation of tree dbh (diameter at the breast height) and tree h (height) measurements represented as the function of dbh and h of the sample trees.

4. Discussion

There is a need to characterize the precision of field measurements, such as individual tree dbh and height, using modern calipers and a clinometer. These measurement devices are routinely used for collecting forest resource information in boreal forests. Currently, close-range remote sensing techniques such as terrestrial and mobile laser scanning techniques are actively being developed in conjunction with photogrammetry to provide fully automatic sample plot or tree measurements. To assess accuracy, these and other remote sensing-based modes of measurement are often compared to traditional caliper and clinometer measurements of dbh and height. Thus, our aim with this study was to characterize the precision afforded by these conventional measures in order to provide baseline information for assessment of these new measurement tools and methods.

In order to describe the possible variation between the mensurationists, both the amount of sample trees and the number of mensurationists used in this study are large. In earlier studies, attention has been paid to one of these factors, e.g., Elzinga et al. [32] studied the observed variation of dbh measurements with 879 sample trees and two mensurationists, whereas McRoberts et al. [35] investigated the variation with 61 sample trees and eight or nine repeat measurements. The aim in this study was to have several mensurationists (4) and a large number of sample trees (319).

Although we found no statistically significant differences in dbh or height as measured by our four mensurationists, the results showed variation. For example, the largest difference between two cross-measured dbh measurements (i.e., dbh_{obs}) from the same tree was 2.1 cm. For tree height, the respective difference was 4.2 m. Despite the relatively large maximum differences, it should be noted that for 80.6% of the trees, the range in the dbh_{obs} measurements was less than 1.0 cm.

In dbh_{obs} measurements, there seemed to be very little variation over the entire data set, since the relative standard deviation was only 1.5% which is in line with the results of Hyppönen and Roiko-Jokela [33] who obtained 1.4% standard deviation in dbh measurements. Päivinen et al. [38] have also studied the precision of dbh measurements performed by several mensurationists; the standard deviation for dbh in their study was 0.69 cm (2.8%). In both of the previous studies mentioned, the number of unique trees in the data set was relatively small when compared to this study (n = 319 trees), since the data consisted of only 38 [33] and 64 [38] unique trees. In those studies, the measurements were then repeated several times, to get the total amount of dbh measurements to 540 and 520 measurements, respectively. In our study, spruce had the lowest relative standard deviation in dbh measurements which is supported by the findings of Päivinen et al. [38]. One possible reason for the greater precision found in this study can be related to the nature of spruce bark, since spruce bark is more homogenous than the outer surface of pine or birch trees. In [37], 95% of measured diameters were within 5 cm of the control measurements when the mean dbh was 52.7 cm, whereas Elzinga et al. [32] reported that in dbh measurements, errors greater than 5% may be expected in 5% of measurements. In general, the design of different studies may restrict the comparison of the results to each other if different kinds of measurement devices are used; if the diameter is acquired with single or cross measurement; or if the cardinal directions for caliper measurements are undefined, as it was in this study; or if the height of 1.3 m is marked on the tree or not.

The dbh values used in the analysis of this study are based on cross measurements, that is, two measurements from each tree from perpendicular directions. The results of this study indicate that even though cross measurements are slightly more time consuming than single caliper measurements, the greater precision they yield is worth the effort and the use of cross measurements is therefore recommended. The standard deviation of dbh measurements was approximately 1.5 times greater when using only one measurement from each tree (1.5% vs. 2.2%), and thus cross measurements improved precision notably.

Laser technology has also been used to measure dbh. In these studies, calipers have been used as a reference data and thus, comparison with these studies is not straightforward. The reported accuracy in dbh measurements has varied from 0.8 cm to 1.6 cm (standard error) with the laser-relascope [47], and from 0.88 cm to 1.43 cm with laser-dendrometers [48,49]. With camera-based systems, the accuracies have varied from 0.7 cm to 2.3 cm [30,50–52]. Terrestrial laser scanning (TLS) is also used for dbh measurements. Several studies have reported measurement accuracies (RMSE) between 0.9 cm and 8.9 cm [27,53–59]. Vastaranta et al. [60] compared several of the methods mentioned earlier and reported standard errors of 0.83 cm, 0.85 cm and 1.43 cm in dbh measurements with the TLS, laser-camera, and laser-relascope, respectively.

The standard deviation of 0.5 m (2.9%) reported for the height measurements is on the same level with the previous studies: Hyppönen and Roiko-Jokela [33] reported standard deviations of 0.8 m and 0.56 m for height measurements done with two different models of Suunto clinometers. Päivinen et al. [38] measured tree heights with Suunto clinometer as well, and reported standard deviation of 0.67 m. The advantage of our study is to report the precision of a modern clinometer, which is rather different from the viewpoint of a user than the clinometer used in earlier studies. The Vertex clinometer used in this study measures distance based on ultrasound, which allows flexible measuring distance. This is in contrast to the Suunto clinometer, which requires a fixed measuring distance. However, the Vertex clinometer requires calibration, and if a user fails to calibrate the instrument, the results may not be reliable and accurate. The laser-relascope has been found to measure the tree height with standard error between 0.49 m to 0.99 m [47,60]. In Omule [37], the variation in height measurements among the crew was reported to be 21.8% with a mean tree height of 32.3 m.

Based on our results, the precision of height measurements varied somewhat more between the three tree species than the dbh measurements did. For pine and spruce, the relative standard deviations in height measurements were within 0.1 percentage points (2.7% and 2.6%, respectively), but for birch, the value was 3.6%. Most probably, the reason for the lower precision of birch measurements is that when measuring the height of birch trees, observing the exact end point of the tree top can be more difficult than with the coniferous trees. In Päivinen et al. [38], the precision of pine, spruce and birch height measurements were 0.71 m, 0.55 m and 0.79 m, but no significant differences between the species were reported in the study.

We found that the size of the tree (dbh or height) did not impact the precision of field measurements (Figure 5). There was little variation in relative standard deviation for tree height between the groups of different sized trees in both dbh- and height-based size classes. However, the variation falls within one percentage point in both classifications and no trend can be seen from the results. Hyppönen and Roiko-Jokela [33] also noticed that the height of the tree had no effect on the standard deviation of tree height measurements.

In general, there was more imprecision in measuring tree height than dbh. According to Larjavaara and Muller-Landau [61], the tree height is difficult to measure accurately for standing trees in the field. When measuring the tree height, the mensurationist needs to operate from a distance, whereas for dbh measurements, the measurement is performed while standing next to the tree stem. When using clinometers for height measurements, it can also be difficult to see and define the exact end point of the tree top—especially in very dense forests or in other areas with restricted visibility. Moreover, there is always a risk of error from optical illusion if a tree is leaning, even if the mensurationist always selects the best possible measurement direction. Furthermore, there is always some randomness and variation when the measurements are performed by using steel calipers and by different mensurationists. Most importantly, for dbh measurements, different methods of defining the measuring height of 1.3 m above the base and the cardinal directions for the caliper cross-measurement cause variation between the observations. A standard measure can be used for determining the exact measuring height, as it was done in this study. In addition, the asymmetrical form of the tree stem and the occasional irregularity of the tree bark may cause differences in measurement results. Non-circular stem form was taken into account by utilizing two dbh measurements from perpendicular directions.

The measurements were carried out independently by the four mensurationist which could have caused bias as the sample trees were not identified together. Nevertheless, they had the stem maps of all sample plots for recognizing the sample trees to be measured. We did not determine one reference dbh and height value to which other measurements were compared. Therefore, the results do not reveal the possible bias in the measurements; however, this is a realistic scenario that is often encountered in operational context.

5. Conclusions

Four individual mensurationists each measured the same 319 trees in different plots using calipers and clinometers in order to characterize the precision of traditional field measurements of dbh and height. We found no statistically significant differences between independent measures of dbh or height. Overall, the standard deviation for dbh was 1.5%, whilst the standard deviation of measured height measures was 2.9%. It was possible to obtain greater precision in dbh than height measurements, while Norway spruce was measured with greater precision than Scots pine or birch. There were no clear differences in the precision of dbh or height measurements for different dbh- or height-based size classes. Based on the results presented herein, it can be stated that the traditional field-based methods tested in the study for tree dbh and height measurements provide precise and reliable information from forests. However, precision should be a stated objective, and must be explicitly considered when planning and implementing a field campaign. For example, in this study, we demonstrated that the use of dbh cross-measurements, instead of single dbh measurements, notably improved the precision of dbh measurements.

In addition, the low variation between repeated measurements from sample trees demonstrated that calipers and clinometers are not prone to outliers, which further supports the use of these traditional measuring tools as a useful benchmark for modern automatic or semi-automatic forest measurement applications and techniques. Calipers and clinometers remain an important source of forest measurement data and can be used to benchmark accuracy of the new forest measurement techniques, such as terrestrial laser scanning.

Acknowledgments: The study was funded by the Academy of Finland through the Centre of Excellence in Laser scanning research (project number 272195). The research leading to these results has also received funding from the European Community's Seventh Framework Programme ([FP7/2007-2013]) under grant agreement no. 606971. The authors would also like to thank Häme University of Applied Science for supporting our research activities at Evo study site.

Author Contributions: Ville Luoma was responsible for analyzing the data. Ninni Saarinen was responsible for the analysis and writing the article together with Ville Luoma. Michael A. Wulder and Joanne C. White wrote the paper together with Ville Luoma and Ninni Saarinen. Mikko Vastaranta designed the experiment together with Ville Luoma, Markus Holopainen and Juha Hyyppä. The article was improved by the contributions of all the co-authors at various stages of the analysis and writing process.

Conflicts of Interest: The authors declare no conflict of interest. The founding sponsors had no role in the design of the study; in the collection, analyses, or interpretation of data; in the writing of the manuscript, and in the decision to publish the results.

References

1. Husch, B.; Beers, T.; Kershaw, J., Jr. *Forest Mensuration*, 4th ed.; John willey & Sons, Inc.: Hoboken, NJ, USA, 2003.
2. Van Laar, A.; Akca, A. *Forest Mensuration*; Cuvillier Verlag: Goettingen, Germany, 1997.
3. Avery, T.E.; Burkhart, H.E. *Forest Measurements*; McGraw-Hill Cop.: Boston, NY, USA, 2002.
4. Clark, N.A.; Wynne, R.H.; Schmoldt, D.L. A review of past research on dendrometers. *For. Sci.* **2000**, *46*, 570–576.
5. Kellogg, R.M.; Barber, F.J. Stem eccentricity in coastal western Hemlock. *Can. J. For. Res.* **1981**, *11*, 715–718. [CrossRef]
6. Matérn, B. *On the Shape of the Cross-Section of a Tree Stem. An Empirical Study of the Geometry of Mensurational Methods*; Sveriges lantbruksuniversitet: Umea, Sweden, 1990.
7. Williamson, R.L. Out-of-roundness in douglas-fir stems. *For. Sci.* **1975**, *21*, 365–370.
8. Binot, J.-M.; Pothier, D.; Lebel, J. Comparison of relative accuracy and time requirement between the caliper, the diameter tape and an electronic tree measuring fork. *For. Chron.* **1995**, *71*, 197–200. [CrossRef]
9. Tallant, B.; Pelkki, M. A comparison of four forest inventory tools in southeast Arkansas. In *Competitiveness of Southern Forest Products Markets in a Global Economy: Trends and Predictions, Proceedings of the Southern Forest Economics Workshop, St. Augustine, FL, USA, 14–16 March 2004*; Alavalapati, J.R.R., Carter, D.R., Eds.; Available online: http://sofew.cfr.msstate.edu/papers/0304tallant.pdf (accessed on 31 January 2017)

10. Tomppo, E.; Kuusinen, N.; Mäkisara, K.; Katila, M.; McRoberts, R.E. Effects of field plot configurations on the uncertainties of ALS-assisted forest resource estimates. *Scand. J. For. Res.* **2016**. [CrossRef]
11. Andersen, H.-E.; Reutebuch, S.E.; McGaughey, R.J. A rigorous assessment of tree height measurements obtained using airborne lidar and conventional field methods. *Can. J. Remote Sens.* **2006**, *32*, 355–366. [CrossRef]
12. Andersen, H.-E.; McGaughey, R.J.; Reutebuch, S.E. Estimating forest canopy fuel parameters using lidar data. *Remote Sens. Environ.* **2005**, *94*, 441–449. [CrossRef]
13. Gobakken, T.; Næsset, E. Assessing effects of positioning errors and sample plot size on biophysical stand properties derived from airborne laser scanner data. *Can. J. For. Res.* **2009**, *39*, 1036–1052. [CrossRef]
14. Holmgren, J.; Nilsson, M.; Olsson, H. Simulating the effects of lidar scanning angle for estimation of mean tree height and canopy closure. *Can. J. Remote Sens.* **2003**, *29*, 623–632. [CrossRef]
15. Næsset, E. Practical large-scale forest stand inventory using a small-footprint airborne scanning laser. *Scand. J. For. Res.* **2004**, *19*, 164–179. [CrossRef]
16. Maltamo, M.; Eerikäinen, K.; Pitkänen, J.; Hyyppä, J.; Vehmas, M. Estimation of timber volume and stem density based on scanning laser altimetry and expected tree size distribution functions. *Remote Sens. Environ.* **2004**, *90*, 319–330. [CrossRef]
17. Popescu, S.C.; Wynne, R.H.; Nelson, R.F. Estimating plot-level tree heights with lidar: Local filtering with a canopy-height based variable window size. *Comput. Electron. Agric.* **2002**, *37*, 71–95. [CrossRef]
18. Vastaranta, M.; Wulder, M.A.; White, J.C.; Pekkarinen, A.; Tuominen, S.; Ginzler, C.; Kankare, V.; Holopainen, M.; Hyyppä, J.; Hyyppä, H. Airborne laser scanning and digital stereo imagery measures of forest structure: Comparative results and implications to forest mapping and inventory update. *Can. J. Remote Sens.* **2013**, *39*, 382–395. [CrossRef]
19. Näslund, M. Skogsförsöksanstaltens Gallringsförsök i Tallskog. *Medd. Stat. Skogsförsöksanst.* **1936**, *29*, 1–169. Available online: http://pub.epsilon.slu.se/10159/1/medd_statens_skogsforskningsanst_029_01.pdf (accessed on 31 January 2017).
20. Richards, F.J. A flexible growth function for empirical use. *J. Exp. Bot.* **1959**, *10*, 290–301. [CrossRef]
21. Sharma, M.; Parton, J. Heightdiameter equations for boreal tree species in ontario using a mixed-effects modeling approach. *For. Ecol. Manag.* **2007**, *249*, 187–198. [CrossRef]
22. Kangas, A.; Maltamo, M. *Forest Inventory: Methodology and Applications*; Springer Science & Business Media: Dordrecht, The Netherlands, 2006.
23. Laasasenaho, J. Taper Curve and Volume Functions for Pine, Spruce and Birch [*Pinus sylvestris, Picea abies, Betula pendula, Betula pubescens*]. *Commun. Inst. For. Fenn.* **1982**, *108*, 1–77. Available online: http://urn.fi/URN:ISBN:951-40-0589-9 (accessed on 31 January 2017).
24. Liang, X.; Hyyppa, J.; Kukko, A.; Kaartinen, H.; Jaakkola, A.; Yu, X. The use of a mobile laser scanning system for mapping large forest plots. *IEEE Geosci. Remote Sens. Lett.* **2014**, *11*, 1504–1508. [CrossRef]
25. Liang, X.; Kankare, V.; Yu, X.; Hyyppa, J.; Holopainen, M. Automated stem curve measurement using terrestrial laser scanning. *IEEE Trans. Geosci. Remote Sens.* **2014**, *52*, 1739–1748. [CrossRef]
26. Raumonen, P.; Kaasalainen, M.; Åkerblom, M.; Kaasalainen, S.; Kaartinen, H.; Vastaranta, M.; Holopainen, M.; Disney, M.; Lewis, P. Fast automatic precision tree models from terrestrial laser scanner data. *Remote Sens.* **2013**, *5*, 491–520. [CrossRef]
27. Hackenberg, J.; Morhart, C.; Sheppard, J.; Spiecker, H.; Disney, M. Highly Accurate Tree Models Derived from Terrestrial Laser Scan Data: A Method Description. *Forests* **2014**, *5*, 1069–1105. [CrossRef]
28. Liang, X.; Jaakkola, A.; Wang, Y.; Hyyppä, J.; Honkavaara, E.; Liu, J.; Kaartinen, H. The use of a hand-held camera for individual tree 3d mapping in forest sample plots. *Remote Sens.* **2014**, *6*, 6587–6603. [CrossRef]
29. Vastaranta, M.; Latorre, E.; Luoma, V.; Saarinen, N.; Holopainen, M.; Hyyppä, J. Evaluation of a smartphone app for forest sample plot measurements. *Forests* **2015**, *6*, 1179–1194. [CrossRef]
30. Molinier, M.; Lopéz-Sánchez, C.A.; Toivanen, T.; Korpela, I.; Corral-Rivas, J.J.; Tergujeff, R.; Häme, T. Relasphone-Mobile and Participative In Situ Forest Biomass Measurements Supporting Satellite Image Mapping. *Remote Sens* **2016**, *8*, 869. [CrossRef]
31. Pratihast, A.K.; DeVries, B.; Avitable, V.; De Bruin, S.; Lammert, K.; Tekle, M.; Herold, M. Combining Satellite Data and Community-Based Observations for Forest Monitoring. *Forests* **2014**, *5*, 2464–2489. [CrossRef]
32. Elzinga, C.; Shearer, R.C.; Elzinga, G. Observer variation in tree diameter measurements. *West J. Appl. For.* **2005**, *20*, 134–137.

33. Hyppönen, M.; Roiko-Jokela, P. Koepuiden Mittauksen Tarkkuus Ja Tehokkuus. *Fol. For.* **1978**, *356*, 1–25. Available online: http://urn.fi/URN:ISBN:951-40-0344-6 (accessed on 31 January 2017).
34. Johnson, J.E.; Haag, C.L. Reliability of height and diameter remeasurements on Red Pine (Pinus resinosa ait.) seedlings. *Tree Plant. Notes* **1985**, *36*, 27–29.
35. McRoberts, R.E.; Hahn, J.T.; Hefty, G.J.; Cleve, J.R.V. Variation in forest inventory field measurements. *Can. J. For. Res.* **1994**, *24*, 1766–1770. [CrossRef]
36. Melson, S.; Azuma, D.; Fried, J.S. A first look at measurement error on fia plots using blind plots in the pacific northwest. In Proceedings of the Third Annual Forest Inventory and Analysis Sumposium, St. Paul, MN, USA, 2002; McRoberts R.E., Reams G.A., van Deusen P.C., Moser J.W., Eds.
37. Omule, S.A.Y. Personal bias in forest measurements. *For. Chron.* **1980**, *56*, 222–224. [CrossRef]
38. Päivinen, R.; Nousiainen, M.; Korhonen, K.T. Puutunnusten Mittaamisen Luotettavuus. Accuracy of Certain Tree Measurements. *Folia For.* **1992**, *787*, 1–18. Available online: http://urn.fi/URN:ISBN:951-40-1197-X (accessed on 31 January 2017).
39. Kitahara, F.; Mizoue, N.; Yoshida, S. Effects of training for inexperienced surveyors on data quality of tree diameter and height measurements. *Silva Fenn.* **2010**, *44*, 657–667. [CrossRef]
40. Liu, S.; Bitterlich, W.; Cieszewski, C.J.; Zasada, M.J. Comparing the use of three dendrometers for measuring diameters at breast height. *South. J. Appl. For.* **2011**, *35*, 136–141.
41. Moran, L.A.; Williams, R. Field note—Comparison of three dendrometers in measuring diameter at breast height field note. *North. J. Appl. For.* **2002**, *19*, 28–33.
42. Williams, M.S.; Bechtold, W.A.; LaBau, V. Five instruments for measuring tree height: An evaluation. *South. J. Appl. For.* **1994**, *18*, 76–82.
43. Guillemette, F.; Lambert, M.-C. Relative effects of dendrometers on the estimation of diameter at breast height, stand basal area and stand volume in uneven-aged northern hardwoods. *For. Chron.* **2009**, *85*, 446–452. [CrossRef]
44. Maltamo, M.; Bollandsas, O.M.; Naesset, E.; Gobakken, T.; Packalen, P. Different plot selection strategies for field training data in als-assisted forest inventory. *Forestry* **2010**, *84*, 23–31. [CrossRef]
45. Saarinen, N.; Vastaranta, M.; Kankare, V.; Tanhuanpää, T.; Holopainen, M.; Hyyppä, J.; Hyyppä, H. Urban-tree-attribute update using multisource single-tree inventory. *Forests* **2014**, *5*, 1032–1052. [CrossRef]
46. Vastaranta, M.; Saarinen, N.; Kankare, V.; Holopainen, M.; Kaartinen, H.; Hyyppä, J.; Hyyppä, H. Multisource single-tree inventory in the prediction of tree quality variables and logging recoveries. *Remote Sens.* **2014**, *6*, 3475–3491. [CrossRef]
47. Kalliovirta, J.; Laasasenaho, J.; Kangas, A. Evaluation of the laser-relascope. *For. Ecol. Manag.* **2005**, *204*, 181–194. [CrossRef]
48. Parker, R.C.; Matney, T.G. Comparison of optical dendrometers for prediction of standing tree volume. *South. J. Appl. For.* **1999**, *23*, 100–107.
49. Skovsgaard, J.P.; Johannsen, V.K.; Vanclay, J.K. Accuracy and precision of two laser dendrometers. *Forestry* **1998**, *71*, 131–139. [CrossRef]
50. Ashley, M.; Roger, R. Tree heights and upper stem diameters. *Photogramm. Eng.* **1969**, *35*, 136.
51. Bradshaw, F. Upper stem diameter measurements with the aid of 35 millimeter photographs. *Aust. For. Res.* **1972**, *6*, 17–20.
52. Varjo, J.; Henttonen, H.; Lappi, J.; Heikkonen, J.; Juujärvi, J. Digital Horizontal Tree Measurements for Forest Inventory. *Working Papers of the Finnish Forest Research Institute* **2006**, *40*, 1–23. Available online: http://www.metla.fi/julkaisut/workingpapers/2006/mwp040.htm (accessed on 31 January 2017).
53. Henning, J.G.; Radtke, P.J. Detailed stem measurements of standing trees from ground-based scanning lidar. *For. Sci.* **2006**, *52*, 67–80.
54. Huang, H.; Li, Z.; Gong, P.; Cheng, X.; Clinton, N.; Cao, C.; Ni, W.; Wang, L. Automated methods for measuring DBH and tree heights with a commercial scanning lidar. *Photogramm. Eng. Remote Sens.* **2011**, *77*, 219–227. [CrossRef]
55. Kankare, V.; Vauhkonen, J.; Tanhuanpää, T.; Holopainen, M.; Vastaranta, M.; Joensuu, M.; Krooks, A.; Hyyppä, J.; Hyyppä, H.; Alho, P.; et al. Accuracy in estimation of timber assortments and stem distribution—A comparison of airborne and terrestrial laser scanning techniques. *ISPRS J. Photogramm. Remote Sens.* **2014**, *97*, 89–97. [CrossRef]

56. Liang, X.; Hyyppä, J. Automatic stem mapping by merging several terrestrial laser scans at the feature and decision levels. *Sensors* **2013**, *13*, 1614–1634. [CrossRef] [PubMed]
57. Maas, H.G.; Bienert, A.; Scheller, S.; Keane, E. Automatic forest inventory parameter determination from terrestrial laser scanner data. *Int. J. Remote Sens.* **2008**, *29*, 1579–1593. [CrossRef]
58. Simonse, M.; Aschoff, T.; Spiecker, H.; Thies, M. Automatic Determination of Forest Inventory Parameters Using Terrestrial Laser Scanning. In Proceedings of the ScandLaser Scientific Workshop on Airborne Laser Scanning of Forests, Umeå, Sweden, 3–4 Sptember 2003; Swedish University of Agricultural Sciences, Department of Forest Resource Management and Geomatics: Umeå, Sweden, 2003; pp. 252–258.
59. Tansey, K.; Selmes, N.; Anstee, A.; Tate, N.J.; Denniss, A. Estimating tree and stand variables in a corsican pine woodland from terrestrial laser scanner data. *Int. J. Remote Sens.* **2009**, *30*, 5195–5209. [CrossRef]
60. Vastaranta, M.; Melkas, T.; Holopainen, M.; Kaartinen, H.; Hyyppä, J.; Hyyppä, H. Laser-Based Field Measurements in Tree-Level Forest Data Acquisition. *Photogramm. J. Finl.* **2009**, *21*, 51–61. Available online: https://pdfs.semanticscholar.org/881a/2577fd02a960b7f2f3dbe825932f79cb198f.pdf (accessed on 31 January 2017).
61. Larjavaara, M.; Muller-Landau, H.C. Measuring tree height: A quantitative comparison of two common field methods in a moist tropical forest. *Method Ecol. Evol.* **2013**, *4*, 793–801. [CrossRef]

8

Soil Microbial Communities in Natural and Managed Cloud Montane Forests

Ed-Haun Chang [1], Guanglong Tian [2] and Chih-Yu Chiu [3,*]

[1] Mackay Junior College of Medicine, Nursing and Management, Beitou, Taipei 112, Taiwan; s108@mail.mkc.edu.tw

[2] Environmental Monitoring and Research Division, Monitoring and Research Department, Metropolitan Water Reclamation District of Greater Chicago, 6001 W. Pershing Road, Cicero, IL 60804, USA; TianG@mwrd.org

[3] Biodiversity Research Center, Academia Sinica, Nangang, Taipei 11529, Taiwan

* Correspondence: bochiu@sinica.edu.tw

Abstract: Forest management often results in changes in soil microbial communities. To understand how forest management can change microbial communities, we studied soil microbial abundance and community structure in a natural *Chamaecyparis* (NCP) forest, a disturbed *Chamaecyparis* (DCP) forest, a secondary (regenerated) *Chamaecyparis* (SCP) forest and a secondary (reforested) *Cryptomeria* (SCD) forest. We analyzed soil microbial abundance by measuring phospholipid fatty acids (PLFAs) and microbial community structure by denaturing gradient gel electrophoresis (DGGE) in the studied forest soils. The content of the soil PLFA fungal biomarker decreased from NCP to SCP, DCP and SCD forest soils, associated with the degree of disturbance of forest management. The ratio of soil Gram positive–to-negative bacteria and the stress index (16:1ω7t to 16:1ω7c) increased from NCP to SCP and DCP soils; thus, disturbed forests except for SCD showed increased soil microbial stress. Principal component analysis of soil microbial groups by PLFAs separated the four forest soils into three clusters: NCP, DCP and SCP, and SCD soil. The DGGE analysis showed no difference in the microbial community structure for NCP, DCP and SCP soils, but the community structure differed between SCD and the three other forest soils. In cloud montane forests, disturbance due to forest management had only a slight influence on the soil microbial community, whereas reforestation with different species largely changed the soil microbial community structure.

Keywords: PLFA; DGGE; reforestation; microbial community; forest management

---

1. Introduction

Soil microorganisms are essential for maintaining soil fertility and plant growth because they play important roles in nutrient cycling and availability [1]. The soil microbial community is vital for a forest ecosystem to maintain long-term sustainability. Forest management such as tree harvesting and replanting can change the soil microbial community and the soil physico-chemical properties, such as the C and N contents, and biochemical activities [2–5]. These effects are attributed to changes in litter quality, root exudates, and nutrient uptake with forest management [6,7].

Chamaecyparis cypress, including *C. formosensis* and *C. obtusa*, can grow to 60 m and live for more than 1000 years; the giant timber ranks at the top in quality. *Chamaecyparis* species are valuable in eastern Asia. Before large-scale logging in Taiwan 100 years ago, *Chamaecyparis* forests were widely distributed in the cloud montane area, about 800 m to 2800 m above sea level. The timber of *Chamaecyparis* is rich in essential oils, and with its rigid intrinsic structure, the wood can resist decay for decades, leading to a large accumulation of organic matter on the soil surface. Large numbers of fallen trees also remain intact on the floor in a natural *Chamaecyparis* forest.

Although soil bacteria in the natural *Chamaecyparis* forest have low diversity because of very high soil acidity and low nutrient fluxes under the perhumid conditions [8], the high soil organic matter (SOM) in this forest supports high soil microbial biomass. Studies have indicated that management of this forest by removing fallen wood or replanting increased soil bacterial diversity [3]. Such changes could be related to the increase in litter decomposition due to disturbance with wood removal or a change in litter quality with the introduced different tree species, possibly leading to an increase in nutrient availability to bacteria. However, little is known about what changes occur in the whole microbial community under such forest management (wood removal and replanting).

Wood harvesting increases the soil exposure, thus increasing the soil temperature fluctuation in addition to the loss of SOM. The reforestation of the natural forest site can lead to the loss of SOM accumulated over the long-term. The soil microbial community might need time to adapt to the litter of new tree species. Therefore, forest management such as tree removal and replanting/reforestation in a cloud montane forest can disturb the forest ecosystem and stress the existing soil microbial community. Thus, we hypothesized that forest management (wood removal and/or replanting) of a natural *Chamaecyparis* forest would decrease the soil microbial abundance and change the soil microbial community structure.

Molecular methods such as DNA-based methods [9,10] and phospholipid fatty acid (PLFA) analysis [2,7] have been used to determine changes in soil microbial community structure resulting from forest management practices on different coniferous species [11]. Particularly, some ratios of PLFAs such as trans/cis (16:1ω7t/16:1ω7c) and cyclo/monounsaturated precursor (cy17:0/16:1ω7c and cy19:0/18:1ω7c) can be used to indicate stress or starvation of microbial communities [12]. Thus, we aimed to examine soil microbial biomass and microbial community structure in forests of *Chamaecyparis* with and without disturbance and forests reforested with *Chamaecyparis* and another species by measuring PLFAs and by denaturing gradient gel electrophoresis (DGGE).

2. Materials and Methods

2.1. Site Description

The study was conducted in the Yuanyang Lake ecosystem (24°35′ N, 121°24′ E) from 1700 m to 2220 m, located in the north-central part of Taiwan. The annual precipitation is about 4000 mm and annual mean temperature approximately 12 °C. The vegetation is dominated by Hinoki cypress (*Chamaecyparis obtusa* Sieb. & Zucc. var. *formosana* (Hayata) Rehder) and Taiwan false cypress trees (*C. formosensis* Matsum.) with understory evergreen broadleaf shrubs predominantly *Rhododendron formosanum*. The natural *Chamaecyparis* forest generally has a large number of fallen trees and a thick layer of undecomposed plant residues.

This study included a natural *Chamaecyparis* (NCP) forest in the lake area and three nearby disturbed forests, located less than 5 km away from the NCP site. The three disturbed forests reflected the increased disturbance along with the forest management for wood harvest and reforestation. Because large numbers of fallen trees remain intact on the floor of the *Chamaecyparis* forest, collecting fallen trees is considered an alternative way to harvest these valuable trees and minimize disturbance to the forest. For establishing one disturbed forest, dead wood and fallen trees were removed about 15 years ago; we call this disturbed *Chamaecyparis* (DCP) forest. The second disturbed forest was the secondary (regenerated) *Chamaecyparis* (SCP) forest after natural *Chamaecyparis* trees were logged 50 years ago. The third disturbed forest was a secondary Japanese cedar-*Cryptomeria japonica* (L. f.) D. Don (SCD) forest reforested 30 years ago after a clear-cut of natural *Chamaecyparis* trees. Japanese cedar was introduced from Japan and widely planted in middle elevation in Taiwan.

Surface soils (0–10 cm) were randomly collected by using a soil corer (3 cm in diameter) from the four forests. Five replicate plots (50 m × 50 m) were selected in each forest. In each plot, three cores were taken. After gentle homogenization and removal of roots and litter, moist soil was sieved through a 2 mm sieve. Soil samples were stored at 4 °C in the dark. Microbial properties, including microbial biomass were analyzed within one month of field collection. Portions of soil samples were freeze-dried at −20 °C immediately after sampling for PLFA analysis and DGGE. Soil samples for chemical analysis were air-dried and milled.

2.2. Biochemical Assays

Soil microbial biomass C and N content (C_{mic} and N_{mic}) was determined in fresh moist soil samples by the chloroform fumigation extraction method [13]. Dissolved organic C in the extracted solution was measured by use of a total organic C analyzer (Model 1010 O.I. Analytical, College Station, TX, USA). Soil microbial N (N_{mic}) was determined by measuring ninhydrin-reactive N as described [14].

Moist soil was oven-dried for 72 h at 105 °C to determine soil moisture content. Soil organic carbon (SOC) and total N contents were determined by use of a Fisons NA1500 elemental analyzer (ThermoQuest Italia, Milan, Italy). Soil pH values were measured in a 1:2.5 soil-to-water suspension.

Extraction and analysis of PLFAs was performed as described [15]. PLFAs were identified and analyzed as described [16]. Fatty acid nomenclature was applied according to Frostegård et al. [15]. The position of the double bond is defined by symbol ω from the methyl end of the molecule. The prefix c refers to cis configuration. Iso and anteiso branching are designated by the prefix i or a, respectively. The prefix cy refers to cyclopropane fatty acids. The fatty acids i15:0, a15:0, 15:0, i16:0, 16:1ω7c, 17:0, i17:0, cy17:0, 18:1ω7c, and cy19:0 represent bacteria; 18:2ω6 represents fungi; 16:1ω5c represents vesicular arbuscular mycorrhizae (VAM) fungi; i15:0, a15:0, i16:0, and i17:0 represent Gram-positive (G+) bacteria; 16:1ω7c, cy17:0, 18:1ω7c, and cy19:0 represent Gram-negative (G−) bacteria; and 10Me18:0 represents actinobacteria [17,18].

Soil DNA was extracted by using the PowerSoil DNA isolation kit (MO BIO lab, Solana Beach, CA, USA). The 16S rDNA and 18S rDNA genes were amplified with bacterial universal primers F968-GC/R1401 [19] and fungal universal primers NS1/GC-Fung, respectively [20]. DGGE involved use of the Dcode Universal Mutation Detection System (Bio-Rad, Hercules, CA, USA) as follows: PCR products were separated on a 6% polyacrylamide gel and 50% to 70% denaturants for bacterial analysis and 7% polyacrylamide and 20% to 45% denaturants for fungal analysis. After running for 16 h at 60 V, the gels were stained with 1:10,000 Gelstar for 30 min and captured by a Gel Doc XR gel imaging system (Bio-Rad).

2.3. Statistical Analysis

Data from biochemical and chemical analysis were converted to an oven-dried basis. Statistical analysis involved ANOVA and Duncan's multiple range test. Principal component analysis (PCA) was used to test relative concentrations (mole percent of total PLFAs) of individual fatty acids. All statistical analyses involved use of SPSS v18.0 (SPSS Inc., Chicago, IL, USA). $p < 0.05$ was considered statistically significant. Hierarchical cluster analysis of the DGGE banding pattern from the forest sites involved use of the unweighted pair-group method by Ward (Quantity One v4.6, Bio-Rad, Hercules, CA, USA).

3. Results

3.1. Soil Characteristics and Microbial Biomass

The soil pH ranged from 3.5 to 3.7, with no significant difference among forests (Table 1). The SOC content was lower in the reforested *Cryptomeria* (SCD) forest than in all *Chamaecyparis* forests, and it did not differ among the three *Chamaecyparis* forests (NCP, DCP and SCP). In addition, the soil total N content was lower in SCD than *Chamaecyparis* soil. The soil C-to-N ratio was lower in SCD and DCP than NCP and SCP forests. The soil microbial biomass C (C_{mic}) content was highest in the NCP and

SCP forests, followed by DCP, then SCD forests. The soil microbial biomass N (N_{mic}) content was highest in SCP and NCP forests, followed by DCP, then SCD forests.

Table 1. Soil chemical properties and microbial biomass under natural and managed forest.

Forest Type [1]	pH	Organic C (%)	Total N (%)	C/N	C_{mic} ($mg \cdot kg^{-1}$)	N_{mic} ($mg \cdot kg^{-1}$)
NCP	3.5 a [2]	48.6 a	1.89 b	23.1 a	4931 a	728 ab
DCP	3.7 a	44.3 a	2.28 a	18.9 b	3601 b	619 b
SCP	3.6 a	46.9 a	1.97 b	25.5 a	4839 a	845 a
SCD	3.7 a	19.7 b	1.16 c	16.5 b	1542 c	264 c

[1] NCP: natural *Chamaecyparis* forest, DCP: disturbed *Chamaecyparis* forest with wood removal, SCP: secondary (regenerated) *Chamaecyparis* forest, SCD: secondary (reforested) *Cryptomeria* forest; [2] Values in each column followed by different letters are significantly different at $p < 0.05$ by Duncan's multiple range test.

3.2. PLFA Analyses

Total PLFA, bacteria, fungi, and actinobacteria content differed among different forests (Table 2). The natural forest (NCP) soil had the highest total PLFA, G− bacteria and VAM fungi content. The NCP soil also had the highest fungi content, followed by SCP, then DCP and SCD soil. The ratio of G+/G− soil bacteria was the highest in DCP soil, followed by SCP and SCD, then NCP soil. The ratio of 16:1ω7t (trans-unsaturated fatty acid) to 16:1ω7c (cis-unsaturated fatty acid) was lower in SCD and NCP than DCP and SCP soil.

Table 2. Content of phospholipid acid biomarkers ($nmol \cdot g^{-1}$ soil) and ratios of biomarkers in natural and managed forest soil.

Forest Type [1]	Total PLFAs	Bacteria	Fungi	VAM Fungi	Actino-Bacteria	G+	G−	G+/G−	16:1ω7t/16:1ω7c
NCP	550 a [2]	240 ab	110 a	21.4 a	34.7 b	71.2 b	57.7a	1.24 c	0.117 b
DCP	511 ab	268 a	64.5 c	15.1 b	42.7 a	97.3 a	46.5 b	2.12 a	0.225 a
SCP	445 c	197 b	80.5 b	15.7 b	32.6 b	68.1 b	39.4 b	1.75 b	0.280 a
SCD	483 bc	232 ab	51.7 c	10.8 c	32.7 b	71.2 b	49.8 ab	1.43 bc	0.099 b

[1] NCP: natural *Chamaecyparis* forest, DCP: disturbed *Chamaecyparis* forest with wood removal, SCP: secondary (regenerated) *Chamaecyparis* forest, SCD: secondary (reforested) *Cryptomeria* forest; [2] Values in each column followed by different letters are significantly different at $p < 0.05$ by Duncan's multiple range test.

3.3. Microbial Community Patterns by Forest Management

The first principal component (PC1) and second principal component (PC2) together accounted for 61.6% of the variation in PLFAs in the soil (Figure 1). The loadings in PC1 were contributed mainly by G+ bacteria and actinobacteria (i15:0, i17:0; 10Me16:0, 10Me18:0; negative loadings) and fungi (18:2ω6C, 18:1ω9c; positive loadings) (Figure 1). The loading in PC2 was contributed mainly by G− bacteria (16:1ω7c, 18:1ω7c; negative loading). Soil microbial communities, as analyzed by the PCA of PLFA levels, significantly differed among the forest types and could be divided into three large clusters: one for NCP soil, another for DCP and SCP soils, and the third for SCD soil. In this study, PC1, which explained 38.1% of the PLFA variability, separated NCP and SCP (positive loadings) from DCP and SCD soil (negative loadings). PC2, which explained 23.5% of the PLFA variability, separated SCD (negative loadings) from DCP and SCP soil (positive loadings) (Figure 1).

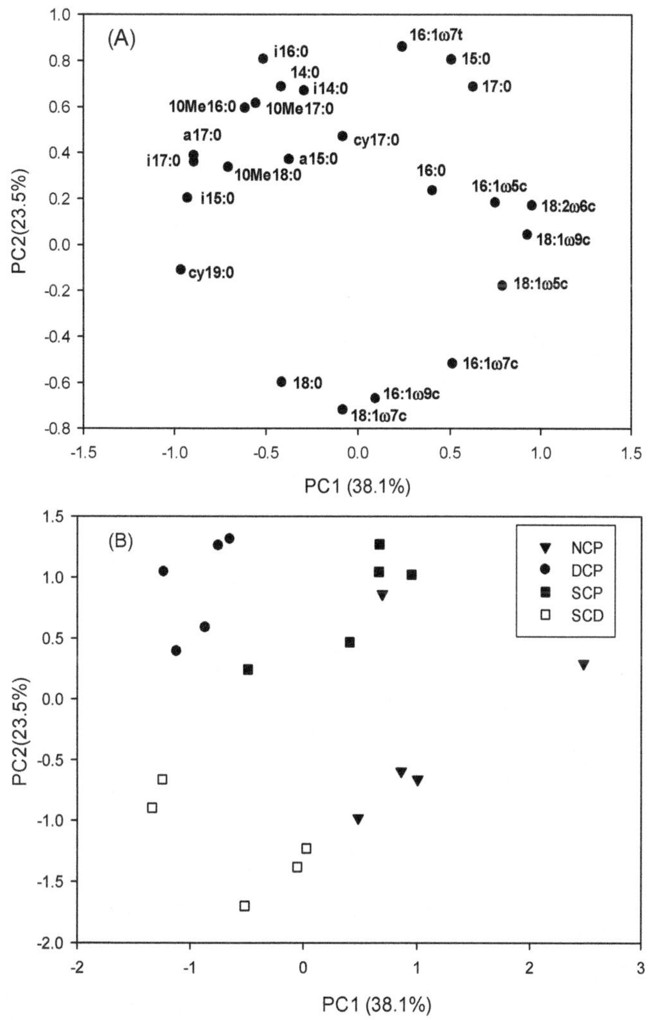

Figure 1. Principle component analysis of microbial phospholipid fatty acids (mole percent of total PLFAs) in natural and managed forest soil. (**A**) Loadings of fatty acid distribution in two PCs; (**B**) Sample distribution by the first two principal components (PC). (NCP, natural *Chamaecyparis* forest; DCP, *Chamaecyparis* forest with wood removal; SCP, secondary (regenerated) *Chamaecyparis* forest; SCD, secondary (reforested) *Cryptomeria* forest).

3.4. DGGE Analysis

Dendrograms of soil bacterial similarity obtained by DGGE analysis showed similarities and differences among the different forest types (Figure 2). The entire samples from the four forest sites could be broadly divided into two groups: SCD soil as one group and NCP, DCP and SCP soils as the other group. Cluster analysis by fungal communities showed the same two major groups, SCD soil versus NCP, DCP and SCP soils (Figure 3). In this study, the similarity between NCP and SCP soil for fungal community was 35%.

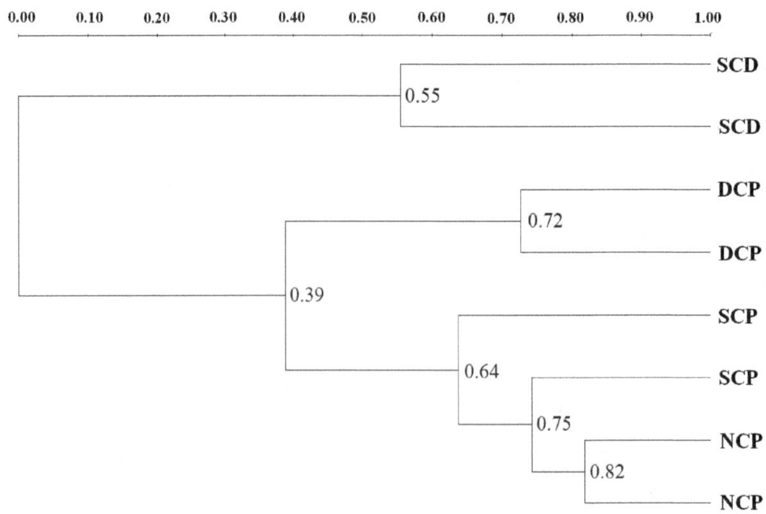

Figure 2. Cluster analysis of forest types by soil bacterial community.

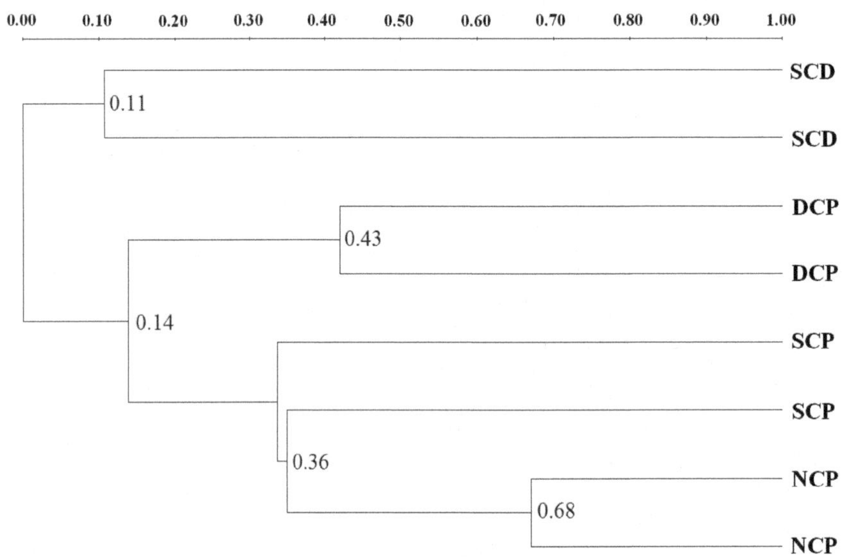

Figure 3. Cluster analysis of forest types by soil fungal community.

4. Discussion

PLFA and DGGE analyses revealed that wood removal and reforestation in *Chamaecyparis* forest ecosystems decreased soil microbial biomass, but only reforestation altered the soil microbial community structure. The natural forest (NCP) had a large number of fallen trees that remained intact on the floor, which reasonably supported the highest soil microbial biomass among the forests studied. The wood removal in the DCP forest might decrease the shading by the overstory and increase the soil temperature, thereby promoting the decomposition of organic matter by microorganisms [21]. The loss of SOM in the DCP forest decreased the soil microbial biomass. The forest clearing caused SOM loss in the regenerated *Chamaecyparis* forest and the reforested cedar forest. Ichikawa et al. [22] reported decreased SOM when a natural broadleaved forest was converted to a *Chamaecyparis obtusa* and *Cryptomeria japonica* forest. The large decrease in SOC by reforestation explained the lowest soil microbial biomass in the SCD forest. The disturbed *Chamaecyparis* forest and reforested cedar forest showed a loss in the soil fungal and VAM fungal abundance, which could be attributed to the decrease in the thickness of the humus layer [23].

The abundance of soil G− bacteria was reported to increase with high SOM content and high substrate availability [24,25]. Thus, the low ratio of G+/G− bacteria in the NCP forest we observed may be due to the better growth of G− bacteria supported by substrate-rich conditions in natural *Chamaecyparis* soils. This result agrees with findings by Lin et al. [3] for the same site that β-Proteobacteria was the most abundant group in the undisturbed NCP forest and Acidobacteria the most abundant in other disturbed forests. The Proteobacteria are a major group of G− bacteria that are copiotrophic and use more C sources [26]. In contrast with Proteobacteria, Acidobacteria are considered an oligotrophic group and exist in nutrient-limited environments [26]. Slow-growing specialists, such as G+ bacteria, feature an effective cell metabolism and effectively use recalcitrant substances such as cellulose and lignin in a coniferous environment [27]. Heipieper et al. [28] showed an increased ratio of 16:1ω7t to 16:1ω7c isomer along with stress and starvation conditions for soil bacteria. The higher ratio of 16:1ω7t to 16:1ω7c in disturbed *Chamaecyparis* forest than in undisturbed NCP forest soils therefore suggests the stress conditions in disturbed *Chamaecyparis* forest and supports our finding of a high ratio of G+/G− bacteria in disturbed *Chamaecyparis* forest soils.

The abundance of G− bacteria was higher in the SCD forest than in other disturbed *Chamaecyparis* forests (SCP and DCP). Sakai et al. [29] indicated that the C accumulation rate was much greater in Japanese cedar than in Hinoki cypress (*Chamaecyparis*) after afforestation. In this study, the SCD forest soil had the lowest C/N ratio (Table 1). Ollinger et al. [30] indicated that a lower C/N ratio of organic matter favors its mineralization in forests. Thus, the 30-year-old SCD forest may provide more easily degradable C to G− bacteria than other *Chamaecyparis* forests.

Some studies have shown that changes in vegetation with replanting of forests may lead to shifts in the soil microbial community structure [31,32]. Our PCA analysis of PLFA content clearly distinguished *Chamaecyparis* from *Cryptomeria* forest soil in PC1. PCR-DGGE dendrograms of soil bacteria and fungi showed primary differences between microbial communities related to the change in plant species. Priha et al. [33] showed that the quality not the quantity of SOM affected the soil microbial community structure. Therefore, the forest type would be an important factor affecting the microbial community structure in these ecosystems. The soil microbial community structure did not differ between disturbed and natural forests in our study, which does not mean that the disturbance in forest management cannot alter the soil microbial community: such a difference might occur with greater disturbance, which is worthy of further investigation.

5. Conclusions

Forest management practices, including wood harvesting and forest conversion, could affect the soil microbial community in the cloud montane forest. The soil microbial abundance was decreased with the conversion of a natural *Chamaecyparis* forest to a secondary *Chamaecyparis* forest, the removal of fallen trees in a *Chamaecyparis* forest, and reforestation, largely due to SOM loss. The disturbance to the *Chamaecyparis* forest led to the increased G+/G− bacteria ratio. However, the change in the soil microbial community structure by forest management in this study was only observed in forest reforested with different tree species. Despite the decrease in the soil microbial abundance due to disturbance, the soil microbial structure in the disturbed and secondary *Chamaecyparis* forest did not differ from the natural *Chamaecyparis* forest. The conversion of natural *Chamaecyparis* forests should be managed carefully to avoid a negative impact on the soil microbial community structure and the need for a long-term restoration. The study implies that the choice of plant species for reforestation is critical to the soil microbial community in the cloud montane forest.

Acknowledgments: This work was funded by the Ministry of Science and Technology, Taiwan (MOST 105-2313-B-436-001) and by Academia Sinica.

Author Contributions: E.H.C. and C.Y.C. conceived and designed the experiments; E.H.C. performed the experiments and analyzed the data; E.H.C. wrote the original draft; G.T. and C.Y.C. reviewed and edited the manuscript.

Conflicts of Interest: The authors declare no conflict of interest.

References

1. Nannipieri, P.; Ascher, J.; Ceccherini, M.T.; Landi, L.; Pietramellara, G.; Renella, G. Microbial diversity and soil functions. *Eur. J. Soil Sci.* **2003**, *54*, 655–670. [CrossRef]
2. Ushio, M.; Kitayama, K.; Balser, T.C. Tree species-mediated spatial patchiness of the composition of microbial community and physicochemical properties in the topsoils of a tropical montane forest. *Soil Biol. Biochem.* **2010**, *42*, 1588–1595. [CrossRef]
3. Lin, Y.T.; Jangid, K.; Whitman, W.B.; Coleman, D.C.; Chiu, C.Y. Soil bacterial communities in native and regenerated perhumid montane forests. *Appl. Soil Ecol.* **2011**, *47*, 111–118. [CrossRef]
4. Lucas-Borja, M.E.; Candel, D.; Jindo, K.; Moreno, J.L.; Andres, M.; Bastida, F. Soil microbial community structure and activity in monospecific and mixed forest stands, under Mediterranean humid conditions. *Plant Soil* **2012**, *354*, 359–370. [CrossRef]
5. Chang, E.H.; Chen, T.H.; Tian, G.; Hsu, C.K.; Chiu, C.Y. Effect of 40 and 80 Years of conifer regrowth on soil microbial activities and community structure in subtropical low mountain forests. *Forests* **2016**, *7*, 244. [CrossRef]
6. Grayston, S.J.; Prescott, C.E. Microbial communities in forest floors under four tree species in coastal British Columbia. *Soil Biol. Biochem.* **2005**, *37*, 1157–1167. [CrossRef]
7. Yin, K.; Zhang, L.; Chen, D.; Tian, Y.; Zhang, F.; Wen, M.; Yuan, C. Understory herb layer exerts strong controls on soil microbial communities in subtropical plantations. *Sci. Rep.* **2016**, *6*, 27066. [CrossRef] [PubMed]
8. Lin, Y.T.; Lin, C.P.; Chaw, S.M.; Whitman, W.B.; Coleman, D.C.; Chiu, C.Y. Bacterial community of very wet and acidic subalpine forest and fire-induced grassland soils. *Plant Soil* **2010**, *332*, 417–427. [CrossRef]
9. Suzuki, C.; Nagaoka, K.; Shimada, A.; Takenaka, M. Bacterial communities are more dependent on soil type than fertilizer type, but the reverse is true for fungal communities. *Soil Sci. Plant Nutr.* **2009**, *55*, 80–90. [CrossRef]
10. Vetrovsky, T.; Baldrian, P. Analysis of soil fungal communities by amplicon pyrosequencing: Current approaches to data analysis and the introduction of the pipeline SEED. *Biol. Fert. Soils* **2013**, *49*, 1027–1037. [CrossRef]
11. Demoling, F.; Nilsson, L.O.; Bååth, E. Bacterial and fungal response to nitrogen fertilization in three coniferous forest soils. *Soil Biol. Biochem.* **2008**, *40*, 370–379. [CrossRef]
12. Frostegård, Å.; Tunlid, A.; Bååth, E. Use and misuse of PLFA measurements in soils. *Soil Biol. Biochem.* **2011**, *43*, 1621–1625. [CrossRef]
13. Vance, E.D.; Brookes, P.C.; Jenkinson, D.S. An extraction method for measuring soil microbial biomass C. *Soil Biol. Biochem.* **1987**, *19*, 703–707. [CrossRef]
14. Amato, M.; Ladd, J.N. Assay for microbial biomass based on ninhydrin-reactive nitrogen in extracts of fumigated soils. *Soil Biol. Biochem.* **1988**, *20*, 107–114. [CrossRef]
15. Frostegård, A.; Tunlid, A.; Bååth, E. Phospholipid fatty acid composition, biomass, and activity of microbial communities from two soil types exposed to different heavy metals. *Appl. Environ. Microbiol.* **1993**, *59*, 3605–3617. [PubMed]
16. Chang, E.H.; Chen, T.H.; Tian, G.; Chiu, C.Y. The effect of altitudinal gradient on soil microbial community activity and structure in moso bamboo plantations. *Appl. Soil Ecol.* **2016**, *98*, 213–220. [CrossRef]
17. Zogg, G.P.; Zak, D.R.; Ringleberg, D.B.; MacDonald, N.W.; Pregitzer, K.S.; White, D.C. Compositional and functional shifts in microbial communities due to soil warming. *Soil Sci. Soc. Am. J.* **1997**, *61*, 475–481. [CrossRef]
18. Zelles, L. Fatty acid patterns of phospholipids and lipopolysaccharides in the characterisation of microbial communities in soil: A review. *Biol. Fert. Soils* **1999**, *29*, 111–129. [CrossRef]
19. Nübel, U.; Engelen, B.; Felske, A.; Snaidr, J.; Wieshuber, A.; Amann, R.I.; Ludwig, W.; Backhaus, H. Sequence heterogeneities of genes encoding 16S rRNAs in Paenibacillus polymyxa detected by temperature gradient gel electrophoresis. *J. Bacteriol.* **1996**, *178*, 5636–5643. [CrossRef] [PubMed]
20. White, T.J.; Bruns, T.; Lee, S.; Taylor, J.W. Amplification and Direct Sequencing of Fungal Ribosomal RNA Genes for Phylogenetics. In *PCR Protocols: A Guide to Methods and Applications*; Innis, M.A., Gelfand, D.H., Sninsky, J.J., White, J.J., Eds.; Academic Press: New York, NY, USA, 1990; pp. 315–322.

21. Williamson, J.R.; Neilsen, W.A. The influence of forest site on rate and extent of soil compaction and profile disturbance of skid trails during ground-based harvesting. *Can. J. For. Res.* **2000**, *30*, 1196–1205. [CrossRef]
22. Ichikawa, T.; Yamaguchi, T.; Takahashi, T.; Asano, Y. Effects of the conversion of the forest management type from natural deciduous broad-leaved forests to artificial Japanese cypress (*Chamaecyparis* obtusa) and Japanese cedar (*Cryptomeria japonica*) forests on soil microbial flora and mineralization characteristics of organic carbon. *Jpn. J. For. Environ.* **2003**, *45*, 81–87.
23. Mahmood, S.; Finlay, R.D.; Erland, S. Effects of repeated harvesting of forest residues on the ectomycorrhizal community in a Swedish spruce forest. *New Phytol.* **1999**, *142*, 577–585. [CrossRef]
24. Zelles, L.; Bai, Q.Y.; Beck, T.; Beese, F. Signature fatty acids in phospholipids and lipopolysaccharides as indicators of microbial biomass and community structure in agricultural soils. *Soil Biol. Biochem.* **1992**, *24*, 317–332. [CrossRef]
25. Bossio, D.A.; Scow, K.M.; Gunapala, N.; Graham, K.J. Determinants of soil microbial communities: Effects of agricultural management, season, and soil type on phospholipid fatty acid profiles. *Microb. Ecol.* **1998**, *36*, 1–12. [CrossRef] [PubMed]
26. Fierer, N.; Mark, A.B.; Robert, B.J. Toward an ecological classification of soil bacteria. *Ecology* **2007**, *88*, 1354–1364. [CrossRef] [PubMed]
27. Kourtev, P.S.; Ehrenfeld, J.G.; Haggblom, M. Experimental analysis of the effect of exotic and native plant species on the structure and function of soil microbial communities. *Soil Biol. Biochem.* **2003**, *35*, 895–905. [CrossRef]
28. Heipieper, H.J.; Meulenbeld, G.; VanOirschot, Q.; de Bont, J.A.M. Effect of environmental factors on the trans/cis ratio of unsaturated fatty acids in Pseudomonas putida S12. *Appl. Environ. Microbiol.* **1996**, *62*, 2773–2777. [PubMed]
29. Sakai, H.; Inagaki, M.; Noguchi, K.; Sakata, T.; Yatskov, M.A.; Tanouchi, H.; Takahashi, M. Changes in soil organic carbon and nitrogen in an area of Andisol following afforestation with Japanese cedar and Hinoki cypress. *Soil Sci. Plant Nutr.* **2010**, *56*, 332–343. [CrossRef]
30. Ollinger, S.V.; Smith, M.L.; Martin, M.E.; Hallett, R.A.; Goodale, C.L.; Aber, J.D. Regional variation in foliar chemistry and N cycling among forests of diverse history and composition. *Ecology* **2002**, *83*, 339–355.
31. Zornoza, R.; Guerrero, C.; Mataix-Solera, J.; Scow, K.M.; Arcenegui, V.; Mataix-Beneyto, J. Changes in soil microbial community structure following the abandonment of agricultural terraces in mountainous areas of Eastern Spain. *Appl. Soil Ecol.* **2009**, *42*, 315–323. [CrossRef] [PubMed]
32. Burns, J.H.; Anacker, B.L.; Strauss, S.Y.; Burke, D.J. Soil microbial community variation correlates most strongly with plant species identity, followed by soil chemistry, spatial location and plant genus. *AoB Plants* **2015**, *7*, plv030. [CrossRef] [PubMed]
33. Priha, O.; Grayston, S.J.; Pennanen, T.; Smolander, A. Microbial activities related to C and N cycling and microbial community structure in the rhizospheres of *Pinus sylvestris*, *Picea abies* and *Betula pendula* seedlings in an organic and mineral soil. *Fems Microb. Ecol.* **1999**, *30*, 187–199. [CrossRef]

Assessing Wildland Fire Risk Transmission to Communities in Northern Spain

Fermín J. Alcasena [1,*], Michele Salis [2], Alan A. Ager [3], Rafael Castell [4] and Cristina Vega-García [1,5]

1. Agriculture and Forest Engineering Department (EAGROF), University of Lleida, Alcalde Rovira Roure 191, Lleida 25198, Spain; cvega@eagrof.udl.cat
2. Euro-Mediterranean Center on Climate Change (CMCC), IAFES Division, Via Enrico De Nicola 9, Sassari 07100, Italy; michele.salis@cmcc.it
3. USDA Forest Service, Rocky Mountain Research Station, 72510 Coyote Road, Pendleton, OR 97801, USA; aager@fs.fed.us
4. Bomberos de Navarra, Agencia Navarra de Emergencias (ANE), Calle Aoiz 35 bis 3°, Pamplona 31004, Spain; rcastels@navarra.es
5. Forest Sciences Centre of Catalonia, Carretera de Sant Llorenç de Morunys km 2, Solsona 25280, Spain

* Correspondence: ferminalcasena@eagrof.udl.cat

Abstract: We assessed potential economic losses and transmission to residential houses from wildland fires in a rural area of central Navarra (Spain). Expected losses were quantified at the individual structure level (n = 306) in 14 rural communities by combining fire model predictions of burn probability and fire intensity with susceptibility functions derived from expert judgement. Fire exposure was estimated by simulating 50,000 fire events that replicated extreme (97th percentile) historical fire weather conditions. Spatial ignition probabilities were used in the simulations to account for non-random ignitions, and were estimated from a fire occurrence model generated with an artificial neural network. The results showed that ignition probability explained most of spatial variation in risk, with economic value of structures having only a minor effect. Average expected loss to residential houses from a single wildfire event in the study area was 7955€, and ranged from a low of 740 to the high of 28,725€. Major fire flow-paths were analyzed to understand fire transmission from surrounding municipalities and showed that incoming fires from the north exhibited strong pathways into the core of the study area, and fires spreading from the south had the highest likelihood of reaching target residential structures from the longest distances (>5 km). Community firesheds revealed the scale of risk to communities and extended well beyond administrative boundaries. The results provided a quantitative risk assessment that can be used by insurance companies and local landscape managers to prioritize and allocate investments to treat wildland fuels and identify clusters of high expected loss within communities. The methodological framework can be extended to other fire-prone southern European Union countries where communities are threatened by large wildland fires.

Keywords: wildland urban interface; wildfire simulation modeling; wildfire risk transmission; community fireshed

---

1. Introduction

Most wildfires that cause human fatalities and losses to property occur in the rapidly expanding interface areas between wildlands and human development [1,2]. This area where residential and other infrastructures intermingle with flammable vegetation is widely known as wildland–urban interface (WUI) or rural–urban interface (RUI) [3,4]. While the former definition is mainly used for predominantly wildland vegetation areas surrounding developed areas, the latter is most commonly used in Mediterranean landscapes where fuels have been influenced by human activities for millennia [5,6]. In areas lacking sharp transitions between development and wildlands, where structures are surrounded by hazardous fuels, the term intermix has been used to describe the juxtaposition of fuels and dwellings [7]. In all cases, a number of factors have contributed to wildfire losses in developed areas (hereafter WUI), including urban expansion, increased fuel loadings from expansion of shrub and forest vegetation into abandoned agricultural lands, and suburban sprawl over metropolitan agricultural belts [8–10]. Likewise, fire suppression policies have contributed to a buildup of fuels in and around developed areas, resulting in higher hazard within developed areas and structure ignition [11]. Wildland fires in the WUI are a growing concern at global scales due to escalating losses to life and property [12,13], and have become a priority for wildfire management policies in many fire-prone areas [14].

Previous efforts on WUI wildfire risk characterization in Mediterranean landscapes have emphasized the importance of flammable vegetation surrounding communities [15], since fuel loadings are directly related to fire intensity and structure loss [16]. Aggregation of dwellings (isolated, grouped and urban center) combined with vegetation types or land covers have been proposed as a WUI classification system to inform risk and vulnerability assessments [17,18]. Other studies have focused on ignition likelihood to measure wildfire risk [19–21]. The vast majority of fires in the Mediterranean basin are caused by humans [22–24], and most fire-occurrence modeling studies include explanatory variables to describe human activities, such as population density, accessibility (e.g., distance to roads, distance to railways, distance to forest tracks) and human activities [25,26]. However, neither of these previous approaches account for the likelihood of loss from large fires (e.g., 5000–50,000 ha) that ignite at some distant location and spread to urban development. Thus, low fire ignition probability close to a WUI area does not necessarily translate to low burn probability, and vice versa. Moreover, fire intensity can substantially vary depending on fire weather and fire front spreading direction [27,28].

To better account for the spatial scale of wildfire risk to human communities, a growing number of researchers have employed wildfire simulation methods [29,30]. Both burn probability and fire intensity in the home ignition zone (HIZ, the immediate 30–60 m-buffer area around dwellings) [11] can be estimated by simulating a large number of fires (e.g., 10^4–10^5) to assess wildfire exposure from large fires [31,32]. These estimates can be then used in risk assessments to quantify the potential socioeconomic impacts, including expected net value change on residential structures [27,28,33]. While it is generally agreed that higher wildfire exposure results in larger losses in the WUI, variability in structure susceptibility and economic valuation can substantially affect risk estimates at the scale of individual dwellings. For instance, high overall exposure levels can be mitigated by construction materials and structure design [34]. These differences in construction can be incorporated into risk assessments using different susceptibility relationships [35,36]. Simulation studies can also be used to understand the scale of risk to communities to help identify responsible landowners [37,38]. For instance, using wildfire transmission analysis, fire effects on valued resources can be traced back to the ignition location [39], and landscape planning to reduce hazardous fuels can then target these areas for fuel treatments [40].

In this paper, we assess potential wildfire economic losses and transmission to residential houses in the rural communities of Juslapeña Valley, northern Spain. We used simulation modeling to map the source of wildfire exposure to communities and estimated the expected financial loss at the scale of individual structures. The simulation modeling incorporated a fine scale ignition probability grid developed from historical fire locations. Simulation outputs were used to estimate a number of exposure metrics, including burning probability and fire intensity. We estimated expected loss in the community using wildfire exposure metrics combined with a structure susceptibility function. The later was generated by a panel of local experts using an interactive structured communication technique. We also conducted a transmission analysis to delineate community firesheds and understand the source of wildfire exposure to communities. The methods provide a number of new ways to examine wildfire exposure to communities that can inform wildfire protection and improve fire resiliency in rural–urban interface areas in the Mediterranean region.

2. Material and Methods

2.1. Study Area

The study area is located in the Juslapeña Valley, central Navarra (Spain), 18 km north of the city of Pamplona (Figure 1A). The Juslapeña Valley is a 31.63 km^2 municipality with 548 inhabitants dispersed among 14 small rural villages or councils (minimum administrative division). The climate is transitional Mediterranean with annual rainfall around 1000 mm, a water shortage period from July to September corresponding to the wildfire season, and average maximum temperatures over 30 °C in the warmest month (meteo.navarra.es). The landscape is a mosaic of dryland cereal crops covering the valley bottom, mesoxerophytic pastures with shrubby edgings on marginal agricultural lands (*Genista scorpius* L., *Juniperus communis* L., *Buxus sempervirens* L., and *Prunus spinosa* L.), downy oak (*Quercus pubescens* Mill.) forests on south facing slopes (replaced by *Quercus ilex* L. in shallow soil foothills), beech (*Fagus sylvatica* L.) forests on high elevation north facing slopes, and scattered stands of black pine (*Pinus nigra* Arn.) [41]. Land management is largely conditioned by ownership. Most forests and natural herbaceous pastures are council common lands and agricultural fields are owned by local inhabitants. Community housing is located at mid-slopes, usually surrounded by agricultural lands and orchards at the front southern side, and forested lands arrive closer at the back (Figure 1B). We focused our analysis on residential houses (n = 306 structures), and we did not consider other structures or constructions such as agricultural warehouses. In the study area, there are no industrial sites or sport-recreational facilities. The largest observed wildfires are characterized as fast-spreading one-day summer events with less than 1000 ha burned (e.g., Juslapeña Fire in 2009 and San Cristobal Fire in 2001). Most fires are caused by humans, while lightning represents only 5% of ignitions (1985 to 2013 fire records; mapama.gob.es).

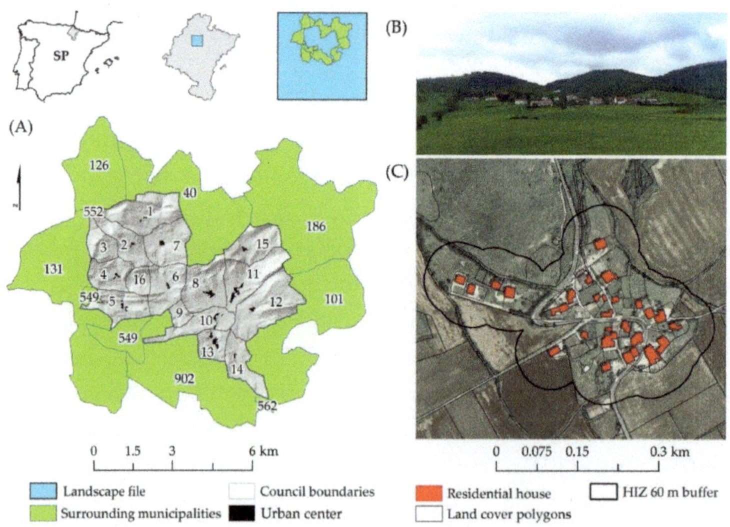

Figure 1. Location of the Juslapeña Valley (3163 ha) in central Navarra (Spain) (**A**). The numbers refer to the regional cadaster council polygon (1 to 16) and municipality code (**B**). The 36,000-ha wildfire modeling domain framed by the landscape file (LCP) encompassing the study area had a wider extension to the south to account for incoming fires from the fire-prone areas of central Navarra. Land covers in the cultural landscapes present sharp edges in vegetation (**B** urban center of the council No. 8). Detailed cartography and cadaster polygons (scale 1/5000) were used to generate surface fuel maps (sigpac.navarra.es) and locate residential houses (catastro.navarra.es) (**C**). The HIZ is the 60-m buffer around structures [11], and was calculated for each residential house to conduct this study. In the figure (**C**) we show the external HIZ contour of the residential houses in urban center of the council No. 8.

2.2. *Wildfire Simulation*

We gathered multiple datasets and geospatial inputs for this modeling approach (Figure 2). We simulated wildfire spread and behavior (fire size, burned area polygons, flame length probabilities, and conditional burn probability) within a 36,000-ha fire modeling domain. Overall, we conducted separated simulations for the most frequent extreme weather conditions of the wildfire season, thus obtaining different sets of modeling outputs. All the output raster grids were obtained at modeling resolution. Details are presented below in the following sections (Table 1).

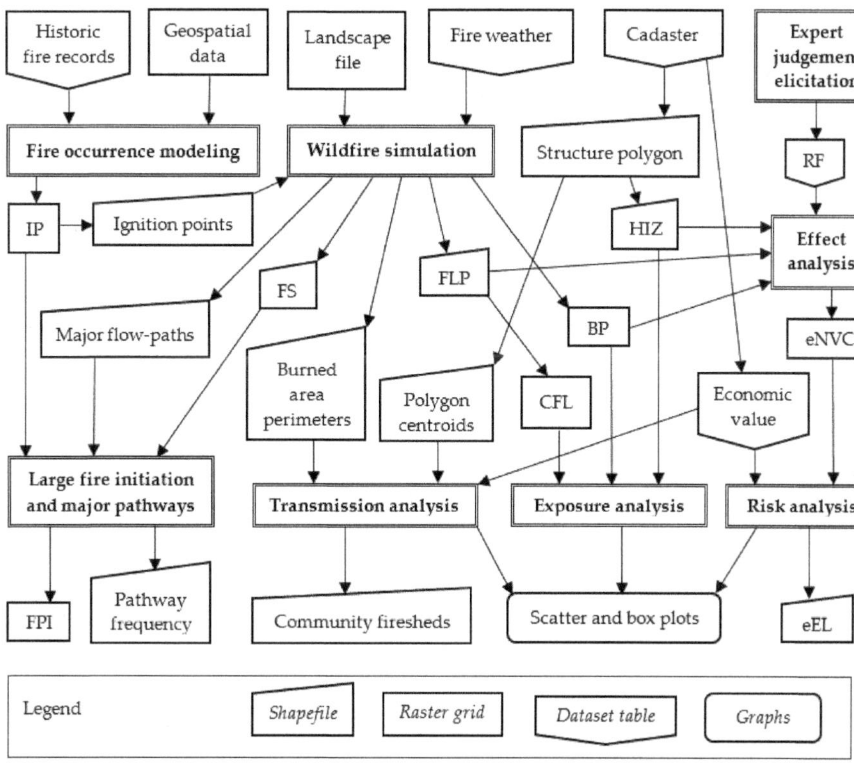

Figure 2. Wildfire simulation and analysis process summary flowchart. Wildfire simulation requires fire weather, landscape and fire ignition input data. Fire initiation, transmission, exposure and risk analysis use different fire modeling outputs. Exposure and risk analyses were conducted at individual structure HIZ level. Results were presented in maps or graphics. See Table 1 for the abbreviations.

Table 1. Summary table with the abbreviations used in this study for the main geospatial inputs, modeling outcomes and analysis results. The terms are described and contextualized for the use in this study. We provide further details in the following sections.

Name (Abbreviation)	Description and Use
Home ignition zone (HIZ)	Area surrounding structures within a 30–60 m-buffer [11]. HIZ was used to assess wildfire exposure and risk on the individual residential houses located in the study area.
Ignition probability (IP)	Fire occurrence probability grid (0–1) generated by artificial neural network analysis [26] of historical ignition locations. It was used to calculate FPI and generate the simulated fire ignition locations.
Fire size (FS)	Fire size (ha) resulting from each individual simulated wildfire. Fire size is output from simulations along with the ignition location. It was combined with IP to generate FPI.

Table 1. *Cont.*

Name (Abbreviation)	Description and Use
Fire potential index (FPI)	Is the grid generated with FS and IP, and it was used to identify large fire initiation areas [31]. The FPI provides spatially explicit valuable information to target anthropic fire ignition prevention priority areas on fire-prone landscapes.
Flame length probability (FLP)	Probability of a fire of a specific flame length given that a pixel burns under the simulated conditions. FLP is output in 0.5-m classes and sums to 1 for a given pixel. A distribution of flame lengths is generated for each pixel since fires can arrive as heading, flanking or backing fires.
Conditional flame length (CFL)	Probability-weighted flame length (m) calculated from the FLP output. CFL was summarized for the HIZ to estimate wildfire hazard and exposure to residential houses [35].
Burn probability (BP)	Number of times a pixel burns as a proportion of the total number of simulated fires (0–1). BP average values for each HIZ were used to estimate wildfire likelihood and assess wildfire exposure to residential houses [35].
Response function (RF)	The susceptibility of structures as a function of flame length represented by percent value loss (%) [42]. It was obtained from expert judgment [35].
Expected net value change (eNVC)	Expectation of *gain* or *loss* in values expressed on a percentage basis (%) [28]. Derived from combining burn probability, intensity, and susceptibility functions to estimate expected change on a percentage basis for structures [27]. Only expected losses were considered in the study.
Expected economic loss (eEL)	Expected loss expressed specifically in economic values (€) given a fire ignition and spread at assumed extreme fire weather conditions. Quantified as the product of the cadaster value of the structures and the average eNVC within the HIZ.

2.2.1. Landscape File and Fire Weather Input Data

We compiled the complete set of input data as required by the FlamMap fire simulator [43], including landscape file (LCP) and wildfire season extreme fire weather data. The LCP is a gridded frame containing the characteristics of the terrain, surface fuels and canopy fuel metrics. The terrain (aspect, slope and elevation) was derived from 5-m resolution digital terrain model raster data (ign.es). Standard fuel models [44,45] were assigned to 1/5000 scale land use land cover considering species composition, shrub cover and forest growth stage (idena.navarra.es and sigpac.navarra.es) (Figure 3). Canopy metrics (canopy height, canopy base height, canopy bulk density and canopy cover), were derived from low density LiDAR data (0.56 returns m^2; ign.es) using FUSION [31,46]. The surface fuel and canopy metric characterization and required raster grid generation were detailed in previous studies [47,48]. The LCP was assembled at 20-m resolution [49] and comprised a 36,000-ha fire modeling domain (Figure 1A). Extreme fire weather conditions were derived using Fire Family Plus [50] from the hourly records of the Pamplona automatic weather station (1999 to 2015 records; meteo.navarra.es), as the 97th percentile ERC-G fuel moisture content [51] and wildfire season dominant winds (Table 2). We generated five wind scenarios considering the most frequent wind directions (frequency >5% in weather records) during wildfire season and the respective 97th percentile wind speeds.

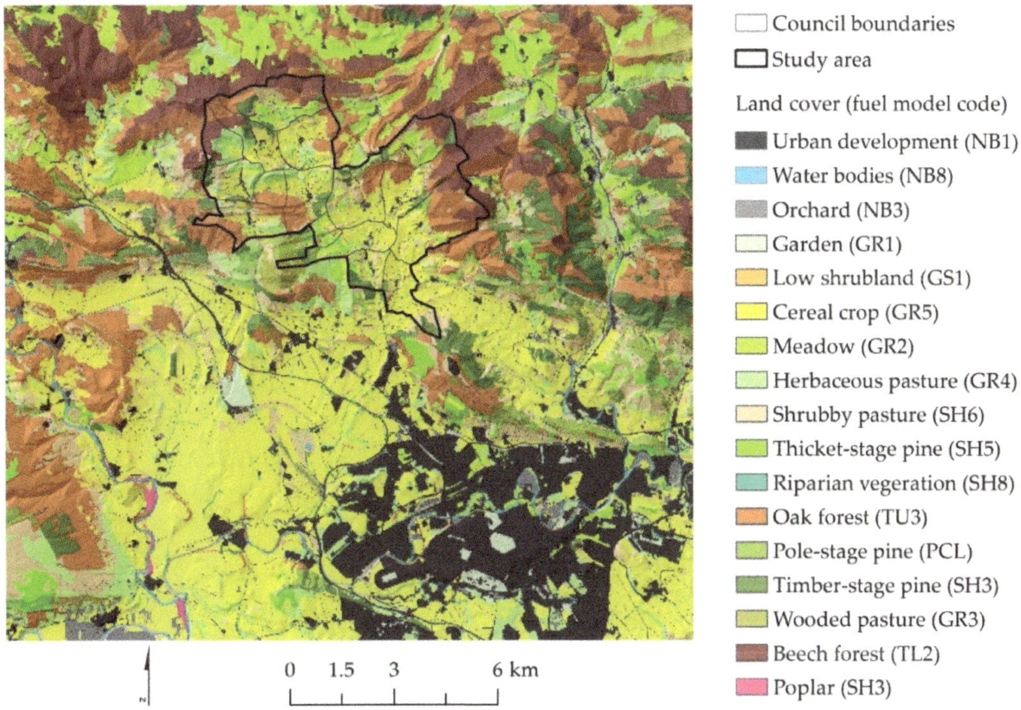

Figure 3. Land cover map (idena.navarra.es) and assigned fuel models [44,45] for the wildfire modeling. The large urban development areas in the southeast correspond to the capital city of Pamplona. Cereal crops occupy all the flat cultivated areas to the south and mountains in the northern part are covered by mosaics of different forest types. See fuel model parameter details in the references [44,45].

Table 2. Fire weather input data, corresponding to the historical 97th percentile conditions, used for wildfire simulations. We considered the most frequent wind directions (frequency >5%) during the last 17 wildfire seasons. Historical weather data were gathered from the meteorological station of Pamplona (meteo.navarra.es). We used standard fuel models for fire modeling, see references [44,45] for further details.

Wind Scenario			Fuel Moisture Content (%)			
				Fuel Model [44,45]		
Direction (°)	Speed ($km \cdot h^{-1}$)	Probability	Fuel Loading Category	GS1, GR5, GR2, GR4, SH6, SH5	TU3, PCL, SH3, GR3	GR1, SH3, TL2, SH8
67.5	32	0.43	1-h	4	6	8
337.5	35	0.28	10-h	5	7	9
45.0	19	0.17	100-h	8	9	12
180.0	31	0.06	Live herbaceous	20	45	70
22.5	23	0.06	Live woody	60	85	100

GS1 = low load, dry climate grass-shrub; GR2 = low load, dry climate grass; GR4 = moderate load, dry climate grass; GR5 = low load, humid climate grass; SH6 = low load, humid climate shrub; SH5 = high load, dry climate shrub; TU3 = moderate load, humid climate timber-grass shrub; PCL = closed and low litter pine stands; SH3 = moderate load, humid climate shrub; GR3 = low load, very coarse, humid climate grass; GR1 = short, sparse dry climate grass; SH3 = Moderate load, humid climate shrub; TL2 = low load broadleaf litter; SH8 = high load, humid climate shrub."

2.2.2. Fire Occurrence Modeling

We used artificial neural networks (ANNs) to construct a fire occurrence model, and ultimately to generate a 20-m resolution ignition probability grid encompassing the modeling domain. A 10,000-fire ignition point input file for wildfire simulation was then created from the ignition probability (IP) grid masked to burnable fuels. ANN models are robust pattern detectors which can approximate mathematical relationships with non-normal distributions and spatially correlated variables where other statistical models could cause multicollinearity [52,53], and have been successfully applied to fire occurrence prediction in previous work [26,54].

The historical fire ignitions within the fire modeling domain (200 ignitions in all, 1985 to 2013 fire records; mapama.gob.es) and the same number of random no-fire observations were matched to topography (elevation, aspect, slope), land cover class, population density, and accessibility (distance to roads, tracks, railways, urban areas and powerlines) 20-m resolution raster grids (ign.es; idena.navarra.es). Ten percent of the fire and no-fire observation variable dataset (40 cases) was set apart for validation purposes before model building. We selected feed-forward, multilayered, non-linear, fully connected, cascade-correlation networks [55], built using Neural Works Predict® v.3.30 software (NeuralWorks Predict®3.30, Serial Number NPSC30-70755, Carnegie, PA, USA) [56] with an adaptive gradient learning rule, a variant of the general algorithm of back-propagation [57], and a weight decay factor which inhibited complexity of the models [58]. The historic fire records of fire and no-fire observations for model building (90%) were further divided in two. One part was used for iterative training (70%, 252 cases) and the other part (30%, 108 cases) for early stopping, the periodic assessment of performance accuracy in order to avoid losing generalization capacity due to overtraining [59]. The cascade-correlation models followed a similar procedure to [60,61], in which the model architecture (number of nodes in the hidden layer) is optimized during training.

The best model found had an 8–6–1 (input–hidden–output) structure, and classification rates of 0.78–0.73–0.69 for training–test–validation datasets (Table 3). When selecting the best ANN classification model, we looked for the highest classification rate on observed and predicted fire/no-fire observations, balanced results between the three datasets and a parsimonious architecture. Variables in the model, by order of importance, were distance to forest tracks (three times input to the model), distance to urban areas (twice input to the model), distance to powerlines (twice input to the model), and population density (once input to the model). Finally, this best fire occurrence model was run at 20-m resolution pixel level to generate the ignition probability grid (IP; values ranging between 0 and 1; Figure 4).

Table 3. Classification table with the results for the best occurrence model. The model was generated with Neural Works Predict® v.3.30 software. This occurrence model was used to generate a 20-m resolution ignition probability grid (Figure 4). Geospatial variables associated with the historical fire ignitions (200 fire ignitions, 1985 to 2013 fire records; mapama.gob.es) (1) and a random sample with the same number of no-fire observations (0) were included within the fire modeling domain. A set of 40 cases (10%) was used for the validation of the model.

Classification Rate		Class	0	1	Total
	0.756	0	99	32	131
Training	0.802	1	24	97	121
	0.779	Total	123	129	252
	0.679	0	36	17	53
Test	0.782	1	12	43	55
	0.731	Total	48	60	108
	0.765	0	13	4	17
Validation	0.609	1	9	14	23
	0.687	Total	22	18	40

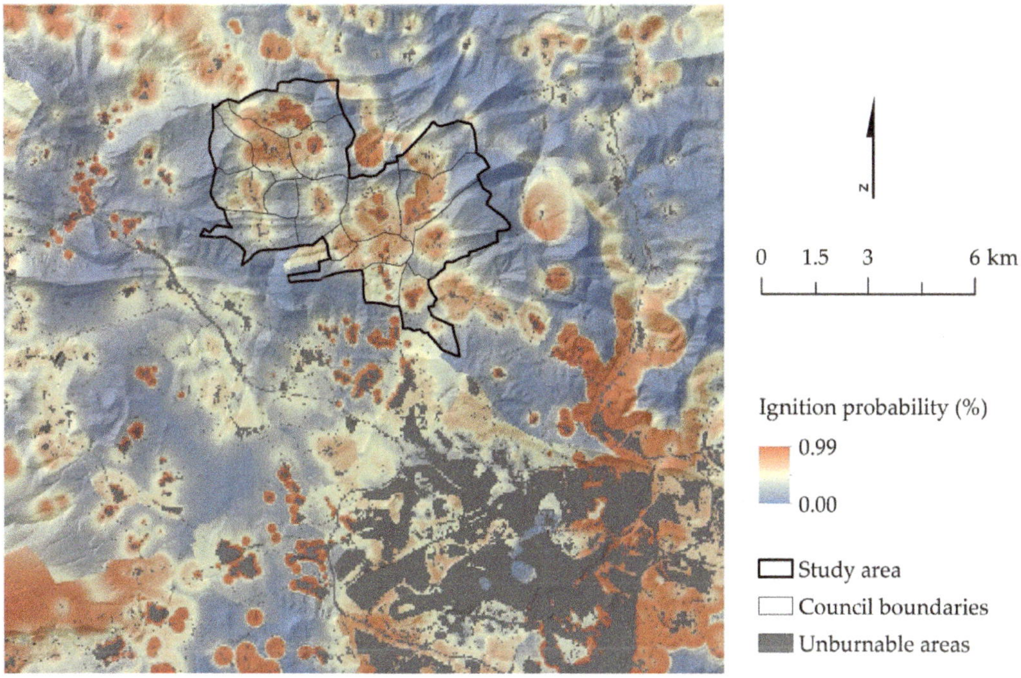

Figure 4. Ignition probability grid generated with an artificial neural network using the geospatial variables associated with the observed ignition data (1985 to 2013 historic fire records; mapama.gob.es). This fire occurrence grid was used to generate the 10,000 fire ignition input data masked to burnable fuels in the wildfire modeling domain. Unburnable areas ($IP = 0$) correspond to urban development, roads and water bodies.

2.2.3. Wildfire Spread and Behavior Simulation

We used FlamMap to simulate wildfires under conditions of constant fuel moisture, wind speed and wind direction [43]. We conducted five different weather scenarios at 20-m resolution, with 10,000 wildfires per scenario (Table 2). FlamMap uses the two-dimensional fire growth minimum travel time algorithm (MTT) [62], which has been widely used worldwide at a broad range of scales with multiple purposes [63–66]. The MTT algorithm replicates fire growth based on the Huygens' principle, where the growth and behavior of the fire edge is modeled as a vector or wavefront [62], and fire spread distance is predicted by the Rothermel's surface fire spread model [67]. Fire duration was set at 6 hour, in agreement with the active fire spread duration of the observed largest wildfire events in the study area (i.e., Juslapeña 2009). We did not consider barriers to fire spread or fire suppression efforts. Overall, modeled fires burned burnable pixels at least once and more than 100 times on average.

FlamMap outputs burn probability (BP) and flame length probability (FLP) grids, as well as a fire size (FS) text file and the fire perimeters (polygons). The burn probability (BP) is the number of times a pixel burns as a proportion of the total number of fires, and is defined as follows:

$$\mathrm{BP} = F/n \qquad (1)$$

where F is the number of times a pixel burns and n is the number of simulated fires per run (n = 10,000 in this study). Specifically, the conditional burn probability in the study area is the BP given that a fire ignites within the fire modeling domain and spreads for 6 hours at assumed fuel moisture and weather conditions (97th percentile fire weather). Fire intensity [68] is first predicted by the MTT algorithm [62] and is converted into flame length as:

$$FL = 0.0775 \times I^{0.46} \qquad (2)$$

where *FL* is flame length (m) and *I* fireline intensity (kW·m^{-1}, kW = kilowatt). Then the program calculates a FLP regular point grid (at the fire simulation resolution) from the multiple burning fires at different flame lengths (i.e., backing, heading and flanking fire spread flame lengths). For every pixel in the FLP output, the probability of flame length is calculated at *i* categories of different fire intensity levels (FILs), given that at least one of the simulated fires has burned the pixel. In this study, FILs were obtained as twenty 0.5-m flame length categories (for FIL_1-FIL_{19} and FIL_{20} >9.5 m).

In the fire size (FS) text file output generated by FlamMap, the simulated burned area (ha) is attributed to each *xy* coordinate fire ignition. Moreover, we also obtained burned-area polygon shapefiles associated with each simulated fire and minimum travel time (MTT) major flow-paths polyline shapefiles for the five fire weather scenarios (Table 2). Travel pathways are straight lines that connect nodes and intersect cells to form segments for which fire behavior is calculated from the input data [43].

2.3. Expert Judgement of Structure Susceptibility

We used a response function (RF) to approximate structure susceptibility (potential losses) using fire intensity level model outputs [36]. To generate a customized RF for residential houses in the study area, we used the Delphi method [69]. The Delphi method is an iterative questionnaire process used to obtain a reliable consensus from a carefully selected expert panel, and it has been used in previous studies to determine wildfire causality from the personnel involved in fire suppression activities [70,71].

We conducted a face-to-face and anonymous two-round questionnaire process with the regional firefighting "Bomberos de Navarra" chiefs, focusing on the most experienced in WUI fire suppression in central Navarra. Fire intensity is the main causative factor of home loss given that a fire reaches a housing structure, and therefore in the questionnaire, potential value loss of structures (as a percentage) was associated to four different fire intensity class response functions (intensity levels of FIL_1, FIL_2-FIL_4, FIL_5-FIL_7, and FIL_8-FIL_{20}). The four fire intensity classes were selected considering previous studies and the capabilities of existing geospatial tools to integrate the fire modeling outputs with potential fire effects [36,49]. In the first round of the questionnaire process, the experts filled the questionnaire anonymously according to their own personal experience to reduce the effect of dominant individuals. Then in the second round, the questionnaire was repeated to the same experts, but included results from the first round (average values and deviation in the fire intensity classes) to meet a higher consensus and refine the final results. The obtained custom RF presented moderate to strong losses in housing structures as fire intensity increased (Table 4), similar to RFs obtained in other studies conducted in Mediterranean areas [35].

Table 4. Custom response function (RF) used to approximate fire effects in terms of value loss (%) on residential houses in the study area [42]. The fire modeling output fire intensity levels (FILs) were grouped into four classes for the geospatial risk assessment [49]. We used the Deplhi method to obtain the susceptibility function from an expert panel composed of the most experienced firefighter chiefs on wildland urban interface fire suppression [69]. The wildfire had negative impacts in structures at all fire intensities.

Valued asset	Relative Net Value Change (%) at Different Fire Intensity Classes			
	Low (FIL_1) FL < 0.5 m	Moderate (FIL_2–FIL_4) 0.5 m < FL < 2 m	High (FIL_5–FIL_7) 2 m < FL <3.5 m	Very high (FIL_8–FIL_{20}) FL > 3.5 m
Residential house	−10	−45	−75	−95

2.4. Residential House Economic Value

We used the official cadaster method described in the Navarra Foral Decree 334/2001 of November 26 to assess the economic value (V) of the individual housing structures in the study area. This Foral Decree approves the procedure for the economic assessment of immovable property in the Foral Community of Navarra throughout the implementation of the Comparison Method of the average market prices, with reference to Inheritance and Gift taxes, and over Property Transfer and Certified Legal Documents (text published in the Boletín Oficial de Navarra No. 155 of 21 December 2001, and the Boletín Oficial de Navarra No. 21 of 18 February 2002; lexnavarra.navarra.es). The method has been updated several times since its first publication, with the Foral Decree 39/2015 of 17 June being the last update. There are specific models to estimate the values for flats, single residential houses, and parking or storage rooms. We used the model for single houses, since most dwellings in the study area were well preserved rural houses or recently built constructions. The main parameters used by the model are the year of the information, type of individual house, location, cadastral category and conducted reforms, year of construction, constructed surface, and the ratio of constructed surface to urban development polygon surface. The residential houses with more than one cadastral sub-division (original building and dwelling expansion) were merged into a single unit. We used market prices from 2015 to obtain the most up-to-date values (Table 5).

Table 5. Summary table of the cadaster economic value (V) for the residential houses in the Juslapeña Valley. Council polygon cadaster codes No. 3 and No. 16 do not have residential houses (Figure 1A). The cadaster value was estimated for the year 2015, considering the model published in the Foral Decree 334/2001 of November 26 (lexnavarra.navarra.es).

Cadaster (Polygon No.)	Council Name	Residential Houses (No.)	Cadaster Economic Value (€)			
			Average	Median	Maximum	Minimum
1	Beorburu	11	108,825	109,767	151,432	52,124
2	Osacar	8	145,249	133,338	223,641	100,226
4	Osinaga	16	130,053	123,993	213,505	50,040
5	Aristregui	23	162,244	191,847	423,310	48,396
6	Larrayoz	17	156,733	138,493	269,403	77,480
7	Nuin	26	142,338	109,033	261,787	52,254
8	Marcalain	31	163,710	141,553	315,990	91,589
9	Iruzkun	1	132,192	132,192	132,192	132,192
10	Garciriain	12	139,194	143,079	199,405	74,682
11	Belzunce	48	168,233	135,984	483,043	64,657
12	Navaz	21	131,148	114,763	223,539	68,061
13	Ollacarizqueta	55	135,030	127,351	319,046	67,402
14	Unzu	16	154,063	174,413	218,129	88,067
15	Usi	21	111,784	108,407	167,685	71,229

2.5. Analysis

Wildfire simulation outputs were used to assess large fire initiation, transmission, exposure and risk to residential houses of rural communities within the study area (Figure 2). We combined five sets of fire simulation outputs (BP, FLP, FS, burned area perimeters and major flow paths), one for each scenario by weighting the relative scenario probability (Table 2).

2.5.1. Large Fire Initiation and Incoming Major Pathways

We estimated fire potential index (FPI) [31], and MTT major flow-paths to spatially analyze where large fires likely initiate and from which surrounding neighboring municipalities do these fires spread to reach the target residential houses. We calculated fire potential index (FPI) as:

$$\text{FPI} = FS \times IP \tag{3}$$

where the FS is the spatially smoothed fire size grid, and IP is the historical-based ignition probability grid generated with the ANN fire occurrence model (Figure 4) used to generate the fire ignition input file. We used a kriging geostatistical analysis method to generate a continuous distribution grid of FS from fire size data contained in the ignition location output point file. MTT flow-paths within surrounding municipality polygons (Figure 1A) were then overlaid and classified in three frequency classes (<33%, 33%–66% and >66%), considering the simulation scenario probability (Table 2), to identify preferential pathways entering to the Juslapeña Valley.

2.5.2. Transmission Analysis

We analyzed how incoming fires are shared among surrounding municipalities (Figure 1A) and mapped the potential impact of each independent fire on dwellings with transmission analysis. We only considered large fires (>100 ha) because small fires do not substantially contribute to total burned area. In the study area observed, large fires (>100 ha) burned about 95% of the total area (1985 to 2012 historic fire records). We quantified (*i*) the number and (*ii*) the economic value of residential houses within fire perimeters. Fire transmission in terms of the number of structures was quantified as:

$$TFS_{ij} = \sum S_j \tag{4}$$

where TFS measures the number of individual S affected structures in the jth municipality (study area) given a large fire (>100 ha) ignited in the ith surrounding municipality (Figure 1A) spreading under 97th percentile fire weather conditions for 6 hours. Correspondingly, the cadastral value of all affected structures contained inside the burned area from transmitted fires was quantified as:

$$TFV_{ij} = \sum V_j \tag{5}$$

where TFV measures the cadaster structure value sum of all houses affected and located in j, given a fire arriving from the ith polygon, and V is the individual structure cadaster value (€). TFV is not the expected economic loss of affected structures, but the value of all affected structures within the burned area polygons. We considered j as the municipality polygon corresponding to the study area (i.e., Juslapeña Valley) containing all the target residential houses, and i as the surrounding municipality polygons (Figure 1A). In total, we analyzed the transmission of 12,515 fires larger than 100 ha ignited from the six different municipality polygons surrounding the study area. Since we focused our analysis only on fires incoming from surrounding polygons, self-burning was not considered (i.e., $i \neq j$).

The TFS transmission results (i.e., the 10,000-fire ignition point file attributed with the number of structures intersected in the fire perimeter polygons) for the five different fire modeling simulation scenarios (Table 2) were separately spatialized into fireshed continuous grids using a 1000-m fixed radius and spherical semivariogram model kriging analysis statistical method. The area estimated within a fireshed is conditional on assumed fire weather and hence we estimated firesheds for each of the scenarios. We also developed contour plots using six different transmission levels to map the internal transmission gradient (0–50, 50–100, 100–150, 150–200, 200–250 and >250 structures).

2.5.3. Exposure Analysis

We analyzed individual residential house wildfire exposure in the home ignition zone (HIZ) (Figure 1C). The HIZ is the 60-m buffer immediately surrounding residential houses that determines structure ignition potential during extreme wildfire events [10]. Fire likelihood and intensity modeling outputs were considered as key causative wildfire risk factors for this analysis. Structure exposure assessment does not account for the fire effects. The geospatial location (polygons) for the individual residential house structures (n = 306) was obtained from cadaster shapefiles (1:5000 scale) of the Regional Government (catastro.navarra.es; Figure 1C).

Wildfire likelihood was estimated as conditional burn probability, and fire intensity as the conditional flame length. We used the pixel-level FIL distribution to calculate the conditional flame length (CFL) as:

$$CFL = \sum_{i=l}^{20} FLP_{\mathrm{i}} \times FL_{\mathrm{i}} \tag{6}$$

where FLP_i is the flame length probability of a fire at the ith flame length category, and FL_i is the flame length (m) midpoint of the ith category FIL. The CFL is the probability-weighted FL assigned to a fire, and is a measure of wildfire hazard [35]. We assessed exposure at individual residential houses from the average values (BP and CFL) within the HIZ.

2.5.4. Risk Analysis

We quantified the expected losses to individual residential houses combining wildfire likelihood and intensity modeling outcomes with expert judgement elicitation response functions [28]. RFs were used to approximate fire effects (losses) to different fire intensity classes. Then, fire effects and respective burning probabilities were considered to estimate the expected net value change [36]. Expected net value change is a risk-neutral measure in terms of *gain* or *loss* expressed on a percentage basis, and allows quantitative wildfire risk assessment for multiple valued resources and human assets [33]. In order to consider the variations between economic values of different houses and quantify economic losses at the individual structure level, we used the latest cadaster reference of economic values (V).

The probabilistic expectation of loss (eNVC) was estimated by combining the customized response function with fire intensity and conditional burning probabilities [28] at the pixel-level on the HIZ:

$$eNVC = \sum_{i=1}^{20} BP \times FLP_i \times RF_i \tag{7}$$

where $eNVC$ is the expected net value change neutral base measure in terms of gain or loss (%) [36], BP is the conditional burn probability, FLP_i is the flame length probability of the ith category FIL, and RF_i is the response function at the ith FIL (Table 4). We assigned the average value within the HIZ to the individual dwellings.

Losses at the individual structure level were monetized using the cadaster value as:

$$eEL_x = eNVC_x \times V_x \tag{8}$$

where eEL is the expected economic loss in the xth residential house (€) given that a fire ignites within the wildfire modeling domain and spreads under extreme fire weather conditions, $eNVC_x$ is the average expected net value change in the xth residential house HIZ, and V_x is the latest cadastral reference value of the xth residential house (€; catastro.navarra.es). Previously, $eNVC$ negative values (fires always produced losses) were transformed into a positive fraction of unity value (e.g., −5% to −0.05).

3. Results

3.1. Large Fire Initiation and Major Pathways

The source location of large fires as quantified by FPI was concentrated around the southwestern and central part of the northern councils (Figure 5). Fires ignited in the south of the study area resulted in larger fire size, and therefore higher FPI values than in the northern and eastern areas. In the northwestern and eastern forested remote areas, the fire ignition probabilities were very low and consequently FPI values were the lowest compared to other areas. Incoming fires exhibited two main paths, either from the northern central part or the southeastern open valleys (Figures 3 and 5). These results highlighted the effect of topography and fuel models in the major flow-paths, especially in the mountainous northern areas of the study area. Fires in the even-aged mature beech forests were largely impenetrable on the northern border within the municipality 126, and most incoming flow-paths were routed through municipality 40, where heading fire spread from the different scenarios' pathways coincided frequently (>66%). On the other hand, in the more fire-prone unmanaged oak and black pine stands, fires arrived from south facing slopes in some cases (e.g., 180° flow-paths). Overall, herbaceous type fuel models located in lowland valley bottom flat areas facilitated the spread of fire and were the preferential fire spread pathways into the study area.

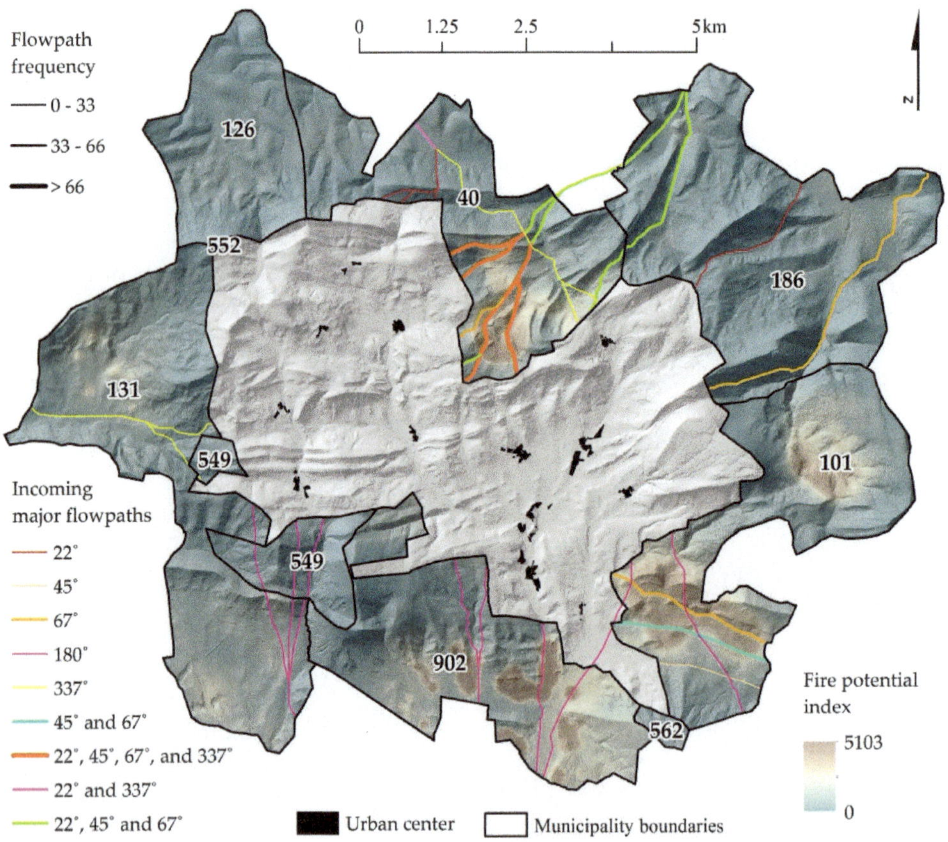

Figure 5. Fire potential index (FPI) and incoming major flow-paths from the surrounding municipalities. FPI was calculated by combining the fire size and ignition probability output grids, and was used to identify the areas where the ignition of a large fire is more likely [31]. Major flow-paths were obtained with the minimum travel time algorithm (MTT) [62] considering the five most recurrent fire weather scenarios (Table 2). The flow-path thickness indicates frequency and color indicates fire scenarios.

3.2. Transmission Analysis

Fires threatening the highest number of residential houses initiated in the southeast 101 and northern 40 municipalities, affecting on average 71 and 80 structures respectively (Figure 6A). The maximum number of structures affected was 188 from a fire ignited in municipality 40. Fires from municipalities 126 and 131 showed limited transmission capability with six or fewer structures burning on average. Although fires in eastern municipality 186 burned on average 15 structures, a few fires (2%) burned more than 100 structures. Due to the limited variability in cadaster economic values of structures within the study area (Table 5), both transmission boxplots depicted similar distributions (Figure 6A,B). Thus, both transmission metrics (TFS and TFV) provide equivalent results in the study area. Given the same response function for all structures (Table 4), economic losses of residential structures from large fires (>100 ha) ignited in surrounding municipalities are highly dependent on HIZ fire intensity and number of structures.

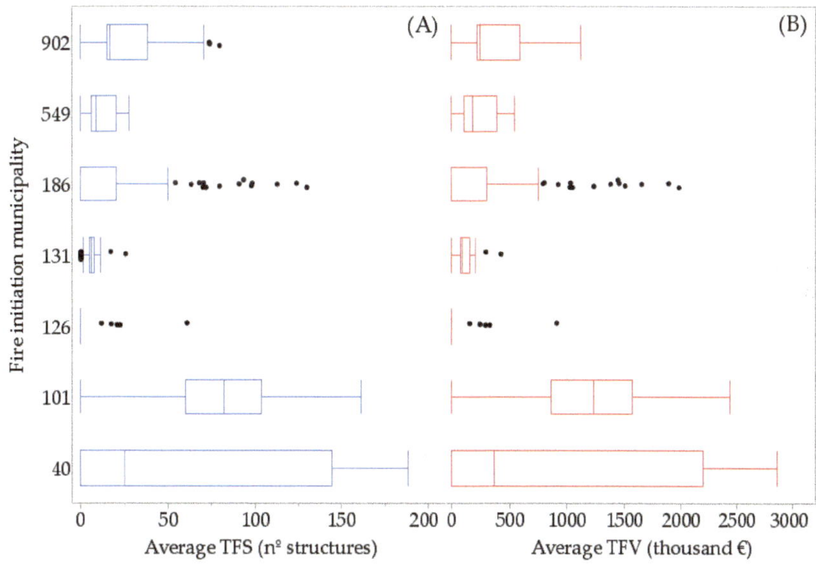

Figure 6. Box plots of average wildfire transmission into the study area from independent ignitions in surrounding municipalities (Figure 1A), in terms of (**A**) number of residential houses and (**B**) cadaster economic value of residential houses affected. For every fire ignition, the number of affected structures and the sum of their economic value was calculated combining the results obtained in the five modeling scenarios (Table 2). Boxes indicate the first/third quartiles, the whiskers indicate 10th/90th percentiles, the black line within the box is the median, and the dots indicate values below the 10th percentile or above the 90th percentile. The municipalities are identified with the cadaster code (Figure 5).

We found a wide variation in predicted community fireshed area for the different scenarios used in the fire simulation (Figure 7A–E). The southern wind direction scenario presented the largest firesheds and smooth gradients, expanding southwards more than 5 km from the study area boundary for the highest >250 structure transmission class. Fires arriving from the south burned through dryland cereal crops and represented the most extreme threat fire scenario to the residential houses in the study area (180°). Firesheds for northwestern to northeastern component wind directions presented the sharpest transitions gradients between transmission classes (337° and 67°). North facing timber litter fuel model beech stands on wind direction perpendicular orientations delineate the fireshed boundaries in northeastern and northwestern wind directions (Figures 3 and 7A,D,E). Highest TFS and TFV values were obtained for fires ignited inside the study area in the majority of cases. Fireshed extension in the north was limited to valley bottom herbaceous fuels on the central part for the scenarios that present similar wind direction and mountain ridge orientation (22° to 67°). Fireshed delineation results agree with the major flow-path results, and overall on larger areas over flow-path influence areas.

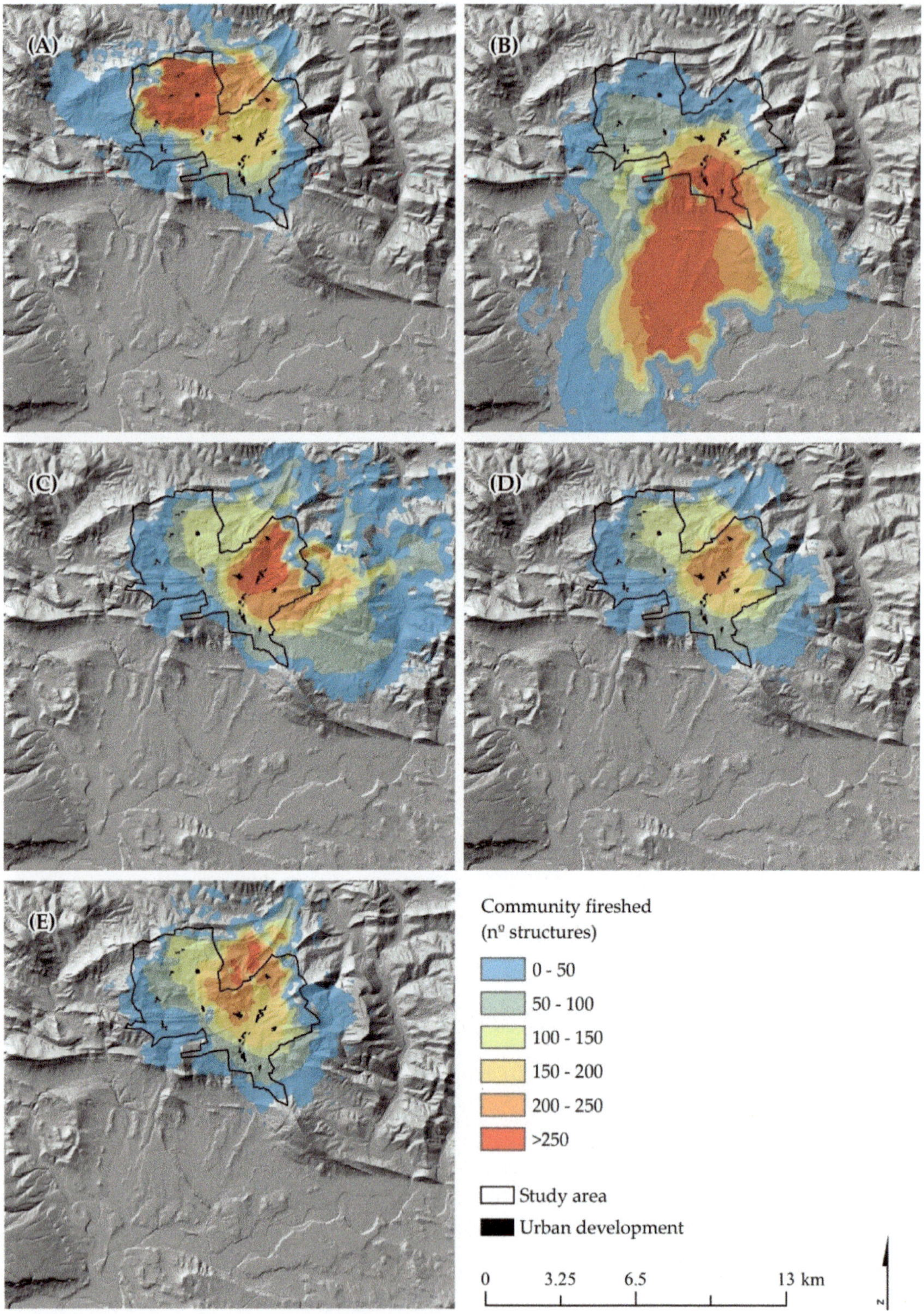

Figure 7. Community fireshed maps corresponding to the number of residential houses burned for the five wildfire scenarios. The letters from **A** to **E** indicate respectively the fire modeling wind direction scenarios of 337°, 180°, 67°, 45° and 22° (Table 2). Fireshed values were generated using a 1-km constant width radius spherical semivariogram model kriging analysis from the transmission values (TFS) assigned to fire ignition locations. Values indicate the number of structures affected by ignitions in a given pixel. Fires were simulated for 97th percentile fire weather conditions and 6-hour duration.

3.3. Exposure Analysis

The burn probability and conditional flame length wildfire modeling outputs showed complex spatial patterns in the study area (Figure 8A,B). As expected, the results highlighted important differences between the fire occurrence IP grid (Figure 4) and conditional burn probability in structure HIZ, since fire occurrence is closely associated with anthropic ignition sources but not necessarily burn probability (Figure 9). While the average IP is usually high on the HIZ (IP >0.8), the BP presents a wide range of values between 0.001 and 0.120 (Figure 8). Southern councils presented the highest ignition probability and burn probability values (e.g., councils No. 13 and No. 14; Figure 1A). Burn probability was higher than 0.1 in most southern areas, ten times higher than values in the northern part of the study area (BP < 0.001; Figure 8A). Highest values were associated in most cases to fast spreading surface fires in herbaceous type fuel models, such as rangelands and cereal crops (the Pamplona Basin northern rim extensive dryland agricultural landscape continuum) that dominate the valley bottom in the southern plain of the study area. On the other hand, the lowest values of the northern mountainous areas corresponded to beech and pine forests on north aspects, both characterized by low biomass understories. The smooth spatial gradients in burn probability were in contrast to the conditional flame length (CFL) (Figure 8B), where CFL highest values did not correspond with high burn probability (Figures 8B and 10A). Low CFL values (<1 m) were obtained in northern areas where the burn probability was the lowest, especially in the low fuel load, timber litter and closed canopy mature forest stands. Mosaics of fuel types, together with wind direction and slope, were the main drivers of fire intensity. High shrubs and dense forests on slopes aligned with the dominant winds (68° and 338° azimuth) showed the highest intensities (CFL > 6 m).

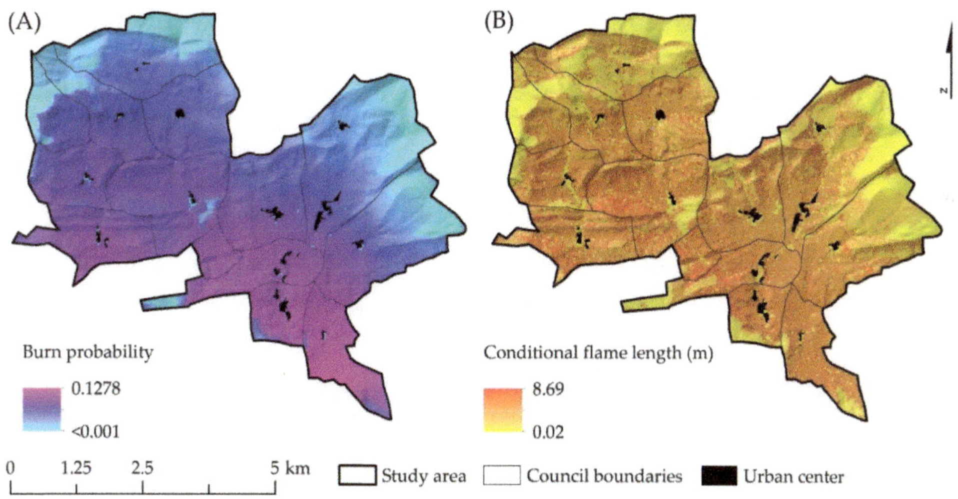

Figure 8. Conditional burn probability (**A**) and conditional flame length (**B**) output maps for the study area. Fires were modeled at 20-m resolution under 97th percentile fire weather conditions. The urban centers containing the bulk of residential structures are indicated with black polygons.

Average burn probability and conditional flame length for pixels within the 60-m circular buffer around individual residential houses varied widely among and within the different councils (Figure 10A,B). Overall, the bulk of houses had average conditional flame length values between 1.5 and 3 m, while the burn probability varied more widely, and was mostly concentrated between 0.4 and 0.11. Around some residential houses located in the central parts of the urban centers, where fuels consisted of managed gardens and orchards, the conditional flame length was the lowest (<0.25 m). Burn probability results showed much wider variations, especially between houses of different councils (Figure 10B). For instance, the residential houses in council No. 15 (located in the northeast, Figure 1A) presented on average four to five times lower burn probability (BP~0.02) compared to the most meridional council No. 13 (BP~0.10). Within the same urban center, residential houses exhibited variations among southern and northern locations, especially in the central parts of the study area (e.g., councils No. 5, No. 6 and No. 8), mainly because upslope spreading fires over cereal crops on the southern sides of urban centers present the fastest spread rates. Therefore, housing aggregation into compact urban centers and the relative structure position in the urban center had a strong effect on HIZ wildfire likelihood. In other words, wildfires were more likely to arrive and impact the southern side, and structures located there were exposed to higher BP. The highest overall exposure was experienced by residential houses nestled within forested and shrubby unmanaged areas with high fuel accumulation.

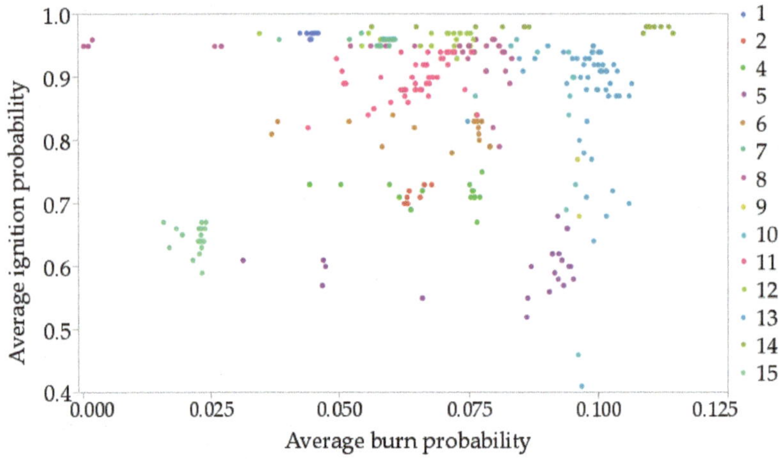

Figure 9. Scatter plot of ignition probability (IP) versus conditional burn probability (BP) for individual residential houses (n = 306 structures). Each dot is related to a different residential house, and values correspond to the mean value in the HIZ [11]. The bubble color indicates the council cadaster polygon (Figure 1A). While BP values showed a wide distribution, most IP values were concentrated above 0.85. Overall, results tended to present clustered aggregations with respect to the council.

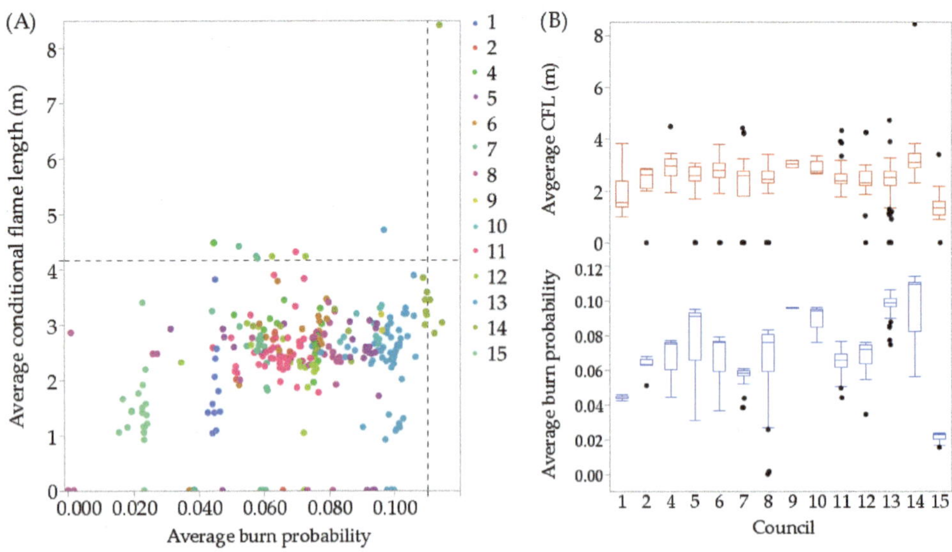

Figure 10. Individual residential house scatter (**A**) and box plots (**B**) for the different councils in the study area. Each point in the scatterplot indicates the average value of burn probability (BP) and conditional flame length (CFL) within the home ignition zone for a single structure. The bubble color indicates the council (Figure 1A), and the dotted lines the 97th percentile values of 0.11 for BP and 4.16 m for CFL. In the box plots, the boxes indicate the first/third quartiles, the whiskers indicate 10th/90th percentiles, the horizontal line within the box is the median, and the black dots indicate values below the 10th percentile or above the 90th percentile.

3.4. Expected Economic Loss

Expected economic loss for individual dwellings (eEL) ranged from a low of 740 to a high of 28,725€ within the study area (mean = 7955€), and also varied widely among the different councils (Figure 11, Table 6). The highest average values were obtained for the southern council No. 14 with 13,323€, followed by councils No. 5 and No. 10 with 12,976€ and 9715€ respectively. On the other hand, the lowest average eEL values were obtained in the low wildfire exposure northern councils No. 1 and No. 15, with 1429€ and 2803€ respectively. Overall, results depicted higher expected economic loss (eEL) for residential houses presenting lower expected net value change (eNVC), that ranged from −1.04% to −11.04%, with an average value of −5.23%. Except for a few cases, most residential houses have cadastral values between 110 and 180 thousand euros (Table 5), and therefore exposure metrics required for risk assessment translated similar patterns into risk outcomes (i.e., higher losses for higher overall exposure). Nonetheless, when the cadastral value varied substantially for the same eNVC (e.g., more than three times), wide differences were observed in terms of eEL. In those cases, the residential house cadastral value influenced the eEL result more than the eNVC (Figure 11).

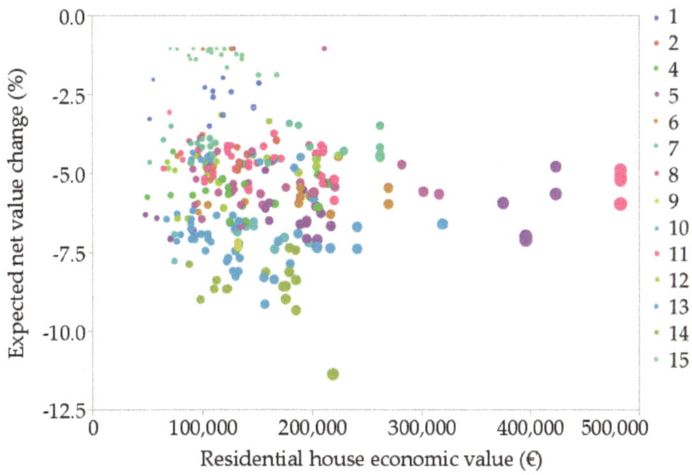

Figure 11. Wildfire risk bubble plot of residential houses in the study area. The expected net value change (eNVC) is the percentage variation with respect to the price of the residential houses. The bubble size indicates the expected economic loss (eEL), which ranged from a low of 3622€ to a high of 28,086€. The color indicates the council.

Table 6. Council level summary table with expected economic loss (eEL) results for the residential houses in the Juslapeña Valley (Figure 1A). Potential expected economic loss was obtained as a result of the implementation of the framework presented in this study (Figure 2). Council polygon cadaster codes No. 3 and No. 16 do not have residential houses (Figure 1A).

Cadaster (Polygon No.)	Council Name	Residential Houses (No.)	Expected Economic Loss (€)			
			Average	Median	Maximum	Minimum
1	Beorburu	11	2803	2826	4248	1120
2	Osacar	8	6378	6263	10,004	3776
4	Osinaga	16	7343	7188	12,459	2870
5	Aristregui	23	12,976	12,480	28,086	3051
6	Larrayoz	17	8347	8152	16,049	1062
7	Nuin	26	5930	4855	11,757	2425
8	Marcalain	31	8167	7502	17,855	1334
9	Iruzkun	1	9592	9592	9707	9478
10	Garciriain	12	9715	9443	14,366	5169
11	Belzunce	48	8044	6232	28,725	2143
12	Navaz	21	6157	5341	9909	3150
13	Ollacarizqueta	55	9105	8219	21,046	3960
14	Unzu	16	13,323	13,419	24,757	6933
15	Usi	21	1429	1222	3131	740

The map of eEL for dwellings in the study area showed how spatial location greatly influenced the result (Figure 12A). Highest economic losses (>10,000€) were located in the southern councils (i.e., No. 14, No. 5 and No. 13) and some individual houses in councils of the central part, while the lowest values (<1000€) were concentrated in the northeastern and northern councils (i.e., No. 1 and No. 15). The highest variation within residential houses of the same council were seen when the urban center tended to present a more scattered linear orientation following the communication corridors (e.g., council No. 11), and greater distances between the most distant houses (>1 km). Spatial patterns in eEL were similar to the gradient observed for the burn probability (Figure 8A), since fire hazard among residential houses (Figure 10A,B) within the study area did not show large differences. Thus, according to the results at individual residential structure level, eEL treatment priorities in the HIZ would be preferentially located in southern councils and structures occluded in hazardous forest fuels.

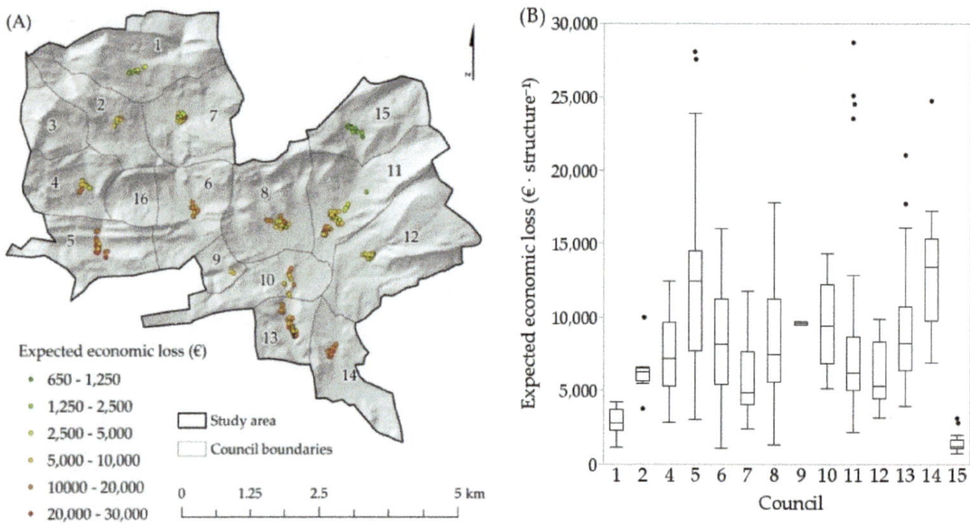

Figure 12. Map (**A**) and box-plots (**B**) of expected economic losses (eEL) for residential houses given that a fire occurs within the fire modeling domain under extreme fire weather conditions. Councils No. 3 and No. 16 do not have residential houses. In the map, every dot corresponds to a single residential house. In the boxplots, boxes indicate the first/third quartiles, the whiskers indicate 10th/90th percentiles, the black line within the box is the median, and the dots indicate values below the 10th percentile or above the 90th percentile (€ structure^{-1}).

4. Discussion

The integration of biophysical fire modeling with susceptibility relationships derived from expert judgement provides a method to calculate expected financial loss to communities from potential wildfire events. Our analysis also demonstrated the tracking of burned areas in the communities to ignition locations, thus providing a linkage between wildland fuels and risk to communities. The results provided useful insights that can inform ignition prevention fuel management programs for reducing risk to communities [72]. Transmission analysis allows the identification of sources of risk in terms of specific landowners within the study area [39,40]. The fireshed mapping defined the scale of risk to rural communities [73] and delineated the area within which fuel treatments could be prioritized to reduce large-fire impacts [37]. Coupling the fireshed with maps of exposure provides a wealth of information to inform the prioritization of wildfire management within the study area [35]. Our fire risk quantitative assessment results showed a very strong structure-level spatial gradient in economic loss within and among the 14 councils in the Juslapeña Valley study area, and provided findings that are potentially useful for insurance companies and local landscape managers.

We identified high probability paths of incoming fire in central north and south east valley bottom flat areas [62,65], which were mainly located in neighboring municipalities 101 and 40. These two areas accounted for the bulk of the transmission as measured by the highest number of residential structures affected. Historically, fires frequently have impacted populated areas after spreading large distances from their ignition location, well beyond community wildfire planning boundaries, underscoring the importance of analyzing firesheds to minimize scale mismatches [41] between the landscape planning and fire risk mitigation efforts [73]. In fact, 67% of large fires (>100 ha) ignited in the surrounding municipalities reached target communities, and each of these fires affected 56 structures on average. Also, we found that the 180° wind direction fire weather scenario in particular resulted in fires spreading the longest distance from ignitions outside the study area administrative boundary (>8 km). Thus, collaborative planning efforts need to involve neighboring administrations and landowners (Figure 7), and the significance of current land management in areas outside of target councils needs to be recognized for its potential to enhance wildfire risk. These practices include grazing, firewood collection in coppice oak forests and thinning in dense conifer plantations.

We contribute several new methods for exposure assessments within the Mediterranean region [47,72]. In particular, we used an ANN fire occurrence model to generate fire ignition input locations, and included an expert-defined response function for structure-scale assessment of potential economic losses. Although wildfire loss or benefit quantification is not possible for many socioeconomic values, a number of important services derived from forests can be represented with market pricing [48]. Specifically, 72% of structures have estimated values ranging from 100 to 250 thousand euros, with relatively few (6%) having values >250 thousand euros. Rather than economic value, we found that spatial patterns of wildfire likelihood were the major causative risk factor, and thus fire occurrence spatiotemporal patterns in Mediterranean environments are especially important for fire prevention. ANN performed well and facilitated the generation of a high-resolution ignition probability grid. Understanding how fire weather and geospatial variables associated with anthropic activities can explain fire occurrence has been conducted in previous works [74,75].

This study highlighted the importance of fire spread modeling for risk assessment in Mediterranean environments where large fires spread through mosaics of fuel type and administrative jurisdiction [32,76,77]. Urban interface classification based on housing density has been considered a key factor in structure loss and risk mitigation in some previous studies [19,78]. Indeed, scattered and occluded houses within wildlands usually present higher exposure levels from catastrophic fire events than densely populated urban development areas [79]. However, structures at the periphery of communities usually incur higher losses since they intercept heading fires and associated embers showers (Figure 13). Although most houses in the study area are built with fire-resistant designs and materials, and have cultivated orchards in the surroundings, exposure to ember showers makes them vulnerable to fire. Isolated dwellings in remote areas are hampered by poor access for ground-based suppression crews, a primary factor contributing to structure loss probability and human fatalities [34]. Urban areas with fire-resistant structures and managed fuels in the HIZ can facilitate fire suppression opportunity, and help relocate residents to save zones during catastrophic fires events. The more typical situation is where developed areas become a resource sink for most of the firefighting resources, creating the potential for entrapment and accidents during mass evacuation during extreme fire events [80].

Figure 13. Close-up view of a residential house level wildfire risk map (eEL) for Marcalain (council cadaster polygon No. 8) expanded from Figure 12, over the June-2015 aerial photograph (idena.navarra.es). Other structures such as farm stores, churches and water deposits were excluded from the analysis. Overall, structures located in the periphery were more exposed to wildfires and presented higher potential losses. On the north side of rural communities, closer to shrublands and forest areas, higher wildfire hazard can enhance potential losses.

Multiple management implications result from this study. First, the results provided useful insights to identify preferential areas for future urban development (e.g., high overall exposure area exclusion criteria) and to inform fire-resistant building design and material requirements. Integrating exposure from other natural hazards such as floods in river basin plains and rock falls or avalanches in mountainous areas is widely accepted as criteria for potential urban development, but fire risk is not accounted in most fire prone areas where catastrophic fires are frequent events (<30 years). In this regard, many southern EU governments concerned with WUI problems are now dictating specific public policies and municipal ordinances to promote community and homeowner involvement in hazardous fuel management. We present structure level risk assessment results that can contribute to risk reduction efforts by identifying where fuel treatment provides the highest benefits at the individual house level. Urban planning and fire managers have limited budgets to cover risk mitigation over thousands of scattered housing communities dispersed throughout fire prone landscapes, and quantitative risk assessment frameworks [28,33,66] can help prioritize planning and investments as well help design specific spatial strategies [81,82].

Reducing structure susceptibility to fire [34] in combination with fuel treatments, both in HIZ [83] and strategically located areas on the landscape [10,35], are the key to mitigating wildfire risk to communities. Fuel treatments reduce potential fire intensity and spread rates by reducing surface and canopy fuel loadings and include a wide range of activities (prescribed burns, low pruning and low fuel load hedges, disrupting tree crown continuity and removing combustible material adjacent to structures) [12,84,85]. Other measures such as the implementation of structure self-protection plans can alleviate extreme fire environments and improve suppression capabilities (e.g., water sprinklers and cannons). Apart from typical treatments in forest fuels and reducing structure susceptibility, other strategies that focus on reducing fire spread over herbaceous land cover could reduce the impacts of long-distance spreading fire events. For instance, we observed long-distance fire events originating in dryland croplands in the southern portion of the study area. By managing herbaceous fuels with extensive grazing in fenced pasture common lands [86,87], and using grass species with patchy growth habit on dryland hay meadows, wildfire spread and intensity could be reduced in these areas. However, implementation of supervised grazing after cereal harvesting that is needed to break fuel beds on the edges between mosaics of cultivated lands is nowadays complicated to implement [88]. Currently, the major risk mitigation effort in agricultural areas is the prevention of ignitions during cereal harvesting operations from equipment, and increasing capabilities for more rapid response to ignitions if they do happen [89].

We assume various sources of error in the models and input data, and results should be viewed as a local approximation of wildfire risk to residential houses in Juslapeña Valley given a large fire event in the study area. Modeling outcomes are conditioned to a specific configuration of extreme fire weather conditions, fuels, topography and the rural–urban interface spatial distribution of the study area. Although fuel models around structures did not differ much from the dominant types in the study area, elsewhere complex interface areas with trimming hedges among structures (e.g., cypress *Cupressus sempervirens* L. hedges) might require a more detailed fuel characterization or various different response functions depending on secondary variables in addition to fire intensity. Community firesheds should be interpreted as a dynamic boundary that changes with assumptions about fire weather, and with existing spatial patterns of fuels as influenced by land management practices. The latter includes forest management practices, grazing practices and agricultural production. Moreover, structure loss is a complex process [12], and is difficult to model at the landscape scale [35]. As in other previous studies, we adopted an expert-defined response function to approximate fire effects at different fire intensities while acknowledging the margin of error [36]. We also did not consider the potential effects of fire suppression that could affect our estimates of structure ignition, especially for low intensity fires with flame lengths <1.2 m [90]. We also understand that structure economic value (conditioned to market changes) might not always be the best way to quantify real risk, due to the lack of correlation among the economic value and the social impact of structure loss on inhabitants. Focusing exclusively on economic criteria when setting treatment priorities might bias results to favor protection of the wealthy neighborhoods at the expense of lower priced homes, although in our study the value of homes did not substantially influence the results.

Further research is needed to better understand not only large fire transmission into the study area, but also the dominant transmission patterns at wider scales (e.g., regional and national), to understand how the study area is integrated into larger scale fire transmission patterns [40]. Understanding major large fire movements would provide a wider perspective to identify the nodes or high priority areas in the landscape requiring investments in treatments. Identification of treatment polygons or stands in priority areas (or firesheds) can be facilitated with optimization models and trade-off analysis to maximize the reduction in risk to multiple values of interest, including structure loss, game species habitat improvement or conifer timber production [91]. The risk assessment in this study should be considered as a preliminary step for mitigation and it does not necessarily reveal the optimal treatment allocation, especially considering that treating fuels at locations far from the urban interface can substantially slow large fire arrival [35]. Analyzing multi-objective treatment strategies in rural–urban intermix fire-prone Mediterranean EU landscapes is challenging, although newer landscape planning tools that allow for integration of fire transmission have opened a wide range of new analytical approaches to analyze trade-offs between local hazard versus large-scale transmitted fire [81].

5. Conclusions

We implemented a fine scale wildfire risk assessment and transmission framework in rural communities of central Navarra (Northern Spain). Potential economic losses were quantified on individual residential houses considering exposure results [42], local expert-defined susceptibility functions, and dwellings cadaster economic values. With the transmission analysis we traced the origin and quantified the potential impacts of large wildfires [40]. Using major flow-paths [62] we identified preferential fire spreading path-ways entering to the study area. We demonstrate that wildfires ignited in neighboring municipalities far beyond human communities can cause substantial economic losses. This work increases the awareness and knowledge on wildfire risk assessment in Southern European fire-prone areas, and highlights the need of a collaborative planning and management among neighboring communities, different landowners and landscape managers to mitigate losses from wildfires.

Acknowledgments: The authors would like to thank the Forest Service of Navarra (DDRMAyAL) and the Fire Service of Navarra (ANE, Bomberos de Navarra) for the collaboration in this study. Fire data records were provided by the Centre for National Information on Forest Fires (Centro de Coordinación de la Información Nacional sobre Incendios Forestales, CCINIF), the agency responsible for coordinating general forest fires statistics in Spain (Estadística General de Incendios Forestales, EGIF) within the Ministry of Agriculture and Fisheries, Food and Environment (MAPAMA). We would also like to thank to the Soil and Climate Affairs of the Department of Rural Development and Local Administration of Navarra for providing weather data. Michele A. Day contributed with very valuable comments on a later draft. This work was funded by a University of Lleida Research training fellowship to Fermín J. Alcasena Urdíroz.

Author Contributions: F.J.A. was the leading author, responsible for the study design and GIS analysis; M.S. contributed in the study design and wrote risk assessment section; A.A.A. contributed in in the study design and wrote sections of the manuscript; R.C. developed the custom susceptibility function and contributed in the discussion of the results; C.V.G. generated the ANN model, contributed in the study design and wrote sections of the manuscript.

Conflicts of Interest: The authors declare no conflict of interest.

Abbreviations

meteo.navarra.es	Meteorología y climatología de Navarra. Nafarroako Gobernua/Gobierno de Navarra
mapama.gob.es	Ministerio de Agricultura y Pesca, Alimentación y Medio Ambiente. Gobierno de España.
sigpac.navarra.es	Sistema de Información Geográfica de Navarra para la Política Agraria Comunitaria. Nafarroako Gobernua/Gobierno de Navarra.
catastro.navarra.es	Servicio de la Riqueza Territorial. Nafarroako Gobernua/Gobierno de Navarra.
ign.es	Instituto Geográfico Nacional. Centro Nacional de Información Geográfica. Ministerio de Fomento. Gobierno de España.
idena.navarra.es	Infraestructura de Datos Espaciales de Navarra. Portal de acceso a la información geográfica de Navarra. Nafarroako Gobernua/Gobierno de Navarra.
lexnavarra.navarra.es	Derecho navarro. Nafarroako Gobernua/Gobierno de Navarra

References

1. Radeloff, V.C.; Hammer, R.B.; Stewart, S.I.; Fried, J.S.; Holcomb, S.S.; McKeefry, J.F. The wildland-urban interface in the United States. *Ecol. Appl.* **2005**, *15*, 799–805. [CrossRef]
2. Price, O.; Bradstock, R. Landscape scale influences of forest area and housing density on house loss in the 2009 Victorian bushfires. *PLoS ONE* **2013**, *8*, e73421. [CrossRef] [PubMed]
3. Martinuzzi, S.; Stewart, S.I.; Helmers, D.P.; Mockrin, M.H.; Hammer, R.B.; Radeloff, V.C. *The 2010 Wildland-Urban Interface of the Conterminous United States—Geospatial Data*; Service U.F.: Fort Collins, CO, USA, 2015.
4. Maillé, E.; Fernández, M.M.; Bouillon, C.; Sirca, C. Wild Fire Risk in the Rural-Urban Interface. In *Forest Fires under Climate, Social and Economic Changes in Europe, the Mediterranean and Other Fire-Affected Areas of the World-Fume-Lessons Learned and Outlook*; Arianoutsou, M., González-Cabán, A., Mouillot, F., Oechel, W.C., Spano, D., Thonicke, K., Vallejo, V.R., Vélez, R., Eds.; Calyptra Pty Ltd.: Adelaide, Australia, 2014; pp. 14–15.
5. Bouillon, C.; Fernandez Ramiro, M.M.; Sirca, C.; Fierro Garcia, B.; Casula, F.; Vila, B.; Long Fournel, M.; Pellizzaro, G.; Arca, B.; Tedim, F.; et al. A Tool for Mapping Rural-Urban Interfaces on Different Scales. In *Advances in Forest Fire Research*; Viegas, Domingos Xavier: Coimbra, Portugal, 2014; pp. 611–625.
6. Pausas, J.G. Simulating mediterranean landscape pattern and vegetation dynamics under different fire regimes. *Plant Ecol.* **2006**, *187*, 249–259. [CrossRef]
7. Theobald, D.M.; Romme, W.H. Expansion of the US wildland–urban interface. *Landsc. Urban Plan.* **2007**, *83*, 340–354. [CrossRef]
8. Badia, A.; Serra, P.; Modugno, S. Identifying dynamics of fire ignition probabilities in two representative mediterranean wildland-urban interface areas. *Appl. Geogr.* **2011**, *31*, 930–940. [CrossRef]

9. Pellizzaro, G.; Arca, B.; Pintus, G.; Ferrara, R.; Duce, P. Wildland Urban Interface Dynamics during the Last 50 Years in North East SARDINIA. In *Modelling Fire Behavior and Risk*; Spano, D., Bacciu, V., Salis, M., Sirca, C., Eds.; Nuova Stampa Color Publishers: Muros, Spain, 2012; pp. 249–254.
10. Costa, P.; Castellnou, M.; Larrañaga, A.; Miralles, M.; Daniel, K. *Prevention of Large Wildfires Using the Fire Types Concept*; Cerdanyola del Vallès: Barcelona, Spain, 2011.
11. Cohen, J. The wildland-urban interface fire problem: A consequence of the fire exclusion paradigm. *For. History Today* **2008**, *3*, 20–26.
12. Mell, W.E.; Manzello, S.L.; Maranghides, A.; Butry, D.; Rehm, D.B. The wildland-urban interface fire problem: Current approaches and research needs. *Int. J. Wildland Fire* **2010**, *19*, 238–251. [CrossRef]
13. Moritz, M.A.; Batllori, E.; Bradstock, R.A.; Gill, A.M.; Handmer, J.; Hessburg, P.F.; Leonard, J.; McCaffrey, S.; Odion, D.C.; Schoennagel, T.; et al. Learning to coexist with wildfire. *Nature* **2014**, *515*, 58–66. [CrossRef] [PubMed]
14. USDA-USDI (2014). *The National Strategy: The Final Phase in the Development of the National Cohesive Wildland Fire Management Strategy*; United States Department of Agriculture: Washington, DC, USA, 2014; p. 93.
15. Calviño-Cancela, M.; has-Amil, M.L.; García-Martínez, E.D.; Touza, J. Wildfire risk associated with different vegetation types within and outside wildland-urban interfaces. *For. Ecol. Manag.* **2016**, *372*, 1–9. [CrossRef]
16. Gibbons, P.; Van Bommel, L.; Gill, A.M.; Cary, G.J.; Driscoll, D.A.; Bradstock, R.A.; Knight, E.; Moritz, M.A.; Stephens, S.L.; Lindenmayer, D.B. Land management practices associated with house loss in wildfires. *PLoS ONE* **2012**, *7*, e29212. [CrossRef] [PubMed]
17. Madrigal, J.; Ruiz, J.A.; Planelles, R.; Hernando, C. Characterization of wildland-urban interfaces for fire prevention in the province of Valencia (Spain). *For. Syst.* **2013**, *22*, 249. [CrossRef]
18. Lampin-Maillet, C.; Jappiot, M.; Long, M.; Christophe, B.; Morge, D.; Ferrier, J.P. Mapping wildland-urban interfaces at large scales integrating housing density and vegetation aggregation for fire prevention in the South of France. *J. Environ. Manag.* **2010**, *91*, 732–741. [CrossRef] [PubMed]
19. Herrero-Corral, G.; Jappiot, M.; Bouillon, C.; Long-Fournel, M. Application of a geographical assessment method for the characterization of wildland-urban interfaces in the context of wildfire prevention: A case study in Western Madrid. *Appl. Geogr.* **2012**, *35*, 60–70. [CrossRef]
20. Conedera, M.; Tonini, M.; Oleggini, L.; Orozco, C.V.; Leuenberger, M.; Pezzatti, G.B. Geospatial approach for defining the wildland-urban interface in the alpine environment. *Comput. Environ. Urban Syst.* **2015**, *52*, 10–20. [CrossRef]
21. Chuvieco, E.; Aguado, I.; Jurdao, S.; Pettinari, M.L.; Yebra, M.; Salas, J.; Hantson, S.; de la Riva, J.; Ibarra, P.; Rodrigues, M.; et al. Integrating geospatial information into fire risk assessment. *Int. J. Wildland Fire* **2014**, *23*, 606. [CrossRef]
22. Ganteaume, A.; Jappiot, M. What causes large fires in Southern France. *For. Ecol. Manag.* **2013**, *294*, 76–85. [CrossRef]
23. Curt, T.; Fréjaville, T.; Lahaye, S. Modelling the spatial patterns of ignition causes and fire regime features in Southern France: Implications for fire prevention policy. *Int. J. Wildland Fire* **2016**, *25*, 785–796. [CrossRef]
24. Salis, M.; Ager, A.A.; Alcasena, F.J.; Arca, B.; Finney, M.A.; Pellizzaro, G.; Spano, D. Analyzing seasonal patterns of wildfire exposure factors in Sardinia, Italy. *Environ. Monit. Assess.* **2015**, *187*, 1–20. [CrossRef] [PubMed]
25. Mann, M.L.; Batllori, E.; Moritz, M.A.; Waller, E.K.; Berck, P.; Flint, A.L.; Flint, L.E.; Dolfi, E. Incorporating anthropogenic influences into fire probability models: Effects of human activity and climate change on fire activity in California. *PLoS ONE* **2016**, *11*, e0153589. [CrossRef] [PubMed]
26. Miller, C.; Ager, A.A. A review of recent advances in risk analysis for wildfire management. *Int. J. Wildland Fire* **2013**, *22*, 1–14. [CrossRef]
27. Finney, M.A. The challenge of quantitative risk analysis for wildland fire. *For. Ecol. Manag.* **2005**, *211*, 97–108. [CrossRef]
28. Bar Massada, A.; Radeloff, V.C.; Stewart, S.I.; Hawbaker, T.J. Wildfire risk in the wildland–urban interface: A simulation study in Northwestern Wisconsin. *For. Ecol. Manag.* **2009**, *258*, 1990–1999. [CrossRef]
29. Haas, J.R.; Calkin, D.E.; Thompson, M.P. A national approach for integrating wildfire simulation modeling into wildland urban interface risk assessments within the United States. *Landsc. Urban Plan.* **2013**, *119*, 44–53. [CrossRef]

30. Salis, M.; Ager, A.; Arca, B.; Finney, M.A.; Bacciu, V.; Duce, P.; Spano, D. Assessing exposure of human and ecological values to wildfire in Sardinia, Italy. *Int. J. Wildland Fire* **2013**, *22*, 549–565. [CrossRef]
31. Alcasena, F.J.; Salis, M.; Vega-García, C. A fire modeling approach to assess wildfire exposure of valued resources in Central Navarra, Spain. *Eur. J. For. Res.* **2015**, *135*, 87–107. [CrossRef]
32. Calkin, D.C.; Finney, M.A.; Ager, A.A.; Thompson, M.P.; Gebert, K.G. Progress towards and barriers to implementation of a risk framework for us federal wildland fire policy and decision making. *For. Policy Econ.* **2011**, *13*, 378–389. [CrossRef]
33. Penman, T.D.; Nicholson, A.E.; Bradstock, R.A.; Collins, L.; Penman, S.H.; Price, O.F. Reducing the risk of house loss due to wildfires. *Environ. Model. Softw.* **2015**, *67*, 12–25. [CrossRef]
34. Mitsopoulos, I.; Mallinis, G.; Arianoutsou, M. Wildfire risk assessment in a typical Mediterranean wildland–urban interface of Greece. *Environ. Manag.* **2014**, *55*, 900–915. [CrossRef] [PubMed]
35. Thompson, M.P.; Calkin, D.E.; Gilbertson-Day, J.W.; Ager, A.A. Advancing effects analysis for integrated, large-scale wildfire risk assessment. *Environ. Monit. Assess.* **2011**, *179*, 217–239. [CrossRef] [PubMed]
36. Ager, A.A.; Day, M.A.; Finney, M.A.; Vance-Borland, K.; Vaillant, N.M. Analyzing the transmission of wildfire exposure on a fire-prone landscape in Oregon, USA. *For. Ecol. Manag.* **2014**, *334*, 377–390. [CrossRef]
37. Scott, J.H.; Thompson, M.P.; Gilbertson-Day, J.W. Exploring how alternative mapping approaches influence fireshed assessment and human community exposure to wildfire. *GeoJournal* **2015**, 1–15. [CrossRef]
38. Haas, J.R.; Calkin, D.E.; Thompson, M.P. Wildfire risk transmission in the Colorado front range, USA. *Risk Anal.* **2015**, *35*, 226–240. [CrossRef] [PubMed]
39. Ager, A.A.; Day, M.A.; Short, K.C.; Evers, C.R. Assessing the impacts of federal forest planning on wildfire risk mitigation in the Pacific Northwest, USA. *Landsc. Urban Plan.* **2016**, *147*, 1–17. [CrossRef]
40. Peralta, J. *Sectorización Fitoclimática de Navarra. Series de Vegetación y Sectorización Fitoclimática de la Comarca Agraria III*; Departamento de Agricultura, Ganadería y Alimentación, Ed.; Trabajos Catastrales S.A.: Pamplona, Spain, 2000; p. 71.
41. Ager, A.A.; Preisler, H.K.; Arca, B.; Spano, D.; Salis, M. Wildfire risk estimation in the Mediterranean area. *Environmetrics* **2014**, *25*, 384–396. [CrossRef]
42. Finney, M.A. An Overview of Flammap Fire Modeling Capabilities; Fuels Management-How to Measure Success. In Proceedings of the RMRS-P-41, Fort Collins, CO, USA, 28–30 March 2006; Andrews, P.L., Butler, B.W., Eds.; USDA Forest Service, Rocky Mountain Research Station: Fort Collins, CO, USA, 2006; pp. 213–220.
43. Burgan, R.E. *Standard Fire Behavior Fuel Models: A Comprehensive Set for Use with Rothermel's Surface Fire Spread Model*; RMRS-GTR-153; USDA Forest Service, Rocky Mountain Research Station: Fort Collins, CO, USA, 2005; p. 72.
44. Fernandes, P.M. Combining forest structure data and fuel modelling to classify fire hazard in Portugal. *Ann. For. Sci.* **2009**, *66*, 1–9. [CrossRef]
45. González-Olabarria, J.-R.; Rodríguez, F.; Fernández-Landa, A.; Mola-Yudego, B. Mapping fire risk in the model forest of Urbión (Spain) based on airborne lidar measurements. *For. Ecol. Manag.* **2012**, *282*, 149–156. [CrossRef]
46. McGaughey, R.J. *Fusion/ldv: Software for Lidar Data Analysis and Visualization*; USDA Forest Service: Portland, OR, USA, 2014; p. 175.
47. Alcasena, F.J.; Salis, M.; Naulsar, N.J.; Aguinaga, A.E.; Vega-García, C. Quantifying economic losses from wildfires in black pine afforestations of Northern Spain. *For. Policy Econ.* **2016**, *73*, 153–167. [CrossRef]
48. Ager, A.A.; Vaillant, N.M.; Finney, M.A. Integrating fire behavior models and geospatial analysis for wildland fire risk assessment and fuel management planning. *J. Combust.* **2011**. [CrossRef]
49. Bradshaw, L.; MacCormick, E. *Fire Family Plus User's Guide, version 2.0*; USDA Forest Service: Ogden, UT, USA, 2000.
50. Nelson, R.M. Prediction of diurnal change in 10-h fuel stick moisture content. *Can. J. For. Res.* **2000**, *30*, 1071–1087. [CrossRef]
51. Hilbert, D.W.; Ostendorf, B. The utility of artificial neural networks for modelling the distribution of vegetation in past, present and future climates. *Ecol. Model.* **2001**, *146*, 311–327. [CrossRef]
52. Scrinzi, G.; Marzullo, L.; Galvagni, D. Development of a neural network model to update forest distribution data for managed alpine stands. *Ecol. Model.* **2007**, *206*, 331–346. [CrossRef]

53. Kalabokidis, K.; Ager, A.; Finney, M.; Athanasis, N.; Palaiologou, P.; Vasilakos, C. Aegis: A wildfire prevention and management information system. *Nat. Hazards Earth Syst. Sci.* **2016**, *16*, 643–661. [CrossRef]
54. Fahlman, S.E.; Lebiere, C. The Cascade-Correlation Learning Architecture. In *Advances in Neural Information Processing Systems*; Touretzky, D.S., Ed.; Morgan Kaufmann: San Francisco, CA, USA, 1990; pp. 524–532.
55. NeuralWare. *Neuralworks Predict, the Complete Solution for Neural Data Modelling*; NeuralWare: Carnegie, PA, USA, 2014; p. 397.
56. Webros, P.J. *The Roots of Backpropagation: From Ordered Derivatives to Neural Networks and Political Forecasting. Adaptive and Learning Systems for Signal Processing Communication and Control*; Wiley: New York, NY, USA, 1994.
57. Hasenauer, H.; Merkl, D.; Weingartner, M. Estimating tree mortality of Norway spruce stands with neural networks. *Adv. Environ. Res.* **2001**, *5*, 405–414. [CrossRef]
58. Corne, S.A.; Carver, S.J.; Kunin, W.E.; Lennon, J.J.; Van Hees, W.W.S. Predicting forest attributes in Southeast Alaska using artificial neural networks. *For. Sci.* **2004**, *50*, 259–276.
59. Alcázar, J.; Palau, A.; Vega-García, C. A neural net model for environmental flow estimation at the Ebro river basin, Spain. *J. Hydrol.* **2008**, *349*, 44–55. [CrossRef]
60. Costafreda-Aumedes, S.; Cardil, A.; Molina, D.M.; Daniel, S.N.; Mavsar, R.; Vega-Garcia, C. Analysis of factors influencing deployment of fire suppression resources in Spain using artificial neural networks. *IForest Biogeosci. For.* **2016**, *9*, 138–145. [CrossRef]
61. Finney, M.A. Fire growth using minimum travel time methods. *Can. J. For. Res.* **2002**, *32*, 1420–1424. [CrossRef]
62. Jahdi, R.; Salis, M.; Darvishsefat, A.A.; Alcasena, F.; Mostafavi, M.A.; Etemad, V.; Lozano, O.M.; Spano, D. Evaluating fire modelling systems in recent wildfires of the Golestan National Park, Iran. *Forestry* **2016**, *89*, 136–149. [CrossRef]
63. Oliveira, T.M.; Barros, A.M.G.; Ager, A.A.; Fernandes, P.M. Assessing the effect of a fuel break network to reduce burnt area and wildfire risk transmission. *Int. J. Wildland Fire* **2016**, *25*, 619–632. [CrossRef]
64. Chung, W.; Jones, G.; Krueger, K.; Bramel, J.; Contreras, M. Optimising fuel treatments over time and space. *Int. J. Wildland Fire* **2013**, *22*, 1118–1133. [CrossRef]
65. Elia, M.; Lovreglio, R.; Ranieri, N.; Sanesi, G.; Lafortezza, R. Cost-effectiveness of fuel removals in mediterranean wildland-urban interfaces threatened by wildfires. *Forests* **2016**, *7*, 149. [CrossRef]
66. Rothermel, R.C. *A Mathematical model For Predicting Fire Spread in Wildland Fuels*; USDA Forest Service: Ogden, UT, USA, 1972; p. 40.
67. Byram, G.M. Combustion of Forest Fuels. In *Forest Fire: Control and use*; Brown, K.P., Ed.; McGraw-Hill: New York, NY, USA, 1959; pp. 61–89.
68. Dalkey, N.C.; Helmer, O. An experimental application of the delphi method to the use of experts. *Manag. Sci.* **1963**, *9*, 458–468. [CrossRef]
69. Lovreglio, R.; Leone, V.; Giaquinto, P.; Notarnicola, A. Wildfire cause analysis: Four case-studies in Southern Italy. *iForest Biogeosci. For.* **2010**, *3*, 8–15. [CrossRef]
70. Meddour-Sahar, O.; Meddour, R.; Leone, V.; Lovreglio, R.; Derridj, A. Analysis of forest fires causes and their motivations in Northern Algeria: The delphi method. *iForest Biogeosci. For.* **2013**, *6*, 247–254. [CrossRef]
71. Salis, M.; Alcasena, F.J.; Ager, A.A.; Casula, F.; Arca, B.; Bacciu, V.; Spano, D. First annual conference of the Italian Society for Climate Sciences. In *Analyzing Changes in Wildfire Likelihood and Intensity in Mediterranean Areas: A case Study From Central Sardinia, Italy*; Italian Society for the Climate Sciences: Lecce, Italy, 2013; pp. 685–706.
72. Ager, A.A.; Kline, J.; Fischer, A.P. Coupling the biophysical and social dimensions of wildfire risk to improve wildfire mitigation planning. *Risk Anal.* **2015**, *35*, 1393–1406. [CrossRef] [PubMed]
73. Cumming, G.S.; Cumming, D.H.M.; Redman, C.L. Scale mismatches in social-ecological systems: Causes, consequences, and solutions. *Ecol. Soc.* **2006**, *11*, 14. [CrossRef]
74. Martínez, J.; Vega-García, C.; Chuvieco, E. Human-caused wildfire risk rating for prevention planning in Spain. *J. Environ. Manag.* **2009**, *90*, 1241–1252. [CrossRef] [PubMed]
75. Salis, M.; Arca, B.; Alcasena, F.; Arianoutsou, M.; Bacciu, V.; Duce, P.; Duguy, B.; Koutsias, N.; Mallinis, G.; Mitsopoulos, I.; et al. Predicting wildfire spread and behaviour in Mediterranean landscapes. *Int. J. Wildland Fire* **2016**. [CrossRef]

76. Mallinis, G.; Mitsopoulos, I.; Beltran, E.; Goldammer, J. Assessing wildfire risk in cultural heritage properties using high spatial and temporal resolution satellite imagery and spatially explicit fire simulations: The case of holy mount Athos, Greece. *Forests* **2016**, *7*, 46. [CrossRef]
77. Lampin-Maillet, C.; Long-Fournel, M.; Ganteaume, A.; Jappiot, M.; Ferrier, J.P. Land cover analysis in wildland-urban interfaces according to wildfire risk: A case study in the South of France. *For. Ecol. Manag.* **2011**, *261*, 2200–2213. [CrossRef]
78. Syphard, A.D.; Keeley, J.E.; Massada, A.B.; Brennan, T.J.; Radeloff, V.C. Housing arrangement and location determine the likelihood of housing loss due to wildfire. *PLoS ONE* **2012**, *7*, e33954. [CrossRef] [PubMed]
79. McCaffrey, S.M.; Rhodes, A. Public response to wildfire: Is the Australian "stay and defend or leave early" approach an option for wildfire management in the United States? *J. For.* **2009**, *107*, 9–15.
80. Salis, M.; Laconi, M.; Ager, A.A.; Alcasena, F.J.; Arca, B.; Lozano, O.; Fernandes de Oliveira, A.; Spano, D. Evaluating alternative fuel treatment strategies to reduce wildfire losses in a Mediterranean area. *For. Ecol. Manag.* **2016**, *368*, 207–221. [CrossRef]
81. Ager, A.A.; Vaillant, N.M.; McMahan, A. Restoration of fire in managed forests: A model to prioritize landscapes and analyze tradeoffs. *Ecosphere* **2013**, *4*, 29. [CrossRef]
82. Price, O.F.; Bradstock, R.A. The efficacy of fuel treatment in mitigating property loss during wildfires: Insights from analysis of the severity of the catastrophic fires in 2009 in Victoria, Australia. *J. Environ. Manag.* **2012**, *113*, 146–157. [CrossRef] [PubMed]
83. Cochrane, M.A.; Moran, C.J.; Wimberly, M.C.; Baer, A.D.; Finney, M.A.; Beckendorf, K.L.; Eidenshink, J.; Zhu, Z. Estimation of wildfire size and risk changes due to fuels treatments. *Int. J. Wildland Fire* **2012**, *21*, 357–367. [CrossRef]
84. Madrigal, J.; Fernández-Migueláñez, I.; Hernando, C.; Guijarro, M.; Vega-Nieva, D.J.; Tolosana, E. Does forest biomass harvesting for energy reduce fire hazard in the Mediterranean basin? A case study in the caroig massif (Eastern Spain). *Eur. J. For. Res.* **2016**. [CrossRef]
85. Casasús, I.; Bernués, A.; Sanz, A.; Villalba, D.; Riedel, J.L.; Revilla, R. Vegetation dynamics in Mediterranean forest pastures as affected by beef cattle grazing. *Agric. Ecosyst. Environ.* **2007**, *121*, 365–370. [CrossRef]
86. Ruiz-Mirazo, J.; Robles, A.B.; González-Rebollar, J.L. Two-year evaluation of fuelbreaks grazed by livestock in the wildfire prevention program in Andalusia (Spain). *Agric. Ecosyst. Environ.* **2011**, *141*, 13–22. [CrossRef]
87. Mena, Y.; Ruiz-Mirazo, J.; Ruiz, F.A.; Castel, J.M. Characterization and typification of small ruminant farms providing fuelbreak grazing services for wildfire prevention in Andalusia (Spain). *Sci. Total Environ.* **2016**, *544*, 211–219. [CrossRef] [PubMed]
88. González, L.A. La maquinaria agrícola y los incendios forestes. *Alerta* **2013**, *2*, 29–34.
89. Andrews, P.L.; Heinsch, F.A.; Schelvan, L. *How to Generate and Interpret Fire Characteristics Charts for Surface and Crown Fire Behavior*; USDA Forest Service: Washington, DC, USA, 2011; p. 40.
90. Vogler, K.C.; Ager, A.A.; Day, M.A.; Jennings, M.; Bailey, J.D. Prioritization of forest restoration projects: Tradeoffs between wildfire protection, ecological restoration and economic objectives. *Forests* **2015**, *6*, 4403–4420. [CrossRef]
91. Ager, A.A.; Day, M.A.; Vogler, K. Production possibility frontiers and socioecological tradeoffs for restoration of fire adapted forests. *J. Environ. Manag.* **2016**, *176*, 157–168. [CrossRef] [PubMed]

10

Sampling Method and Tree-Age Affect Soil Organic C and N Contents in Larch Plantations

Huimei Wang [1], Wenjie Wang [1,2,*] and Scott X. Chang [3]

[1] Key Laboratory of Forest Plant Ecology, Northeast Forestry University, Harbin 150040, China; whm0709@hotmail.com
[2] Northeast Institute of Geography and Agricultural Ecology, Chinese Academy of Sciences, Changchun 130102, China
[3] Department of Renewable Resources, University of Alberta, 442 Earth Science Building, Edmonton, AB T6G 2E3, Canada; scott.chang@ualberta.ca
* Correspondence: wjwang225@hotmail.com

Abstract: We currently have a poor understanding of how different soil sampling methods (pedogenetic horizon versus fixed-depth) influence the evaluation of soil properties. Here, 159 soil profiles were sampled from larch (*Larix gmelinii*) plantations in northeast China using both the pedogenetic horizon and fixed-depth sampling methods. Analysis of variance was used to determine how sampling method influences the assessment of the spatial variation in the concentration and storage of soil organic C (SOC) and N (SON), as well as how these properties are affected by tree age-group (<20, 20–40, and >40 years). In both the 20 cm (surface) and 80 cm (whole profile) sampling depths, pedogenetic sampling resulted in 1.2- to 1.4-fold higher SOC and SON concentrations than fixed-depth sampling. Surface soil nutrient storage between the two sampling methods was not significantly different, but was it was 1.2-fold higher ($p < 0.05$) with pedogenetic sampling than with fixed-depth sampling in the whole soil profile. For a given error limit in SOC and SON assessments, fixed-depth sampling had a 60%~90% minimum sampling intensity requirement compared with pedogenetic horizon sampling. Additionally, SOC was 1.1- to 1.3-fold greater in the >40 years age-group than in the <20 years age-group ($p < 0.05$), while SON was the highest in the 20–40 years age-group ($p < 0.05$). The total amount of SOC and nutrients in soil is fixed regardless how you sample, it is the different assumptions and different ways of extrapolation from samples to the population that cause sampling by horizon versus fixed depth to lead to different conclusions. Our findings highlight that soil sampling method and tree age-group affect the determination of the spatial variation of SOC and SON and future soil assessments should control for methodological differences.

Keywords: *Larix gmelinii*; tree-age effect; degraded farmland; Northeast China; soil nutrient variation; soil heterogeneity; pedogenetic horizon; fixed-depth sampling

---

1. Introduction

Accurate evaluation of soil properties is important for species-to-location matching during afforestation, precision agroforestry [1–3], and the investigation of forest ecological functions such as carbon (C) sequestration and nutrient dynamics [4]. Soil heterogeneity occurs horizontally and vertically [5]; thus, the lack of a standardized sampling procedure to assess soil fertility and C inventory could affect conclusions drawn from different studies. Therefore, research is needed to understand how sampling method influences the evaluation of the inter-site variations (i.e., interaction between sampling method and site variation) of soil nutrients. Accounting for the effect of sampling method could improve comparability among studies [6,7], particularly in terms of between-site soil fertility assessments [8].

Forest development with age can increase C storage and recover nutrient levels in soils degraded by long-term agricultural cultivation [9–12]. Soils can range from being a C sink (11 to 238 $g \cdot C \cdot m^{-2} \cdot year^{-1}$) [4,13–15] to a source of atmospheric CO_2 (0.1 to 14.1 $g \cdot C \cdot m^{-2} \cdot year^{-1}$) [11,16]. Additionally, studies examining post-afforestation conditions have reported both the depletion of soil nutrients [4,10,17–19] and the accumulation of soil organic N (SON) with increasing soil organic C (SOC) [9,12,20] over time (e.g., tree age effect). Previous land use, the tree species planted, and soil nutrient supply are thought to explain the contradictions in tree age effects on soil properties after afforestation [9–12,17–20]. Furthermore, forest age itself [21], soil depth, and other site-related variations [22–25] have been used to describe the spatial and temporal changes in nutrient accumulation. However, differences in sampling methods (fixed-depth [4,11,13] versus pedogenetic horizon [26–28] sampling) are a potentially powerful explanation for between-study differences in soil properties [29,30] that has not been fully studied. Indeed, comprehensive research including soil sampling method, sampling site, sampling depth, and tree age would provide a more holistic view of how different sampling methods alter evaluations of SOC and SON.

The objective of this study was to determine how soil sampling method, sampling site, sampling depth, and tree age independently and interactively influence SOC and SON, particularly variation across the soil profile in different soil depths and the effect of tree age. The investigation took advantage of widespread larch plantations (4.5 million hectares, or 70% of the total plantation area) available in northeastern China [11,31]. We hypothesized that soil sampling methods could greatly affect the evaluations of spatial changes of and tree-age effects on SOC and N, independent of soil depth inclusion and unit expression of storage or concentration. Our results may contribute to improving inter-study comparisons of SOC and SON measurements, leading to a broader study applicability and improved interpretation of conclusions drawn across disparate reports.

2. Materials and Methods

2.1. Study Site and Soil Sampling

All sampling was conducted in *Larix gmelinii* plantations located in northeast China. This region (45°20′–47°14′ N, 127°30′–128°55′ E) has a continental monsoon climate with mean annual temperatures ranging from −0.3 to 2.6 °C and mean annual precipitation ranging from 676 to 724 mm. The soil is generally a typical Dark-Brown Forest Soil, with occasional eluviation features. The soil in Chinese Soil Taxonomy is Mollic Bori-udic Cambosols or Albic Bori-udic Cambosols [32]; it is a Eutroboralf in the US Soil Taxonomy and a Halic Luvisol in the FAO (Food and Agriculture Organization) system [33]. Soil pH ranges from 4.9 to 6.5. The details of the study sites have been previously described [11].

A total of 159 plots were sampled from four sites: the Laoshan experimental station (LS; 57 plots), the Maoershan experimental farm (MES; 37 plots), the Dongshan forest farm (DS; 30 plots), and the Daqingchuan forest farm (DQC; 35 plots).

Fixed-depth and pedogenetic horizon sampling were employed for soil sampling. Soil cutting rings with a 100 cm^3 volume were used for sampling soil from a soil pit. A soil pit was excavated in each of four subplots established in each of the 159 20 m× 20 m plots (for a total of 636 soil pits), and the samples from the four pits from the same horizon/layer of the same sampling method were mixed to form a composite sample for that layer and plot. Soil samples were collected from one or two sides of the pit for sampling the same layer or depth. Those composite samples from a plot make up the samples for a soil profile for each plot. Such composite samples better represent the soil from a plot. Under fixed-depth sampling, the soil was collected from four depths: 0–20, 20–40, 40–60, and 60–80 cm. Under pedogenetic horizon sampling, soil samples were collected from the A, B, and C horizons. The first horizon contains partially humified organic matter, giving the soil a darker color than soils from lower horizons. The second horizon has undergone sufficient changes during soil genesis, with maximum accumulation of materials such as silicate clay, metal (e.g., Fe and Al) oxides, and organic material. The third horizon comprises unconsolidated material below the solum (A and B horizons) that is little affected by soil-forming processes and lacks pedogenic development.

A total of 1067 composite soil samples were collected (pedogenetic: 113 plots × 3 samples with one each from horizons A, B, and C, plus 46 plots with only A and B horizons, for a total of 431 samples; fixed-depth: 159 plots × 4 samples from the four depths = 636 samples). At the time of sampling, the soil samples were placed in cloth soil-sampling bags for transportation back to the laboratory. To compare the effect of sampling method, the surface soil layer (0–20 cm) and the whole profile soil (0–80 cm) as two sampling depths were examined in this study. In other words, within the 0–20 cm and 0–80 cm sampling depths, we assessed the impact of sampling method on the assessment of SOC and SON.

Tree age was determined using an increment borer (Zhonglinweiye, Beijing, China) on at least five trees in each plantation plot. Fifty-six plots contained trees younger than 20 years (<20 years age-group), with an average tree age of 8.1 ± 7.1 years (SD, standard deviation). The second age-group (20–40 years) had 73 plots, with an average tree age of 25.5 ± 5.3 years. The third age-group (>40 years) had 30 plots and an average tree age of 44.8 ± 2.3 years.

2.2. *Bulk Density, SOC, and SON Measurements*

Once the soil samples were brought back to the laboratory, they were air-dried in a ventilated room until they reached a constant weight. Cutting rings with a 100 cm^3 volume were used for intact bulk density measurement, and bulk density was determined based on the oven-dried mass and volume (400 cm^3 of soil from four cutting rings as a composite sample) of each site. In the fixed-depth sampling, we measured one soil bulk density at each soil depth (0–20, 20–40, 40–60, and 60–80 cm) of each site. In the pedogenetic sampling, we measured one soil bulk density in each horizon (A, B, and C) of each site, and in some places, the soil bulk density at some transitional horizons (e.g., AB horizon) was also measured. Soil was ground to <0.25 mm for measuring SOC and SON concentrations with a heated dichromate/titration and the semimicro-Kjeldahl method, respectively. Details about those methods are available in a previous publication of our group [11].

2.3. *Storage Calculation*

The storage of SOC and SON was calculated with the following equation:

$$\text{Storage } (g{\cdot}m^{-2}) = a \times \rho_b \times depth \times (1- V_{\text{gravel}}), \quad (1)$$

where a is SOC or SON concentration ($g{\cdot}kg^{-1}$), ρ_b is soil bulk density ($g{\cdot}cm^{-3} = Mg{\cdot}m^{-3}$, Mg: million gram), and V_{gravel} is the fraction of gravel volume (determined via water displacement) in the soil.

2.4. *Assessing the Effect of Sampling Method on Tree Age Effect and Spatial Variation in SOC and SON*

Analysis of variance (ANOVA) was used to determine whether soil sampling method affected the spatial variation (among study sites) of SOC and SON and differences among tree age-groups. Specifically, we examined whether study site and age-group interacted with sampling method (all were included in the model as fixed factors). The concentration and storage of SOC and SON at 0–20 and 0–80 cm soil were analyzed. Duncan post-hoc tests were then conducted to identify the specific parameter (SOC concentration, SOC storage, SON concentration, and SON storage) that was affected by site, age-group, and sampling method. Marginal means were estimated and pairwise comparisons (LSD test) were used to separate the means across predictor variables, and their interactions with sampling method (site × method and age-group × method). All of these analyses were performed in SPSS 17.0 (SPSS Inc., Armonk, NY, USA).

2.5. *Number of Samples Required for Estimating SOC and SON at Specific Error Limits*

As a way of quantifying differences between the two soil sampling methods, statistical analyses [34,35] were conducted to compare the number of sampling points (n) required for estimating mean SOC and SON at a site within 5%, 10%, 20%, and 30% of its actual value, at the 95% probability level for both datasets (pedogenetic and fixed-depth). The following equation was used:

$$n = \frac{t_\alpha^2 s^2}{D^2}, \tag{2}$$

where t_α is the Student's t statistic with degrees of freedom at the α probability level, s is standard deviation, and D is the specified error limit. Only the SOC and SON concentration data were used as patterns in the storage data were similar.

3. Results

3.1. Differences between Sites, Sampling Methods, and Age-Groups

The SOC and SON concentrations in both the surface and whole profile soil differed significantly among sites ($p < 0.001$) and age-groups ($p < 0.05$), and between sampling methods ($p < 0.001$). However, SOC and SON storage did not follow the same pattern. For example, no difference between sampling methods was observed in surface soil SOC storage ($p > 0.05$; Table 1).

The highest SOC and SON concentrations were usually observed at DQC or DS, while the lowest values were found in LS (Table 2). Inter-site variation (the highest/lowest ratio) was 2.8 and 3.5 for surface soil SOC and SON, respectively, whereas inter-site variation was 2.9 and 3.2 for SOC and SON, respectively, in the whole profile soil. The highest SOC and SON storage in surface and whole profile soils were found at DS and DQC, respectively, with the lowest values at LS. The ratios of the highest to the lowest values for SOC and SON storage in the surface soil were 2.3 and 2.6, respectively, while those for the whole profile soil were 2.5 and 2.7, respectively (Table 2).

Pedogenetic sampling generally yielded higher SOC and SON concentrations (Table 2). Surface and whole profile soil SOC and SON concentrations were 1.2- to 1.4-fold higher in pedogenetic than in fixed-depth sampling (Table 2). Different patterns were found in their storage. Whole profile soil storage of both SOC and SON was 1.2-fold higher under pedogenetic than under fixed-depth sampling ($p < 0.05$), but the difference was only 1.05- to 1.07-fold ($p > 0.05$) for the surface soil (Table 2).

Age-group differences increased linearly from <20, to 20–40, and >40 years groups in SOC concentration and storage in the surface soil, SON concentration in the surface soil, and SOC storage in the whole profile soil (Table 2). Forests in the 20–40 years age-group exhibited peak whole profile soil.

Table 1. ANOVA result on the influence of sampling method, age-group, and study site on soil organic C (SOC) and N (SON) concentrations, storage, in surface (0–20 cm), and the whole profile soil (0–80 cm). Bold fonts indicate statistical significance ($p < 0.05$).

	Concentration ($g{\cdot}kg^{-1}$)				Storage ($kg{\cdot}m^{-2}$)			
	SOC		SON		SOC		SON	
	F	*p*-Value	*F*	*p*-Value	*F*	*p*-Value	*F*	*p*-Value
Surface soil (0–20 cm)								
Site	104.31	**<0.001**	119.5	**<0.001**	131.26	**<0.001**	118.77	**<0.001**
Age-group	8.50	**<0.001**	5.51	**0.004**	11.17	**<0.001**	4.13	**0.017**
Sampling method	15.29	**<0.001**	15.50	**<0.001**	1.65	0.200	1.83	0.177
Site × age-group	0.51	0.771	1.76	0.122	0.62	0.682	2.38	**0.039**
Site × method	8.08	**<0.001**	6.18	**<0.001**	4.71	**0.003**	2.12	0.098
Age-group × method	2.68	0.070	0.35	0.708	2.46	0.087	0.06	0.946
Site × age-group × method	0.75	0.586	0.66	0.656	0.67	0.647	0.41	0.839
Whole profile soil (0–80 cm)								
Site	87.56	**<0.001**	95.16	**<0.001**	135.95	**<0.001**	122.43	**<0.001**
Age-group	9.74	**<0.001**	11.32	**<0.001**	9.39	**<0.001**	9.25	**<0.001**
Sampling method	37.45	**<0.001**	25.57	**<0.001**	28.60	**<0.001**	10.42	**0.001**
Site × age-group	1.30	0.262	1.64	0.149	1.46	0.203	1.94	0.088
Site × method	6.37	**<0.001**	3.25	**0.022**	13.76	**<0.001**	4.89	**0.002**
Age-group × method	0.18	0.839	0.11	0.892	0.18	0.832	0.47	0.625
Site × age-group × method	0.33	0.897	0.49	0.782	0.21	0.958	0.48	0.790

Table 2. Differences in soil organic C (SOC) and N (SON) concentration and storage across four sites, three age-groups, and two soil sampling methods.

Item	Site	Concentration ($g \cdot kg^{-1}$)	Storage ($kg \cdot m^{-2}$)	Age-Group (Year)	Concentration ($g \cdot kg^{-1}$)	Storage ($kg \cdot m^{-2}$)	Method	Concentration ($g \cdot kg^{-1}$)	Storage ($kg \cdot m^{-2}$)
	DQC	**50.3 c**	**8.96 c**	<20	**35.5 a**	**6.83 a**	FD	**36.3 a**	7.35 a
	DS	**54.5 c**	**9.95 c**	20–40	**40.7 b**	**7.65 b**	PE	**43.8 b**	7.71 a
SOC 0–20 cm	LS	**19.4 a**	**4.29 a**	>40	**45.2 b**	**8.30 b**			
	MES	**39.9 b**	**6.61 b**						
	Highest/lowest	2.8	2.3	>40/<20	1.3	1.2	PE/FD	1.2	1.05
	DQC	**23.8 c**	**21.1 c**	<20	**18.5 a**	**16.6 a**	FD	**17.5 a**	**16.2 a**
	DS	**28.6 d**	**24.1 d**	20–40	**22.02 b**	**18.7 b**	PE	**23.6 b**	**19.7 b**
SOC 0–80 cm	LS	**10.0 a**	**9.71 a**	>40	**21.4 ab**	**18.8 b**			
	MES	**19.4 b**	**16.4 b**						
	Highest/lowest	2.9	2.5	>40/<20	1.2	1.1	PE/FD	1.4	1.2
	DQC	**3.80 c**	**0.68 c**	<20	**2.58 a**	**0.51 a**	FD	**2.6 a**	0.52 a
	DS	**3.70 c**	**0.67 c**	20–40	**3.02 b**	**0.57 b**	PE	**3.2 b**	0.56 a
SON 0–20 cm	LS	**1.10 a**	**0.26 a**	>40	**3.05 b**	**0.55 ab**			
	MES	**2.80 b**	**0.55 b**						
	Highest/lowest	3.5	2.6	20–40/<20	1.2	1.1	PE/FD	1.2	1.08
	DQC	**1.80 c**	**1.66 c**	<20	**1.35 a**	**1.21 a**	FD	**1.30 a**	**1.19 a**
	DS	**1.90 c**	**1.51 bc**	20–40	**1.69 b**	**1.42 b**	PE	**1.70 b**	**1.37 b**
SON 0–80 cm	LS	**0.60 a**	**0.61 a**	>40	**1.37 a**	**1.19 a**			
	MES	**1.60 b**	**1.39 b**						
	Highest/lowest	3.2	2.7	20–40/<20	1.3	1.2	PE/FD	1.3	1.2

Note: DQC, Daqingchuan site; DS, Dongshan site; LS, Laoshan site; MES, Maoershan site; FD, fixed-depth sampling; PE, pedogenetic horizon sampling. Different letters in the same column and sampling depth indicate significant differences. Means that are significantly different among the treatment under comparison are highlighted.

SOC and SON concentrations, as well as whole profile soil and surface soil SON storage (Table 2). In general, the >40 years group had 1.1- to 1.3-fold higher SOC than the <20 years age-group, while the 20–40 years group tended to have higher SON (Table 2).

3.2. SOC and SON Concentrations Across Study Sites: Interaction with Sampling Method

Marked site by sampling method interaction was found for SOC and SON concentrations in the surface and whole profile soil (Table 1; Figure 1). Both surface and whole profile soil SOC concentrations were higher with pedogenetic than with fixed-depth sampling. Inter-site differences were also found (Figure 1). The highest increases from one sampling method to another in surface soil SOC concentrations generally occurred in DQC (52%), while moderate increases (7%–10%) were observed in DS, LS, and MES (Figure 1). The whole profile soil (0–80 cm) typically accompanied higher SOC in pedogenetic sampling than in fixed-depth sampling. For example, SOC concentration at DS was 8% higher with pedogenetic sampling than with fixed-depth sampling in the surface soil, whereas it was 38% higher when the whole profile soil was evaluated. Furthermore, at LS, SOC in the surface soil was 10% higher, but that in the whole profile soil was 20% higher in pedogenetic than in fixed-depth sampling (Figure 1).

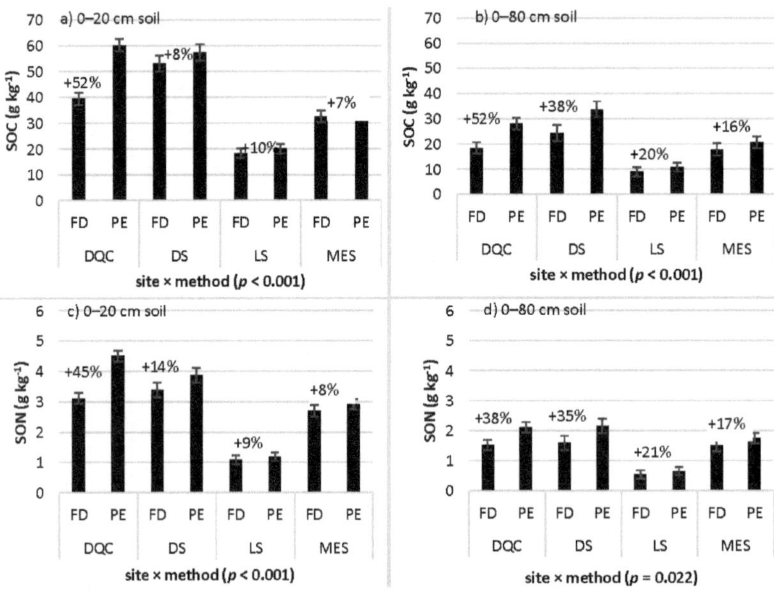

Figure 1. Sampling method influences soil organic C (SOC) and N (SON) concentrations depending on the sampling site (site × sampling method interaction). Data are presented for the two different sampling depths (surface soil: 0–20 cm and whole profile soil: 0–80 cm). A plus sign (+) indicates that the pedogenetic method resulted in higher values than the fixed-depth method, while a minus sign (−) indicates the opposite result. DQC, Daqingchuan site; DS, Dongshan site; LS, Laoshan site; MES, Maoershan site; FD, fixed-depth sampling; PE, pedogenetic horizon sampling. The error bars represent standard error of the mean. 0–20 cm soil for SOC (**a**); 0–80 cm soil for SOC (**b**); 0–20 cm soil for SON (**c**); 0–80 cm soil for SON (**d**).

Sampling influence on SON concentration was similar to that on SOC. Pedogenetic sampling led to higher SON in both surface and whole profile soil as compared with fixed-depth sampling, again with differences among sites (Figure 1). The largest increases (38%–45%) of pedogenetic over fixed-depth sampling occurred at DQC, whereas the other sites had relatively smaller sampling-induced changes (14%–35% at DS, 9%–21% at LS, and 8%–17% at MES; Figure 1). In the surface soil, pedogenetic sampling resulted in 14% (DS), 9% (LS), and 8% (MES) increases in SON concentration as compared with fixed-depth sampling, whereas the increases were 35% (DS), 21% (LS), and 17% (MES) when the whole profile soil was evaluated (Figure 1).

3.3. SOC and SON Storage across Study Sites: Interaction with Sampling Method

The interaction between sampling method and study site significantly affected SOC and SON storage in the two sampling depths, except SON storage in the surface soil (Table 1; Figure 2). At DQC, pedogenetic sampling resulted in 16%–22% higher SOC and SON storage compared with fixed-depth sampling in the surface soil; and 30%–44% higher SOC and SON storage when the whole profile soil was considered. At DS, there was a <6% difference in SOC and SON storage in the surface soil, and a 14%–27% increase in the whole profile soil from pedogenetic to fixed-depth sampling. Additionally, SOC and SON storage values were similar under both methods at LS and MES (Figure 2).

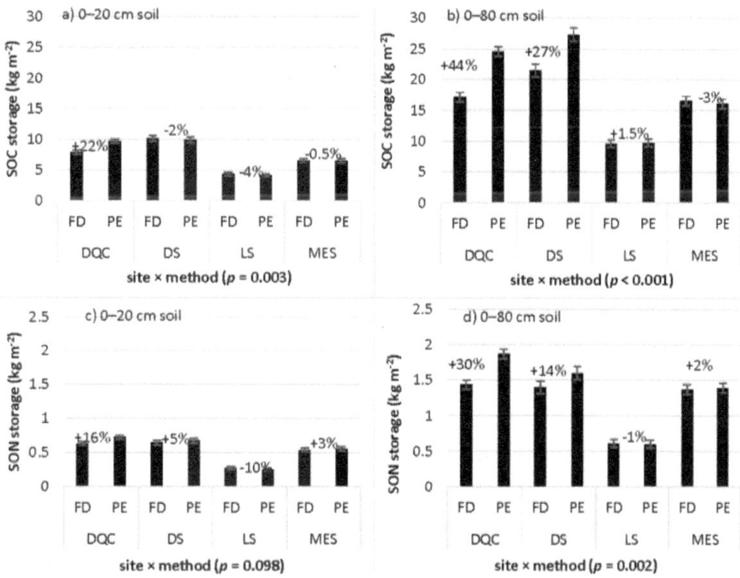

Figure 2. The influence of sampling method on soil organic C (SOC) and N (SON) storage across study sites (site×method interaction) and sampling depth (surface soil: 0–20 cm, whole profile soil: 0–80 cm). A plus sign (+) indicates that the pedogenetic method resulted in higher values than the fixed-depth method, while a minus sign (−) indicates the opposite result. DQC, Daqingchuan site; DS, Dongshan site; LS, Laoshan site; MES, Maoershan site; FD, fixed-depth sampling; PE, pedogenetic horizon sampling. The error bars represent standard error of the data. 0–20 cm soil for SOC (**a**); 0–80 cm soil for SOC (**b**); 0–20 cm soil for SON (**c**); 0–80 cm soil for SON (**d**).

3.4. Age-Group Differences in SOC and SON Concentrations: Interaction with Sampling Method

Age-group effects were similar between the two sampling methods for both SOC and SON concentrations (age-group* method interaction: $p > 0.05$; Table 1 and Figure 3). Moreover, age-group related differences were apparent in surface versus whole profile soil (Figure 3).

Both pedogenetic and fixed-depth sampling found sharp increases (6%–34%) in surface soil SOC concentrations from the <20 years to the 20–40 years age-group, which then moderately increased (9%–10%) from the latter to the >40 years age-group. Similarly, both fixed-depth and pedogenetic sampling found sharp increases (11%–22%) in SON concentration in the surface soil from the <20 years to the 20–40 years age-group, also followed by a small increase (2%) to the >40 years age-group (Figure 3).

Unlike in the surface soil, SOC and SON concentrations in the whole profile soil were lower in the >40 years age-group compared with the 20–40 years age-group (Figure 3). Both sampling methods found sharp increases (19%–28%) in SOC concentration in the 20–40 years age-group from the <20 years age-group, but 4%–5% decreases from the 20–40 years age-group to the >40 years group. Both sampling methods also found sharp increases (25%–27%) in SON concentration in the 20–40 years age-group from the <20 years age-group, followed by 17%–21% decreases in the >40 years age-group from the 20–40 years age-group (Figure 3).

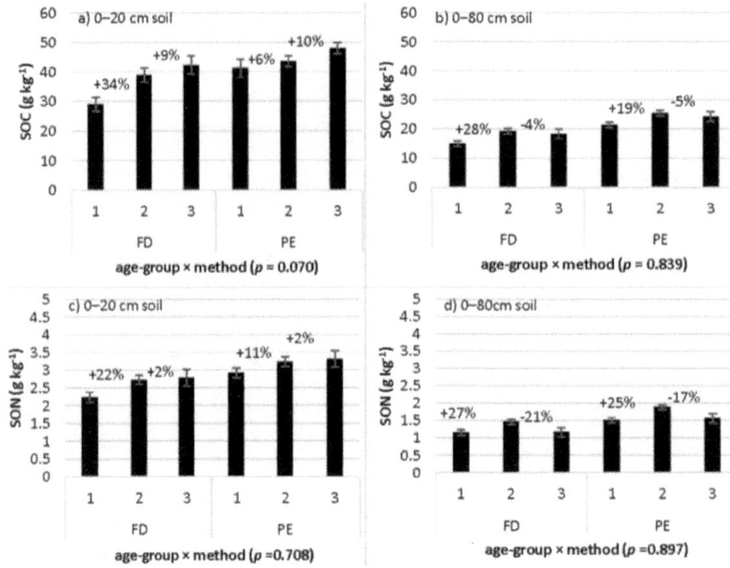

Figure 3. The interaction of age-group and sampling method affected soil organic C (SOC) and N (SON) concentrations, as well as their variation between surface (0–20 cm) and whole profile soil (0–80 cm). A plus sign (+) indicates that the pedogenetic method resulted in higher values than the fixed-depth method, while a minus sign (−) indicates the opposite result. DQC, Daqingchuan site; DS, Dongshan site; LS, Laoshan site; MES, Maoershan site; FD, fixed-depth sampling; PE, pedogenetic horizon sampling. The error bars represent standard error of the data. 0–20 cm soil for SOC (**a**);. 0–80 cm soil for SOC (**b**); 0–20 cm soil for SON (**c**); 0–80 cm soil for SON (**d**).

Owing to non-significant interactions between the age-group and sampling method (Figure 3), the age-group effect could be expressed as group means of the two sampling methods (Table 2). For the surface soil, marked increases in SOC and SON concentrations were found in the 20–40 years age-group from the <20 years age-group, while no marked changes were found between the 20–40 years and >40 years age-groups. However, different patterns were found in the 0–80 cm soil, with marked decreases of SON concentrtion from the 20–40 years age-group to the >40 years age-group (Table 2).

3.5. Age-Group Differences in SOC and SON Storage: Interaction with Sampling Method

Similar to SOC and SON concentrations, the sampling method did not alter age-group effects on SOC and SON storage (non-significant age-group × method interaction; Table 1 and Figure 4). Both fixed-depth and pedogenetic sampling revealed increases (11%–13%) in SON storage from the <20 years to the 20–40 years age-group, then small decreases (2%–3%) from the 20–40 years to the >40 years age-group. In addition, both sampling methods found sharp increases (7%–24%) in SOC storage from the <20 years to the 20–40 years age-group, then small increases (7%) from the 20–40 years to the >40 years age-group. Age-group differences in the whole profile soil SOC and SON storage clearly contrasted those in the surface soil (Figure 4). For example, there were sharp increases (15%–17%) in SOC storage from the <20 years to the 20–40 years age-group, but then no obvious changes (0%–2% decreases) between the 20–40 years and the >40 years age-groups regardless of the sampling method used. Furthermore, there were sharp increases (15%–20%) in SON storage from the <20 years to the 20–40 years age-group, and 13%–20% decreases from the 20–40 years to the >40 years age-group regardless of the sampling method used (Figure 4).

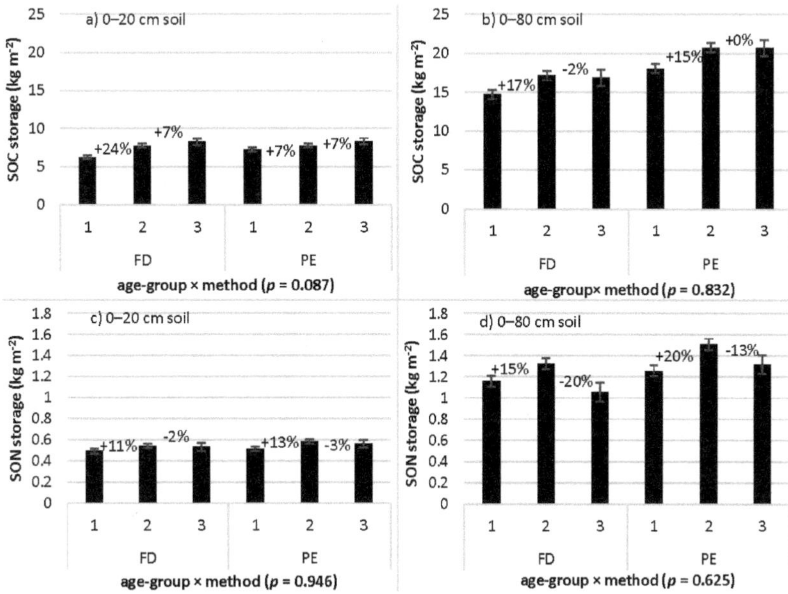

Figure 4. The interaction of age-group and sampling method on variation in soil organic C (SOC; upper two panels) and N (SON; lower two panels) storage, and differences between surface (0–20 cm; **a** and **c**) and whole profile soil (0–80 cm; **b** and **d**). A plus sign (+) indicates higher values in an older group than in the immediately younger one (e.g., >40 years versus 20–40 years), while a minus sign (−) indicates the opposite. 1 = <20 years age-group; 2 = 20–40 years age-group; 3 = >40 years age-group. FD, fixed-depth sampling; PE, pedogenetic horizon sampling. The error bars represent standard error of the data. 0–20 cm soil for SOC (**a**); 0–80 cm soil for SOC (**b**); 0–20 cm soil for SON (**c**); 0–80 cm soil for SON (**d**).

There were moderate increases in SOC storage and sharper decreases in SON storage from the 20–40 years to the >40 years age-group when the whole profile soil was considered (Figure 4). For example, surface soil SOC storage in the >40 years age-group was 7% higher than that in the 20–40 years age-group, whereas the percentage difference was a decrease of 0%–2% in the whole profile soil. Moreover, surface soil SON storage in the >40 years age-group was 2%–3% lower than that in the 20–40 years age-group, but the percentage difference increases to 13%–20% in the whole profile soil (Figure 4).

Owing to non-significant interactions (Figure 4), the age-group effect could be expressed as pooled means of two sampling methods (Table 2). For the surface soil, marked increases in SOC and SON storage were usualy found from the <20 years to the 20–40 years age-group, while no marked changes were found between the 20–40 years and >40 years age-groups. However, different patterns were found in the 0–80 cm soil, with marked decreases of SON storage in the >40 years age-group (Table 2).

3.6. Minimum Sample Number Required for Estimating SOC and SON at Specific Error Limits: Effects of Sampling Method

Fewer sampling points are required with fixed-depth sampling than with pedogenetic sampling, across multiple error limits (Figure 5). At least 538 samples should be collected for analyzing whole profile soil SOC concentration under pedogenetic sampling at the 5% error limit, whereas 37% fewer points are required with fixed-depth sampling to secure the same precision. This pattern of requiring more data points for pedogenetic sampling holds as the error limit increases (Figure 5). Similarly, under the same error limit, at least 543 pedogenetic samples are needed to analyze whole profile soil SON concentrations, compared with 480 samples for in fixed-depth sampling (12% fewer sampling points required) (Figure 5).

Similar results were observed for whole profile soil SOC and SON storage. At the 5% limit, pedogenetic sampling and fixed-depth required 387 and 233 samples, respectively, to estimate SOC storage, as well as 395 and 364 samples for SON storage estimation. At other error limits, these general patterns again held (Figure 5). Outcomes for surface-soil SOC and SON storage and concentration data were similar to the whole profile soil results (data not shown).

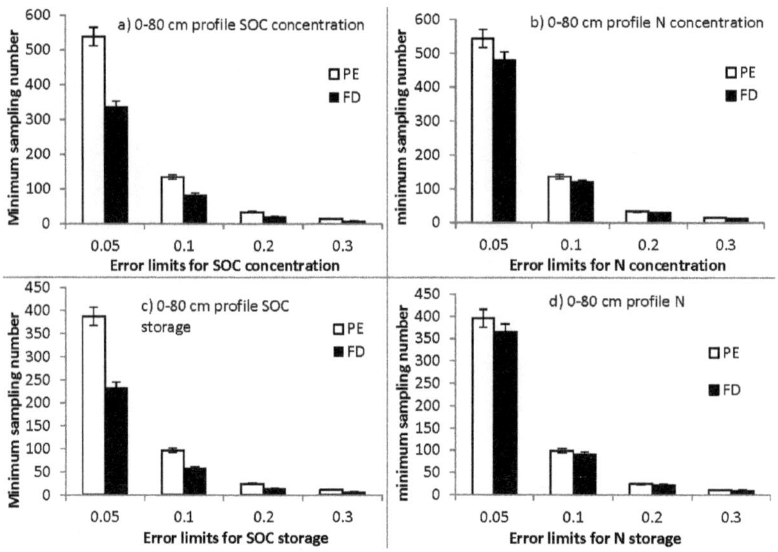

Figure 5. Effects of sampling method on minimum sampling number at different error limits to estimate differences in soil organic C (SOC) and soil organic N (SON) concentration and storage. PE, pedogenetic sampling; FD, fixed-depth sampling. Vertical bars represent a 5% error. 0–80 cm soil profile SOC concentration (**a**); 0–80 cm soil profile SON concentration(**b**); 0–80 cm soil profile SOC storage (**c**); 0–80 cm soil profile SON storage (**d**).

4. Discussion

4.1. Age-Group Effect on SOC and SON Is Dependent on Sampling Depth but Not Sampling Method

Compared with younger forests, older forests can have both higher and lower SOC and SON [11,14–16], but few have examined how sampling methods may affect such evaluations. Here, we demonstrated that sampling methods did not alter the estimate of the age-group effect on SOC and SON in larch plantations (Table 1, Figures 3 and 4). Pedogenetic horizon sampling in previous studies has revealed large increases (31.3 to 48.2 $g \cdot kg^{-1}$) in soil organic matter (SOM) concentration due to thinning [36], as well as large SOM variations (184 to 510 $g \cdot kg^{-1}$) within different larch forests [37]. Using fixed-depth sampling, SOM was found to increase from 47.7 to 51.6 $g \cdot kg^{-1}$ after a second rotation of a 36-years old larch forest at the same site [38]. Therefore, differences in sampling methods do not explain contradictory patterns of SOC and SON variation over time in the studied plantations. As seen in previous reports [9–12,17,18,20], prior land use, tree species planted, and soil nutrient supply may all affect SOC and SON dynamics after afforestation.

The patterns of tree age effect on SOC and SON was strongly dependent on soil sampling depth under both sampling methods. Under a deeper soil sampling depth, the SOC and SON estimates (storage and concentration) were accompanied by moderate SOC increases or sharper SON decreases from the 20–40 years to the >40 years age-group. Previous soil studies in larch forests were focused on the more dynamic surface soil (0–20 cm), which also contains the highest root density [39,40]. However, this focus may have overlooked the importance of deep soil, which is significant given that SOM and nutrient storage in deep soils are important for forest vegetation with deep root systems [41,42], such as the larch trees studied in this paper. A recent, 12 years chronosequence study in larch plantations showed that SOC initially decreased before increasing with stand age [43]. Larch plantations in northeast China could accumulate SOC at a rate that ranges from 57.9 to 139.4 $g{\cdot}m^{-2}{\cdot}year^{-1}$ in the top 20 cm of soil [11]. Moreover, researchers also noted a slight increase in surface soil N storage (~0.33 $kg{\cdot}m^{-2}$), which decreased from 0.11–0.16 to 0.06–0.11 $kg{\cdot}m^{-2}$ in the 20–60 cm soil during larch forest development [11]. Larch tree growth could cause divergent changes in SIC (soil inorganic carbon) and SOC levels, particularly in terms of their vertical distribution, and therefore, these effects should be fully considered in SIC-rich calcareous soils [44]. Larch reforestation could markedly affect the temporal dynamics in the concentration, storage, and vertical distribution of most soil nutrients, as indicated by findings showing eight SOC, SON, P, and K related parameters with differences between surface and subsurface soils [21]. These studies suggest that deeper soil C and nutrient depletion was possible. In contrast, this study suggests that larch plantation establishment in northeast China could sequester C in the mineral soil without depleting SON, at least in the surface soil of stands in the >40 years age-group. However, the inclusion of deeper soil layers could strongly modify these patterns to indicate possible depletion of SOC and SON with stand age.

4.2. Sampling Method Effects on SOC and SON Differ between Their Concentration and Storage

Understanding SOC and SON spatial variation is important in order to improve intensive forest management [2], but the influence of pedogenetic versus fixed-depth sampling on the assessment of SOC and SON spatial variation has not been statistically assessed [29,30,45,46]. This study demonstrated that sampling method significantly affected ($p < 0.05$) the identified size of inter-site variations of SOC and SON. Moreover, such sampling effects differed between concentration and storage of SOC and SON, as well as across the two soil sampling depths. These effects suggest that the effect of sampling method and its interaction with soil sampling depth on nutrient storage or concentration should be considered in future research, particularly when comparing results across different studies.

The A and B horizons in this study were generally thicker than 20 cm, with the A horizon at LS, MES, DS, and DQC averaging 17.8, 27.0, 40.5, and 25.0 cm, respectively, whereas the corresponding B horizon averaged 55.2, 66.1, 69.2, and 49.6 cm. Thus, fixed-depth sampling with a 20-cm increment may collect mixed A and B horizon samples. This corroborates with the literature that if soil horizons with different SOC concentrations are mixed during fixed-depth sampling, SOC and SON concentrations could be much higher in fixed-depth sampling than in pedogenetic sampling [47]. Others have also reported that SOC stock variability decreases under fixed-depth sampling as compared with pedogenetic horizon sampling [29]. Therefore, fixed-depth sampling could result in higher SOC and SON estimates than pedogenetic sampling depending on the depth of the soil sampled and also sampling intensities at different horizons [29,47].

Effects of sampling method on SOC and SON have been reported previously [29,30,45], but quantitative assessment on the effect of sampling method on inter-site variation and differences between sampling depths is lacking. In the upper 30 cm of ploughed Gleysols, VandenBygaart et al. found that pedogenetic horizon sampling reduced the variability in SOC stock [45]. Sampling by pedogenetic horizon is also recommended over fixed-depth when monitoring SOC stock variation in hydromorphic soils of the agricultural landscape [46], as well as when studying pedogenetic processes controlling SOC stocks. In contrast, fixed-depth sampling is preferred for determining regional SOC stock [29]. Yet another report [30] indicated that fixed-depth sampling tends to overestimate cultivation-induced SOC depletion, and thus they recommended the use of pedogenetic soil information (e.g., soil classes) in SOC stock calculation and C sequestration when assessing the impact of land-use change. Our findings differed from these reports; we showed that pedogenetic sampling typically resulted in much higher SOC and SON than fixed-depth sampling (Table 2), which was 1.05- to 1.2-fold higher for SOC and SON storage and 1.2- to 1.4-fold higher for SOC and SON concentration. Pedogenetic sampling was also used in several major datasets in China (e.g., the 1979–1982 national survey) and has since become the main sampling method for nationwide evaluation of SOC and SON budgets [48–50]. Currently, however, more studies use the fixed-depth method for estimating SOC and SON storage in China [11,21,51]. This study provides a statistical assessment of differences in SOC and SON between two major sampling methods, which should be useful for meta-analysis and soil fertility diagnoses across multiple locations.

4.3. Lower SOC and SON Spatial Variations Requiring Lower Sampling Intensity in Fixed-Depth Sampling

Many studies have addressed the effect of forest development on SOC and SON dynamics [11,21], inter-site and vertical soil variations [29,52], as well as the minimum sample number required for accurate assessment [20–23]. The minimum number of samples required should be determined for detecting a specified change in soil properties [34,53]. For designing efficient soil sampling methods, Conant et al. found that differences in the order of 2.0 Mg·C·ha^{-1} could be detected through collecting and analyzing samples from at least five (tilled) or two (forest) microplots in Tennessee [5]. Compared with the large number required as was identified in this paper (300–600 samples), a proper sampling procedure, such as the microplot averaging method, could strongly decrease the number of samples required for accurate evaluation [5]. Moreover, the smaller variation with the fixed-depth sampling indicates that this method requires a lower sampling intensity (10%–40% lower) for a given precision as compared with pedogenetic horizon sampling in assessing changes in SOC and SON over time. Average SOC and SON changing rates are smaller relative to their total amount in the soil, resulting in a small signal to noise ratio [5]. The methodological effect on the minimum sample number required, together with previous findings on proper soil sampling design [5], may increase the precision of SON and SOC evaluations for a given error limit.

5. Conclusions

Larch plantation establishment on degraded farmland increased SOC and SON storage in the 20–40 years age-group, and sampling method (fixed-depth versus pedogenetic horizon sampling) did not alter this pattern ($p > 0.05$). Furthermore, pedogenetic horizon sampling resulted in higher SOC and SON storage and inter-site variations than fixed-depth sampling. At a given error limit for SOC and SON estimation, the minimum number of sampling points required for fixed-depth sampling was 60%~90% of that required for pedogenetic horizon sampling. These sampling method effects can help interpret differences in data collected in disparate studies (e.g., for meta-analysis), and improve our understanding of the assessment of spatial variations of SOC and SON.

Acknowledgments: This study was supported financially by the National Key Research and Development Program (2016YFA0600802), Fundamental Research Funds for the Central Universities (2572016EAJ1&DL13EA03-03), China's National Foundation of Natural Sciences (41373075, 31100457), and the "111" project (B16010). Thanks are due to De'an Xia from Forestry College, Northeast Forestry University for his kind support with statistical analysis and data explanation.

Author Contributions: W.W. conceived and designed the experiment; H.W. performed the experiment and analyzed the data; and H.W., W.W. and SC wrote the manuscript.

Conflicts of Interest: The authors declare no conflict of interest.

References

1. Rushton, B.T. Matching tree species to site conditions in reclamation. In *Evaluation of Alternatives for Restoration of Soil and Vegetation on Phophatic Clay Settling Ponds*; Odum, H.T., Rushton, B.T., Paulic, M., Everett, S., McClanahan, T.R., Munroe, M., Wolfe, R.W., Eds.; Florida Institute of Phosphate Research: Bartow, FL, USA, 1991.
2. Kravchenko, A.N. Influence of spatial structure on accuracy of interpolation methods. *Soil Sci. Soc. Am. J.* **2003**, *67*, 1564–1571. [CrossRef]
3. Franzen, D.W.; Hopkins, D.H.; Sweeney, M.D.; Ulmer, M.K.; Halvorson, A.D. Evaluation of soil survey scale for zone development of site-specific nitrogen management. *Agron. J.* **2002**, *94*, 381–389. [CrossRef]

4. Wang, H.-M.; Wang, W.-J.; Chen, H.; Zhang, Z.; Mao, Z.; Zu, Y.-G. Temporal changes of soil physic-chemical properties at different soil depths during larch afforestation by multivariate analysis of covariance. *Ecol. Evol.* **2014**, *4*, 1039–1048. [CrossRef] [PubMed]
5. Conant, R.T.; Smith, G.R.; Paustian, K. Spatial variability of soil carbon in forested and cultivated sites: Implications for change detection. *J. Environ. Qual.* **2003**, *32*, 278–286. [CrossRef] [PubMed]
6. Guo, L.B.; Gifford, R.M. Soil carbon stocks and land use change: A meta analysis. *Glob. Chang. Biol.* **2002**, *8*, 345–360. [CrossRef]
7. Laganiere, J.; Angers, D.A.; Pare, D. Carbon accumulation in agricultural soils after afforestation: A meta-analysis. *Glob. Chang. Biol.* **2010**, *16*, 439–453. [CrossRef]
8. Zinkevičius, R. Influence of soil sampling for precision fertilizing. *Agron. Res.* **2008**, *6*, 423–429.
9. Morris, S.J.; Bohm, S.; Haile-Mariam, S.; Paul, E.A. Evaluation of carbon accrual in afforested agricultural soils. *Glob. Chang. Biol.* **2007**, *13*, 1145–1156. [CrossRef]
10. Berthrong, S.T.; Jobbagy, E.G.; Jackson, R.B. A global meta-analysis of soil exchangeable cations, pH, carbon, and nitrogen with afforestation. *Ecol. Appl.* **2009**, *19*, 2228–2241. [CrossRef] [PubMed]
11. Wang, W.J.; Qiu, L.; Zu, Y.G.; Su, D.X.; An, J.; Wang, H.Y.; Zheng, G.Y.; Sun, W.; Chen, X.Q. Changes in soil organic carbon, nitrogen, pH and bulk density with the development of larch (*Larix gmelinii*) plantations in china. *Glob. Chang. Biol.* **2011**, *17*, 2657–2676.
12. Li, D.; Niu, S.; Luo, Y. Global patterns of the dynamics of soil carbon and nitrogen stocks following afforestation: A meta-analysis. *New Phytol.* **2012**, *195*, 172–181. [CrossRef] [PubMed]
13. Springsteen, A.; Loya, W.; Liebig, M.; Hendrickson, J. Soil carbon and nitrogen across a chronosequence of woody plant expansion in North Dakota. *Plant Soil* **2010**, *328*, 369–379. [CrossRef]
14. Covington, W. Changes in forest floor organic matter and nutrient content following clear cutting in northern hardwoods. *Ecology* **1981**, *62*, 41–48. [CrossRef]
15. Garten, J. Soil carbon storage beneath recently established tree plantations in Tennessee and South Carolina, USA. *Biomass Bioenergy* **2002**, *23*, 93–102. [CrossRef]
16. Wirth, C.; Czimczik, C.J.; Schulze, E.D. Beyond annual budgets: Carbon flux at different temporal scales in fire-prone siberian scots pine forests. *Tellus B* **2002**, *54*, 611–630. [CrossRef]
17. Kueffer, C.; Klingler, G.; Zirfass, K.; Schumacher, E.; Edwards, P.J.; Gusewell, S. Invasive trees show only weak potential to impact nutrient dynamics in phosphorus-poor tropical forests in the seychelles. *Funct. Ecol.* **2008**, *22*, 359–366. [CrossRef]
18. Cavard, X.; Bergeron, Y.; Chen, H.Y.H.; Pare, D. Effect of forest canopy composition on soil nutrients and dynamics of the understorey: Mixed canopies serve neither vascular nor bryophyte strata. *J. Veg. Sci.* **2011**, *22*, 1105–1119. [CrossRef]
19. Wei, X.; Shao, M.; Gale, W.; Li, L. Global pattern of soil carbon losses due to the conversion of forests to agricultural land. *Sci. Rep.* **2014**, *4*, 4062. [CrossRef] [PubMed]
20. Sauer, T.J.; James, D.E.; Cambardella, C.A.; Hernandez-Ramirez, G. Soil properties following reforestation or afforestation of marginal cropland. *Plant Soil* **2012**, *360*, 375–390. [CrossRef]
21. Wang, W.; Wang, H.; Zu, Y. Temporal changes in SOM, N, P, K, and their stoichiometric ratios during reforestation in China and interactions with soil depths: Importance of deep-layer soil and management implications. *For. Ecol. Manag.* **2014**, *325*, 8–17. [CrossRef]
22. Yim, M.H.; Joo, S.J.; Shutou, K.; Nakane, K. Spatial variability of soil respiration in a larch plantation: Estimation of the number of sampling points required. *For. Ecol. Manag.* **2003**, *175*, 585–588. [CrossRef]
23. Adachi, M.; Bekku, Y.S.; Konuma, A.; Kadir, W.R.; Okuda, T.; Koizumi, H. Required sample size for estimating soil respiration rates in large areas of two tropical forests and of two types of plantation in malaysia. *For. Ecol. Manag.* **2005**, *210*, 455–459. [CrossRef]
24. Laiho, R.; Penttilä, T.; Laine, J. Variation in soil nutrient concentrations and bulk density within Pearland forest sites. *Silva Fenn.* **2004**, *38*, 29–41. [CrossRef]
25. Kulmatiski, A.; Vogt, D.J.; Siccama, T.G.; Beard, K.H. Detecting nutrient pool changes in rocky forest soils. *Soil Sci. Soc. Am. J.* **2003**, *67*, 1282–1286. [CrossRef]
26. Verboom, W.H.; Pate, J.S. Evidence of active biotic influences in pedogenetic processes. Case studies from semiarid ecosystems of south-west Western Australia. *Plant Soil* **2006**, *289*, 103–121. [CrossRef]

27. Funakawa, S.; Hirooka, K.; Yonebayashi, K. Temporary storage of soil organic matter and acid neutralizing capacity during the process of pedogenetic acidification of forest soils in Kinki District, Japan. *Soil Sci. Plant Nutr.* **2008**, *54*, 434–448. [CrossRef]
28. Eger, A.; Almond, P.C.; Condron, L.M. Pedogenesis, soil mass balance, phosphorus dynamics and vegetation communities across a Holocene soil chronosequence in a super-humid climate, South Westland, New Zealand. *Geoderma* **2011**, *163*, 185–196. [CrossRef]
29. Grueneberg, E.; Schoening, I.; Kalko, E.K.V.; Weisser, W.W. Regional organic carbon stock variability: A comparison between depth increments and soil horizons. *Geoderma* **2010**, *155*, 426–433. [CrossRef]
30. Wiesmeier, M.; Sporlein, P.; Geuss, U.; Hangen, E.; Haug, S.; Reischl, A.; Schilling, B.; von Lutzow, M.; Kogel-Knabner, I. Soil organic carbon stocks in Southeast Germany (Bavaria) as affected by land use, soil type and sampling depth. *Glob. Chang. Biol.* **2012**, *18*, 2233–2245. [CrossRef]
31. Sun, Z.H.; Jin, G.Z.; Mu, C.C. *Study on the Keeping Long-Term Productivity of Larix Olgensis Plantation*; Science Press: Beijing, China, 2009.
32. Chen, Z.; Gong, Z.; Zhang, G.; Zhao, W. Correlation of soil taxa between Chinese Soil Genetic Classification and Chinese Soil Taxonomy on various scales. *Soils* **2004**, *36*, 584–595.
33. Gong, Z.; Chen, Z.; Luo, G.; Zhang, G.; Zhao, W. China Soil Classification Reference System. *Soils* **1999**, 57–63.
34. Petersen, R.; Calvin, L. Sampling. In *Methods of Soil Analysis. Part 1. Physical and Mineralogical Methods. Agronomy Monograph 9*, 2nd ed.; Klute, A., Ed.; American Society of Agronomy, Soil Science Society of America: Madison, WI, USA, 1986; pp. 33–51.
35. Robertson, G.P.; Coleman, D.C.; Bledsoe, C.S.; Sollins, P. *Standard Soil Methods for Long-Term Ecological Research*; Oxford University Press: Oxford, UK, 1999; p. 480.
36. Chen, X.Q. The effect of community structure of artificial larch wood on physic-chemical properties of Baijing soil. *J. Northeast For. Univ.* **1986**, *14*, 113–116.
37. Niu, X.; Wei, J.S.; Zhou, M.; Liu, B. Study on the soil organic matter under *Larix gmelinii* forest. In *Soil Fertilizer Society of Inner Mongolia 2007 Conference on Soil Fertilizer and Sustainable Development*; Inner Mogolica Publisher: Hohhot, China, 2007; Volume 2007, pp. 1145–1149.
38. Yan, D.; Wang, J.; Yang, M. Tendency of Soil Degradation in the Pure Larch Plantations. *Chin. J. Ecol.* **1997**, *16*, 62–66.
39. Kajimoto, T.; Osawa, A.; Matsuura, Y.; Abaimov, A.P.; Zyryanova, O.A.; Kondo, K.; Tokuchi, N.; Hirobe, M. Individual-based measurement and analysis of root system development: Case studies for *Larix gmelinii* trees growing on the permafrost region in Siberia. *J. For. Res.* **2007**, *12*, 103–112. [CrossRef]
40. Liu, S.R.; Wang, W.Z.; Wang, M.Q. The characteristics of energy in the formative process of net primary productivity of larch artificial forest ecosystem. *Acta Phytoecol. Geobot. Sin.* **1992**, *16*, 209–219.
41. Jobbágy, E.G.; Jackson, R.B. The vertical distribution of soil organic carbon and its relation to climate and vegetation. *Ecol. Appl.* **2000**, *10*, 423–436. [CrossRef]
42. Rumpel, C.; Kögel-Knabner, I. Deep soil organic matter—A key but poorly understood component of terrestrial c cycle. *Plant Soil* **2011**, *338*, 143–158. [CrossRef]
43. Wang, C.M.; Shao, B.; Wang, R. Carbon sequestration potential of ecosystem of two main tree species in Northeast China. *Acta Ecol. Sin.* **2010**, *30*, 1764–1772.
44. Wang, W.; Su, D.; Qiu, L.; Wang, H.; An, J.; Zheng, G.; Zu, Y. Concurrent changes in soil inorganic and organic carbon during the development of larch, *Larix gmelinii*, plantations and their effects on soil physicochemical properties. *Environ. Earth Sci.* **2012**, *69*, 1559–1570. [CrossRef]
45. VandenBygaart, A.J.; Gregorich, E.G.; Angers, D.A.; McConkey, B.G. Assessment of the lateral and vertical variability of soil organic carbon. *Can. J. Soil Sci.* **2007**, *87*, 433–444. [CrossRef]
46. VandenBygaart, A.J. Monitoring soil organic carbon stock changes in agricultural landscapes: Issues and a proposed approach. *Can. J. Soil Sci.* **2006**, *86*, 451–463. [CrossRef]
47. Palmer, C.J.; Smith, W.D.; Conkling, B.L. Development of a protocol for monitoring status and trends in forest soil carbon at a national level. *Environ. Pollut.* **2002**, *116* (Suppl. 1), S209–S219. [CrossRef]
48. Wu, H.B.; Guo, Z.T.; Peng, C.H. Land use induced changes of organic carbon storage in soils of China. *Glob. Chang. Biol.* **2003**, *9*, 305–315. [CrossRef]
49. Yan, X.Y.; Cai, Z.C.; Wang, S.W.; Smith, P. Direct measurement of soil organic carbon content change in the croplands of China. *Glob. Chang. Biol.* **2011**, *17*, 1487–1496. [CrossRef]

50. Yang, Y.H.; Fang, J.Y.; Ji, C.J.; Ma, W.H.; Mohammat, A.; Wang, S.F.; Wang, S.P.; Datta, A.; Robinson, D.; Smith, P. Widespread decreases in topsoil inorganic carbon stocks across China's grasslands during 1980s-2000s. *Glob. Chang. Biol.* **2012**, *18*, 3672–3680. [CrossRef]
51. Lv, H.; Wang, W.; He, X.; Xiao, L.; Zhou, W.; Zhang, B. Quantifying tree and soil carbon stocks in a temperate urban forest in Northeast China. *Forests* **2016**, *7*, 200. [CrossRef]
52. Wang, Q.; Wang, W. Grsp amount and compositions: Importance for soil functional regulation. In *Fulvic and Humic Acids: Chemical Composition, Soil Applications and Ecological Effects*; Barrett, K.D., Ed.; Nova Science Publishers, Inc.: Hauppauge, NY, USA, 2015.
53. Johnson, C.E.; Johnson, A.H.; Huntington, T.G. Sample size requirements for the determination of changes in soil nutrient pools. *Soil Sci.* **1990**, *150*, 637–644. [CrossRef]

11

Influence of Forest Harvest on Nitrate Concentration in Temperate Streams

Anne-Christine Mupepele *,† and Carsten F. Dormann †

Biometry and Environmental System Analysis, University of Freiburg, 79106 Freiburg, Germany; carsten.dormann@biom.uni-freiburg.de

* Correspondence: anne-christine.mupepele@biom.uni-freiburg.de

† Current address: Tennenbacherstr. 4, 79106 Freiburg, Germany

Abstract: Forest harvest alters natural nutrient cycles, which is reflected in stream water run-off from harvested catchments. Nitrate is an essential nutrient for plant growth, but increased concentrations in rivers, lakes, and oceans have contributed to eutrophication and anoxic conditions. Based on a literature review, we assessed the impact of three different harvest methods—clearcut, patchcut, and selective harvest—on nitrate concentrations in temperate forest streams. In a meta-analysis, the influence of harvest methods and additional environmental variables was analysed. Nitrate concentrations are significantly influenced by harvest methods, forest composition, site altitude, and time passed after the harvesting. The remaining unexplained between-site variability is small compared to the between-site variability explained by the model, indicating the model's validity. The effect of forest harvest is most pronounced in coniferous and deciduous forests, where clearcuts and patchcuts result in high nitrate run-off three to five years after harvest. Mixed forest plots can compensate for clearcut and patchcut, and do not show a significantly increased nitrate concentration after harvest. Selective harvest at low intensities succeeded in maintaining nitrate levels similar to control or pre-harvest levels in coniferous and mixed forests, and showed a positive but not significant trend in deciduous forests. Coniferous and deciduous monocultures clearly face the problem that nitrate wash-out cannot be minimized by reducing clearcut to patchcut harvest, whereas mixed forests are more suitable to diminish nitrate wash-out in both clearcut and patchcut.

Keywords: forest management; water quality; nitrogen; water chemistry

---

1. Introduction

Stream water quality is influenced by the surrounding landscape, with minerals and salts from the soil leaching into the water [1]. Nitrogen in particular plays a major role in the eutrophication of rivers and lakes and anoxia in coastal oceans [2,3]. It is the main component of agricultural fertilizers, and the increased use of fertilizers is strongly related to an increase in nitrate leaching into nearby rivers [4,5]. Regulations such as the EU Water Framework Directive and the US Clean Water Act attempt to protect water quality and ensure a good status of surface water and groundwater [6–9]. Compared to agricultural areas, water draining from forested watersheds has lower nitrate concentrations [10–12]. Nevertheless, forest harvest can result in an increased run-off and nutrient loss [13–15]. Streamside management zones can reduce nutrient loss [9,16,17], whereas clearcut leads to an increased stream-nitrate concentration, which has been documented in several study sites (e.g., Plynlimon in the UK, Hubbard Brook and Coweeta Experimental Forests in the USA [18–20]). The duration of nitrate run-off and the long-term consequences have received less attention, and so far there has been no endeavour to systematically synthesize forest harvest effects on stream nitrate concentration and their long-term consequences [13,21,22]). Furthermore, forest harvest practices beyond clearcutting—but which are of increasing importance, especially in Central Europe—are less well investigated [13]. In the light of increasing demand for bioenergy and the increased pressure on harvest, forest management is likely to change in the near future [23].

Nitrate concentrations in streams are potentially influenced by the type of forest harvest and other environmental variables accounting for forest management, temporal changes, and site-specific characteristics [24]. Leaves from deciduous forests have another decomposition rate than needles from conifers, and may influence the nitrate leaching from soils to streams [25,26]. N deposition can change absolute values for nitrate output in streams [27], but may have no impact on the difference between harvested and non-harvested control sites adjacent to each other. We hypothesized that nitrate patterns follow a temporal trend, and differences between harvested and control sites are largest in the year following the intervention. In a meta-analysis of the effects of forest harvesting on stream nitrate concentrations, we quantitatively synthesize knowledge gained in individual studies to extract patterns, independent from individual features of research sites, as well as highlighting research gaps to encourage future efforts. Our objective was to find out whether different forest harvest types (clearcut, patchcut, and selective harvest) influence the nitrate concentrations in streams in comparison to unharvested control sites. We further investigated whether other environmental variables could predict differences in nitrate concentrations between harvested and control sites.

2. Methods

2.1. Literature Search and Data Extraction

A scoping literature search according to the Collaboration for Environmental Evidence [28] evaluated relevant search terms, and resulted in a Web of Science search on 25 March 2015 at the University of Freiburg with the topic search terms: ('water quality' OR nitrogen OR nitrate OR nitrite OR 'nonpoint source pollution' OR 'nonpoint-source pollution') AND (stream* OR river*) AND *forest* AND (manage* OR liming OR clear-cut* OR clearcut* OR logging OR harvest* OR deforestation OR thinning OR selective) NOT (*tropic* OR boreal OR mangrove OR South*Asia OR Africa* OR Brazil* OR Marine). Articles with agricultur* in the title were excluded with the option NOT agricultur* in 'Title'. Excluding articles with agriculture in the title was necessary to avoid hitting a large amount of literature comparing run-off from agricultural fields and forest, which was beyond the scope of this study.

- The site was situated in a temperate region, defined according to the nemoral zonobiom in Walter and Breckle [29], covering Central and Eastern Europe, the United Kingdom, large parts of Eastern North America, and smaller parts in Western North America. It equally covers New Zealand, Korea, and Eastern China.
- At least half of the catchment had to be covered by the treated forest to minimize the risk of being influenced by a second unmeasured change in the catchment.
- Forest management needed to be clearly described. Forest should be a main subject in the abstract, so that we could assume that the management is described in detail in the main text.
- Water quality had to be measured in streams or rivers. No soil or lake water quality was considered.
- Nitrogen was given in the form of nitrate or nitrate-nitrogen as response variable to forest management. Studies were excluded if the abstract contained no hint that water chemistry was measured.
- The study design needed to provide a control, and was either a control–impact or a before–after experimental design.

Box 1. Inclusion criteria.

Title and abstract of the articles were screened according to a set of inclusion criteria (Box 1). Two assessors screened a subset of first 100 and then another 500 studies, and differences in their assessment were discussed. From then on, studies were screened by one of the two assessors. Articles that went into full-text reading were checked for additional inclusion criteria:

- Forest harvest could be classed into one of the following categories: clearcut, selective harvest, or patchcut.
- Time of nitrate measurement was given in relation to the forest harvest (e.g., measured in the first year after clearcut).

Data were extracted from publications that met the inclusion criteria. Data from long-term research sites happened to be used several times in different publications. Special care was taken to include these duplicated data only once. Annual mean nitrate concentration, standard deviation, and sample size were extracted from text, tables, and figures for all harvested and control sites. If no control site was available, before-harvest values were taken as comparison. Aside from before–after control–impact (BACI) studies, which have a preharvest calibration phase [12,30], we also included 16 before–after and 13 control–impact studies, as their results did not systematically differ from the 23 BACI studies. WebPlotDigitizer was used to extract data points from figures [31]. Nitrogen-nitrate was converted to NO_3^- in µg/L. Two studies [32,33] did not provide enough data to extract the annual standard deviation, and we estimated standard deviation according to Bracken [34] and Lajeunesse [35], considering the relationship between effect size and variance in all other studies.

Environmental variables were recorded to explain variability in the effect sizes and account for site-to-site differences. In total, nine variables were extracted from every study: harvest year, time after harvest (years after harvest intervention), catchment size of the harvested watershed (ha), mean annual temperature (°C), annual precipitation (mm/annum), altitude (m a.s.l.), and three components of forest management: forest harvest, forest age (years), and forest composition (Table 1). Forest composition was classified as coniferous, mixed, and deciduous stands. Coniferous stands were covered at least 90% with coniferous trees, and deciduous stands had less than 10% coniferous trees.

Forest harvest was categorized as clearcut, patchcut, or selective harvest. Clearcut was defined as tree removal of at least 90% of the forest. Patchcut meant the removal trees in patches of variable sizes, and we further distinguished patchcut into two different intensities: high intensity (meaning that at least 45% of all trees were removed in patches) and low intensity (including forests with less than 45% of forest removal). Selective harvest implies harvesting of individual trees selected according to different criteria (e.g., shelterwood forestry), and we equally distinguished into high intensity with at least 45% of trees removed and low intensity reflecting less than 45% tree removal (classification adapted from [26]).

2.2. Statistical Analysis

All statistical analyses were conducted in R [36], using packages nlme [37] and metafor [38]. To investigate changes in nitrate concentration, the standardized mean difference Hedges' d and the corresponding variance were used to compute effect sizes [39]. $Hedges'\,d = \frac{\bar{X}_t - \bar{X}_c}{S_{\text{pooled}}} \cdot J$; where X_t is the nitrate concentration in the harvested and X_c the concentration in the control forest or before the harvest intervention (Table S1); S_{pooled} is the within-study variance, and J a correction factor for small sample size.

Effect sizes were used in a meta-analysis to quantify and test the relationship between forest harvest, temporal trends, and other environmental predictors which cannot be accomplished by a narrative review. The meta-analysis was realized in the form of an additive mixed model (Equation (1)) to assess the influence of nine environmental predictors on the effect size reflecting changes in nitrate concentrations. Subsequent model selection was based on the F-statistic, and parameters were estimated with maximum likelihood. We assumed that the time after intervention influences the effect size, but not necessarily linearly, and eventually gets back to pre-harvest level after a certain time. Splines on the "time after harvest" allowed flexible functions of time to be fit . Most sites were measured over several years after the forest harvest, resulting in several non-independent effect sizes.

The model included the non-independence of effect sizes, taken in subsequent years on the same site, in the form of a temporal autocorrelation structure with time lag one [40] and nested in a random effect for sites.

$$y_i = \beta_0 + \beta_1 x_{1i} + \beta_2 x_{2i} + \beta_3 x_{3i} + f(t_i) + \gamma_{0j} + \gamma_{1j} \cdot t_i + \epsilon_i; \quad \text{for} \quad \epsilon_i \sim N(0, \Omega) \tag{1}$$

where y_i are effects sizes per site and year, x_i are environmental variables, $f(t_i)$ represents the regression spline on the time after harvest, γ_{0j} & γ_{1j} are random intercept and random slope for time in the j-th study, and Ω is the variance–covariance matrix fitting a continuous time AR(1) structure [41]. Subgroup analysis was done for each harvest type and forest composition ([42] [S2]) .

2.2.1. Between-Site Variability

In linear regressions, we are usually interested in how much of the variance is explained by the model, mainly computed as R^2, a ratio of explained to total variance. In a meta-analysis, the total variance is partitioned into between-study variability—also called heterogeneity—and within-study variability. The between-study variability T^2 is given by the variance of the random effect in linear-mixed models. In our meta-analysis, there are several sites per study, and we are actually interested in the between-*site* variability. The between-site variability of the effect sizes may have changed over the years. We expected a lower variability after a few years due to a diminishing management effect. Models with random slope on "time after harvest" nested in sites were used and performance compared to models with random intercept on sites using the Akaike Information Criterion (AIC).

We further included predictors in our model that explained part of the between-site variability resulting in an estimated between-site variability of $T^2_{\text{total}} = T^2_{\text{explained}} + T^2_{\text{unexplained}}$. Total between-site variability (T^2_{total}) was estimated as the variance of the random effect of a model without predictors, and unexplained between-site variability ($T^2_{\text{unexplained}}$) was given by the variance of the random effect of the model with predictors ([42], p. 200).

In meta-analysis, I^2 is the proportion of the between-site variance to the total variance [43,44]. In meta-regression, part of the between-site variance is explained by predictors. Of main interest is the contribution of the *unexplained* between-site variability to the total unexplained variance, which is the unexplained between-site variability plus within-site variance. Thus, we calculated $I^2_{\text{unexpl}} = \frac{T^2_{\text{unexplained}}}{(T^2_{\text{unexplained}} + S^2)}$; S^2 is the estimated within-site variance according to Nakagawa and Santos [45]. In addition, the between-site variance changed over the years (random slope on time after harvest), and we were therefore interested in the annual ratio of unexplained between-site variability and total unexplained variance (S2).

2.2.2. Study Weights

Meta-analysis has been criticized for its "garbage in, garbage out" problem; i.e., combining weak studies and carrying over fundamental errors of primary studies to the meta-analysis [42]. Studies are commonly weighted according to their inverse variance, with the assumption that the study variance was known. Study variance alone may not reflect reliability of results and sufficiently down-weigh unreliable studies. To avoid giving a strong weight to poor-quality studies, we decreased the weight of weak studies by multiplying the variance of each site with the level of evidence of the study. Levels of evidence were assessed by combining study design and study quality criteria to derive the reliability of underlying study results (sensu [46]). We investigate the consequences of this approach below.

2.2.3. Publication Bias

Publication bias occurs because not all results are published, particularly non-significant results may remain in the drawer [45,47,48]. Egger's regression test detects publication bias, which is reflected in an asymmetric distribution of effect sizes [49]. As the distribution of effect sizes may be explained by covariates, a modified version of the Egger's regression test was used, testing the weighted normalized residuals from the additive mixed model against their standard error [45,50].

3. Results

3.1. Data Extraction and Description

The literature search resulted in 2180 publications in Web of Science. Title and abstract screening by two assessors agreed in 74 of the first 100 studies, and after discussing the divergences, the level of agreement could be increased to 82% for the next 500 studies. Title and abstract screening resulted in 216 studies that were retained for full-text reading. Full-text reading resulted in 21 peer-reviewed papers and 3 raw-data sets. References of papers led to one additional study. The extracted 52 paired-watershed sites were measured over several years, resulting in 348 effects sizes (one effect size per year per site, Table S1).

Most of the 25 studies were conducted in the United States (15), followed by Japan (4), the UK (3), and one study each in Germany, Ireland, and New Zealand. They cover time series up to 40 years, and there was one study that measured the impact of forest harvest in the 53rd and 54th year after harvest [51]. Most data are available for the first ten years after harvest (Figure 1).

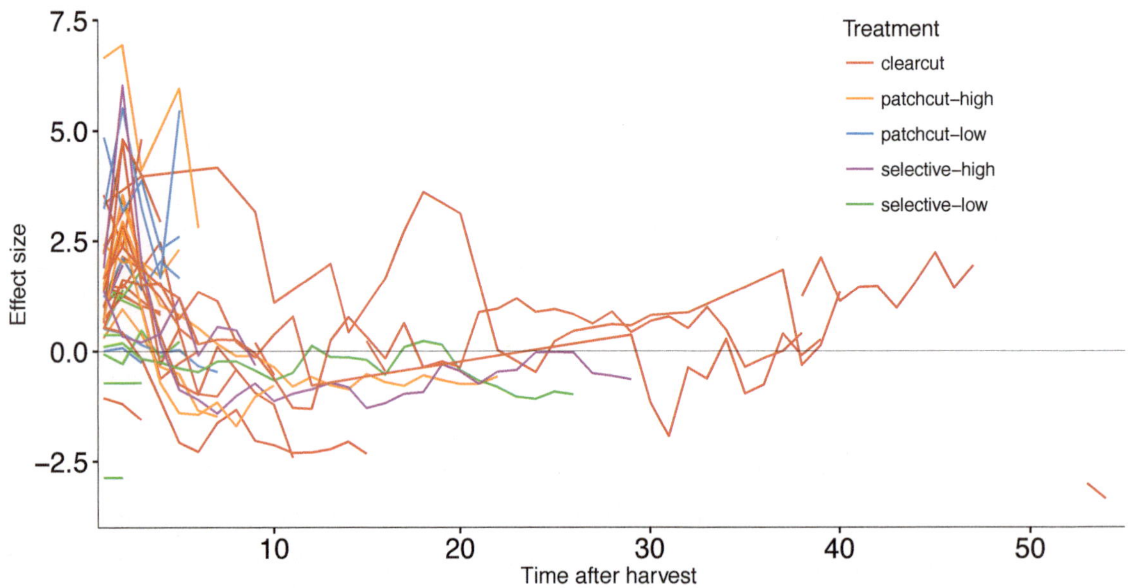

Figure 1. Standardized mean difference of stream nitrate concentrations plotted against post-harvest time series. Colours reflect the different harvest types. Most long-term studies were clearcut treatments. Seven studies had records for more than fifteen years after harvest.

3.2. Assessing the Influence of Forest Harvest on Nitrate Concentrations

Nitrate concentrations in streams are potentially influenced by forest harvest and other environmental variables accounting for forest management, temporal changes, and site-specific characteristics (Table 1).

Table 1. Environmental variables used as predictors in the additive mixed model before model selection.

Environmental Variable	Description
Harvest treatment	Clearcut: 26 sites Patchcut at high intensity: nine sites, and low intensity: six sites Selective harvest at high intensity: three sites, and low intensity: eight sites.
Forest age	20 to 100 years, with four exceptionally old sites of 450 and 600 years.
Forest composition	Coniferous forests: 27 sites Mixed forests: 14 sites Deciduous forests: 11 sites
Time after harvest	The time of NO_3^--measurements in relation to forest harvest ranged from 1 to 54 years. Forty-five effect sizes were from the first year after intervention, decreasing to 39 in the second and 33 in the third year. Two thirds of effect sizes were from the first 12 years.
Harvest year	Earliest harvest took place in 1926, but measurements on this site were taken only 50 years later [51]. First forest harvests with measurements directly after the intervention were from the 1960s. Most studies (15) investigated harvesting between 2000 and 2006. The last forest harvests investigated took place in 2006, and were published in 2011 [52–55]. The 10 sites harvested in the 1980s were those with the most long-term research, and provided 117 data points around 12 years of investigation after harvest.
Catchment size	Sites covering small catchments with less than 100 ha were most abundant (37). One catchment was as large as 2031 ha [56], and the remaining sites all ranged between 100 and 600 ha.
Mean annual temperature	Ranged from 4.6 to 14 °C, whereas most sites were between 5 to 13 °C. Catskill, USA had the lowest mean temperature (4.6 to 5.2 °C). The warmest sites were in Hubbard Brook, USA and Japan with 14 °C.
Annual precipitation	Ranged from 1156 mm/annum in Cloosh, Ireland, to 3333 mm/annum in the Hoh river valley, USA.
Altitude	The lowest place was at 63 m a.s.l, the same catchment that obtained lowest precipitation in Cloosh, Ireland [57]. Great Smoky Mountains National, USA had the most elevated site (1162 m).

An additive mixed effects model with a regression spline on "time after harvest" [58] was used to assess the influence of environmental variables on stream nitrate. Model selection resulted in four predictors (Table 2).

Table 2. F-test for the parameters of the selected model with temporal autocorrelation and a regression spline on time.

Predictor	Degrees of Freedom	F-Statistic	*p*-Value
Time after harvest (spline)	11.46	11.07	<0.001
Harvest treatment	4	7.07	<0.001
Forest composition	2	8.82	<0.001
Altitude	1	25.32	<0.001

In the first three to five years, clearcut and patchcut significantly increased nitrate concentration in coniferous and deciduous forests (Figure 2). Selective harvest at high intensity was significantly increased in the first year after intervention in coniferous forests and in deciduous forests. Mixed forests were influenced by selective harvest at high intensity, and did not significantly differ under all other forest regimes. Selective harvest at low intensity did not alter nitrate concentration, but showed a positive trend in deciduous forests. In some treatments and sites, we found a trend for overcompensation: in mixed forests with high intensities of patchcut and selective harvest and in clearcut and patchcut coniferous forests, the concentration of nitrate drops below the concentration of control streams after four to five years.

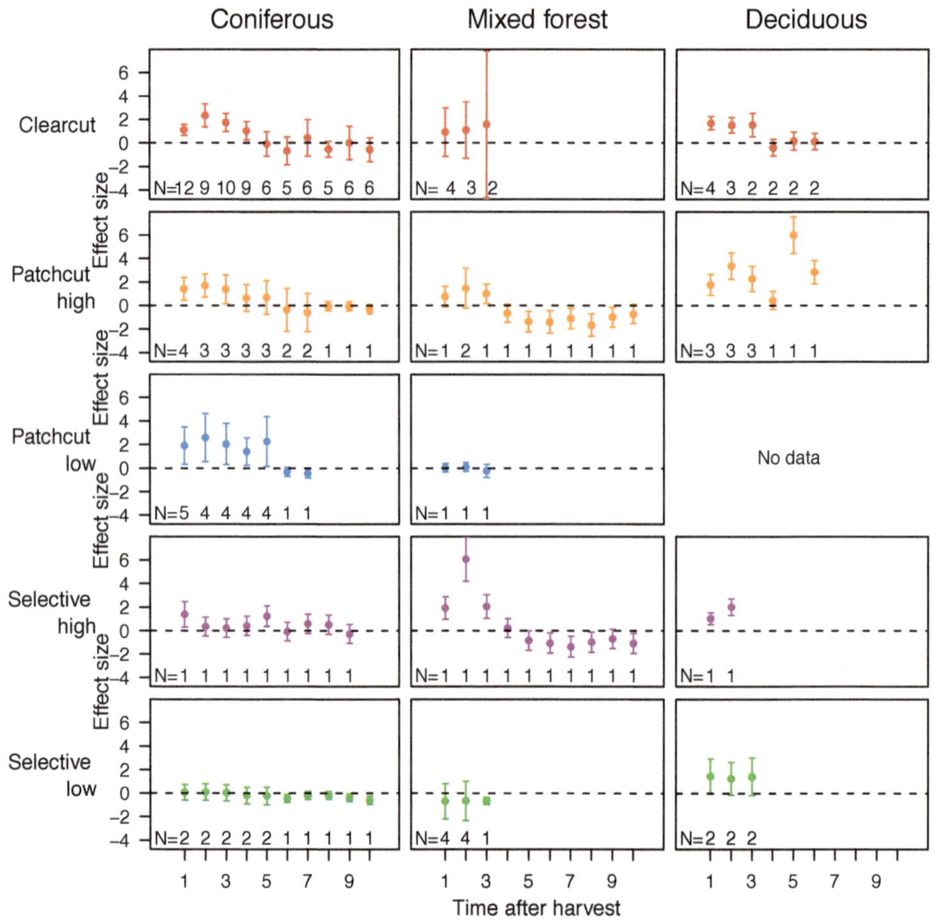

Figure 2. Subgroup analysis for every combination of forest harvest and forest composition. Summary effect sizes are plotted with confidence intervals against time for the first ten years. Number of studies entering the subgroup analysis is given by "N =" in each plot. Coniferous and deciduous forests show a significantly elevated nitrate concentration three to five years after clearcut and patchcut harvests, contrary to mixed forests. Selective harvest at low intensity does not significantly influence nitrate concentrations.

3.3. Between-Site Variability

Models with random slopes on the time after harvest nested in sites attained a better model fit (AIC = 1758) than random intercept models (AIC = 2603; likelihood ratio 837.6; $p < 0.001$). The unexplained between-site variability was small compared to the explained between-site variability (Figure 3a). The ratio of unexplained heterogeneity toward unaccounted variability indicates that between-site variability makes up the largest proportion in the unexplained variance (Figure 3b), with an increasing trend. The increasing trend is related to the increasing unexplained between-site variability with close-to-unchanged within-site variability (S^2, between 0.21 and 0.27).

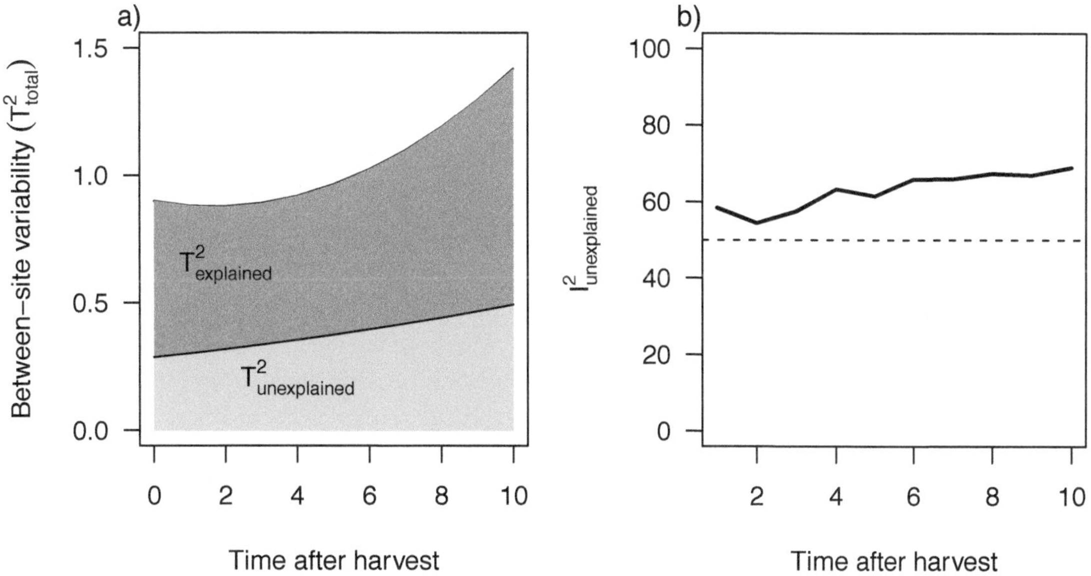

Figure 3. (**a**) Annual between-site variability is partitioned into explained and unexplained heterogeneity. While unexplained heterogeneity linearly increases over the years, heterogeneity explained by the meta-analytical model decreases in the first two years before it increases again after the fourth year. (**b**) Total unexplained variability consists of unexplained between-site and within-site variability. The ratio I^2_{unexpl} is larger than 50 (dotted line), which indicates that between-site variability is larger than within-site variability ($I^2_{unexpl} = T^2_{unexplained}/(T^2_{unexplained} + S^2)$).

3.4. Study Weights and Evidence Assessments

Re-weighting studies by their level of evidence led to no change in model selection, and the magnitude of the effect changes only slightly without any different interpretation. The level-of-evidence score was significantly positively correlated with variance ($r = 0.17, t = 3.12, p = 0.002$). Low variability in the level-of-evidence score (from 3.2 to 4, see also S2) resulted in a low correlation coefficient ($r = 0.17$) that was still significantly different from zero, indicating that studies penalized by their higher variances also scored less well on the level-of-evidence scale. Down-weighting of studies providing weak evidence did not influence our results.

3.5. Publication Bias

The distribution of the standard error against normalized residuals was symmetrically distributed around zero (S2), and Egger's regression test was not significant (t-statistic $= 0.18$, $p = 0.86$). There is thus no evidence of a publication bias.

4. Discussion

The present meta-analysis shows that forest harvesting practices influence nitrate concentration in temperate streams with elevated concentrations three to five years after harvest.

The influence of harvest on stream nitrate was mediated by forest type. Coniferous and deciduous forests induced a similar nitrate pattern, in contrast to mixed forests. Although differences in soil-water nitrate concentrations in deciduous and coniferous forests are known—such as higher N deposition in coniferous and higher topsoil N storage of nitrogen in deciduous forests [13,25,59–61]—they are not reflected in the results of our meta-analysis. Mixed forests do not show an increase in stream nitrate concentration after clearcut and patchcut harvest. Higher species diversity—favoured by microhabitat diversity—may buffer effects of nitrate leaching and lead to a more stable ecosystem [62–64]. Nitrate levels in mixed forests dropped even *below* the pre-harvest level in patchcut and high-intensity selective-harvest forests. This may be caused by rapidly regrowing vegetation with high nitrogen demand [65].Although sample size for mixed forests was low, our analysis points towards the important role that mixed forests can play in minimizing nitrate efflux.

Clearcut and patchcut harvest increased nitrate concentration for the first three-to-five years before returning to control-site level in coniferous and deciduous stands, which is consistent with previous findings [12,25,66]. Clearcuts expose soils to sunlight and associated temperature increase, which can accelerate mineralization and nitrification [67]. Together with rain and no more interception loss, run-off and nutrient wash-out increase [13,68]. An additional source of nitrogen comes from decomposing logging residue releasing nutrients, and may equally increase nitrate leaching to adjacent streams [30]. Reduction of nitrate in leachate after a few years was found to be correlated with regrowing vegetation that took up nitrate [13,69]. In addition, topsoil may have by then lost most nutrients through wash-out [70].

Selective harvest management at low intensity did not alter nitrate concentration. Impact was small and only limited wash-out of nutrients took place [71]. Soil flora faced less severe changes in selective harvest than in clearcut areas. Its impact is probably smallest, due to not opening up larger areas of bare soil. Selective harvest plays an important role in today's forestry practice recommendations [72–76], often aiming at close-to-nature forest management [77]. Selective harvest was represented only by a low number of samples in the meta-analysis, and future research on selective harvest intervention with various forest compositions is needed to verify the reliability of the findings.

The harvest year was included in the set of predictors as proxy for environmental context that may have changed over the years, such as atmospheric N deposition. Nitrogen compounds increased in natural systems through the N deposition derived from anthropogenic industrial nitrogen oxides, which was a major problem in the 1980s and decreased at the beginning of this century [25,78–80]. The time of harvest was not significant in the meta-analytical model, and based on the assumption that N deposition is correlated with year, changing deposition rates did not influence nitrate run-off from harvested forest streams in comparison to control sites. This result was consistent with results of a review by Campbell *et al.* [81].

Between-site differences were likely to be further mediated by soil composition [82–84]. N storage capacity and mineralization rate influence the amount of nitrogen retained in the soil [13,59]. The presence of ammonium and oxygen favour nitrification resulting in nitrate, as well as higher decomposition rates of organic material, favouring nitrate release [85,86]. Water and nitrate saturation of soils both contribute to the amount of nitrate washed out [87–89]. Due to a lack of data, it was not possible to include soil components in our meta-analysis.

Post-harvest management (such as burning and replanting) is known to influence nitrate concentration [90]. In almost half of the studies, information on post-harvest treatment was not available and could not be used as a predictor for differences between paired catchments. Leaving harvest residuals can increase nitrogen [13,22]. Most studies did not mention whether logging residue was left on site to avoid loss of nutrients from the system, but there was a tendency to leave logging residues (12 sites). Regrowing N fixing plants may also influence the nitrate output, although N fixing plants may as well profit from the high nitrate concentrations available in the soil after harvest and not necessarily fix more nitrogen from the air than they need themselves for regrowth [91]. In other forests where nitrogen-fixing plants such as alder—present in two studies of our meta-analysis [54,92]—were harvested, nitrate concentration even decreased compared to pre-harvest level [25], which may indicate that we underestimated the effect of harvest. We have accounted for these site-specific unknown circumstances by including a random effect for "site" in the meta-analysis.

Between-site variability reflects the differences of effect sizes between sites estimated by the meta-analytical model. We expected that harvest has a large impact on between-site variability in the first years, resulting in a high variability reflecting the differences in nitrate concentrations caused by the different harvest regimes. Contrary to our expectation, total between-site variability increased with time after harvest (Figure 3a). We interpret the increasing variability as idiosyncratic trajectories of regrowth in each site, making predictions for unobserved sites more difficult. We can still interpret the first years in comparison to the proportion that remained unexplained by the model. The part of unexplained variances equally increases over the years, but not to the same degree as the total variance. Forest composition, forest harvest, altitude, and the time after harvest explained parts of between-site variability in the model.

Combining weak studies can bias results, and we need to make sure to down-weigh weak studies in meta-analysis. Usually, studies are weighted according to their precision (i.e., the inverse of variance), assuming that studies with low variance contribute higher amounts of information to the analysis [48]. Study variance does not reflect quality shortcomings such as non-randomness or unrepresentative sampling. The evidence assessment tool of Mupepele et al. [46] accounts for study quality, and the resulting level of evidence was multiplied with study variance to down-weigh weak studies. This did not change the results in our meta-analysis, because levels of evidence were similar in all studies, and were positively correlated with variance. Furthermore, the random-effects model with relatively large between-site variability compared to within-site variance (Figure 3b) minimized the importance of weights ([93], p. 104). Nevertheless, quality shortcomings of individual studies in meta-analysis need to be revealed and incorporated in the analysis, although this did not influence the results in our study.

5. Conclusions

Nitrate leaching from harvested forests was generally low and below a harmful level for aquatic ecosystems [12,94,95]. Nevertheless, forest harvest influences stream nitrate concentration up to five years after harvesting, depending on harvest type and forest composition. Clearcut and patchcut lead to significantly increased nitrate concentrations in coniferous and deciduous forests, but not in mixed forests. Selective harvest has the lowest influence on nitrate concentration. A mixed forest and selective harvest at low intensity was the management option keeping nitrate concentrations in streams close to pre-harvest levels. While managed forests—even with occasional clearcuts—may still be beneficial compared to agriculture, effects should be considered on a landscape scale, avoiding high densities of several small equally-managed adjacent catchments draining to the same river.

Acknowledgments: We thank the participants of the German Forest Science Conference (FoWiTa) 2016 and their input during the discussion of this article. This work was supported by the 7th framework programme of the European Commission in the project OPERAs (grant number 308393, www.operas-project.eu).

Author Contributions: ACM collected the data. ACM and CFD did the analysis and wrote the manuscript.

Conflicts of Interest: The authors declare no conflict of interest.

References

1. Gundersen, P.; Laurén, A.; Finér, L.; Ring, E.; Koivusalo, H.; Sætersdal, M.; Weslien, J.O.; Sigurdsson, B.D.; Högbom, L.; Laine, J.; et al. Environmental services provided from riparian forests in the nordic countries. *Ambio* **2010**, *39*, 555–566.
2. Diaz, R.; Rosenberg, R. Spreading dead zones and consequences for marine ecosystems. *Science* **2008**, *321*, 926–929.
3. Howarth, R.W. Coastal nitrogen pollution: A review of sources and trends globally and regionally. *Harmful Algae* **2008**, *8*, 14–20.
4. Mosier, A.; Syers, J.K.; Freney, J.R. *Agriculture and the Nitrogen Cycle—Assessing the Impacts of Fertilizer Use on Food Production and the Environment*; Island Press: Washington, DC, USA, 2004; p. 344.
5. Sebilo, M.; Mayer, B.; Nicolardot, B.; Pinay, G.; Mariotti, A. Long-term fate of nitrate fertilizer in agricultural soils. *Proc. Natl. Acad. Sci. USA* **2013**, *110*, 18185–18189.
6. European Parliament. Directive 2000/60/EC of the European Parliament. *Off. J. Eur. Communities* **2000**, *327*, 1–72.

7. Twery, M.J.; Hornbeck, J.W. Incorporating water goals into forest management decisions at a local level. *For. Ecol. Manag.* **2001**, *143*, 87–93.
8. Borja, Á.; Galparsoro, I.; Solaun, O.; Muxika, I.; Tello, E.M.; Uriarte, A.; Valencia, V. The European Water Framework Directive and the DPSIR, a methodological approach to assess the risk of failing to achieve good ecological status. *Estuar. Coast. Shelf Sci.* **2006**, *66*, 84–96.
9. Sweeney, B.W.; Newbold, J.D. Streamside forest buffer width needed to protect stream water quality, habitat, and organisms: A literature review. *J. Am. Water Resour. Assoc.* **2014**, *50*, 560–584.
10. Binkley, D. *Patterns and Processes of Variation in Nitrogen and Phosphorus Concentrations in Forested Streams.* Technical Report Tech. Bull. 838; Research Triangle Park, National Council for Air and Stream Improvement: Cary, NC, USA, 2001.
11. Sun, G.; Riedel, M.; Jackson, R.; Kolka, R.; Amatya, D.; Shepard, J. Influences of management of Southern forests on water quantity and quality. In *Southern Forest Science: Past, Present, and Future*; USDA For. Serv., Gen. Tech. Rep: Asheville, NC, USA, 2004; Volume GTR SRS-75, pp. 195–234.
12. Boggs, J.; Sun, G.; McNulty, S. Effects of timber harvest on water quantity and quality in small watersheds in the Piedmont of North Carolina. *J. For.* **2016**, *114*, 27–40.
13. Hegg, C.; Jeisy, M.; Waldner, P. *Wald und Trinkwasser—Eine Literaturstudie*; Technical Report; Eidg. Forschungsanstalt für Wald, Schnee und Landschaft: Birmensdorf, Switzerland, 2004.
14. Keenan, R.J.; van Dijk, A.I. Planted forests and water. In *Ecosystem Goods and Services from Plantation Forests*; Bauhus, J., van der Meer, P., Kanninen, M., Eds.; Earthscan: London, UK, 2010; Chapter 4, pp. 77–95.
15. Clarke, J.; Kelly-Quinn, M.; Blacklocke, S.; Bruen, M. The effect of forest windrowing on physico-chemical water quality in Ireland. *Sci. Total Environ.* **2015**, *514*, 155–169.
16. Mayer, P.M.; Reynolds, S.K.; McCutchen, M.D.; Canfield, T.J. Meta-analysis of nitrogen removal in riparian buffers. *J. Environ. Qual.* **2007**, *36*, 1172–1180.
17. Dosskey, M.G.; Vidon, P.; Gurwick, N.P.; Allan, C.J.; Duval, T.P.; Lowrance, R. The role of riparian vegetation in protecting and improving water qulity in streams. *J. Am. Water Resour. Assoc.* **2010**, *46*, 1–18.
18. Scott, N.A.; Likens, G.E.; Eaton, J.S.; Siccama, T.G. Trace metal loss following whole-tree harvest of a northeastern deciduous forest, U.S.A. *Biogeochemistry* **2001**, *54*, 197–217.
19. Worrall, F. Changes in stream nitrate concentrations due to land management practices, ecological succession, and climate: Developing a systems approach to integrated catchment response. *Water Resour. Res.* **2003**, *39*, 1–14.
20. Neal, C.; Reynolds, B.; Neal, M.; Wickham, H.; Hill, L.; Williams, B. The impact of conifer harvesting on stream water quality: the Afon Hafren, mid-Wales. *Hydrol. Earth Syst. Sci.* **2004**, *8*, 503–520.
21. Neal, C.; Reynolds, B.; Wilkinson, J.; Hill, T.; Neal, M.; Hill, S.; Harrow, M. The impacts of conifer harvesting on runoff water quality: A regional survey for Wales. *Hydrol. Earth Syst. Sci.* **1998**, *2*, 323–344.
22. Dissmeyer, G. *Drinking Water from Forests and Grasslands*; Technical Report; USDA Forest Service Southern Research Station: Asheville, NC, USA 2000.
23. Forsell, N.; Korosuo, A.; Havlík, P.; Valin, H.; Lauri, P.; Gusti, M.; Kindermann, G.; Obersteiner, M. *Study on Impacts on Resource Efficiency of Future EU Demand For Bioenergy (ReceBio)*; Technical Report; Publications Office of the European Union: Luxembourg, 2016.
24. Passeport, E.; Vidon, P.; Forshay, K.J.; Harris, L.; Kaushal, S.S.; Kellogg, D.Q.; Lazar, J.; Mayer, P.; Stander, E.K. Ecological engineering practices for the reduction of excess nitrogen in human-influenced landscapes: A guide for watershed managers. *Environ. Manag.* **2013**, *51*, 392–413.
25. Gundersen, P.; Schmidt, I.K.; Raulund-Rasmussen, K. Leaching of nitrate from temperate forests effects of air pollution and forest management. *Environ. Rev.* **2006**, *14*, 1–57.
26. Jerabkova, L.; Prescott, C.E.; Titus, B.D.; Hope, G.D.; Walters, M.B. A meta-analysis of the effects of clearcut and variable-retention harvesting on soil nitrogen fluxes in boreal and temperate forests. *Can. J. For. Res.* **2011**, *41*, 1852–1870.
27. Dise, N.; Wright, R. Nitrogen leaching from European forests in relation to nitrogen deposition. *For. Ecol. Manag.* **1995**, *71*, 153–161.
28. Collaboration for Environmental Evidence. *Guidlines for Systematic Review and Evidence Synthesis in Environmental Management*; Technical Report; Centre for Evidence-Based Conservation: Bangor, UK, 2013.
29. Walter, H.; Breckle, S.W. *Vegetation und Klimazonen*; Ulmer: Stuttgart, Germany, 1999.

30. Swank, W.T.; Vose, J.M.; Elliott, K.J. Long-term hydrologic and water quality responses following commercial clearcutting of mixed hardwoods on a southern Appalachian catchment. *For. Ecol. Manag.* **2001**, *143*, 163–178.
31. Rohatgi, A. *WebPlotDigitizer*, 3.10 ed.; WebPlotDigitizer: Austin, TX, USA, 2016. Available online: http://arohatgi.info/ (accessed on 1 September 2015).
32. Reynolds, B.; Stevens, P.A.; Hughes, S.; Parkinson, J.A.; Weatherley, N.S. Stream chemistry impacts of conifer harvesting in Welsh catchments. *Water Air Soil Pollut.* **1995**, *79*, 147–170.
33. Arthur, M.A.; Coltharp, G.B.; Brown, D.L. Effects of best management practices on forest streamwater quality in eastern Kentucky. *J. Am. Water Resour. Assoc.* **1998**, *34*, 481–495.
34. Bracken, M.B. Statistical methods fo analysis of effects of treatment in overviews of randomized trials. In *Effective Care of the Newborn Infant*; Sinclair, J.C., Bracken, M.B., Eds.; Oxford University Press: Oxford, UK, 1992; pp. 13–20.
35. Lajeunesse, M.J. Recovering Missing or Partial Data from Studies: A Survey of Conversions and Imputations for Meta-analysis. In *Handbook of Meta-Analysis in Ecology and Evolution*; Koricheva, J., Gurevitch, J., Mengersen, K., Eds.; Princeton University Press: Princeton, NJ, USA, 2013; pp. 195–206.
36. R Core Team. *R: A Language and Environment for Statistical Computing*; R Foundation for Statistical Computing: Vienna, Austria, 2016.
37. Pinheiro, J.; Bates, D.; DebRoy, S.; Sarkar, D.; R Core Team. *nlme: Linear and Nonlinear Mixed Effects Models; R package version 3.1-128*; R Foundation for Statistical Computing: Vienna, Austria, 2016.
38. Viechtbauer, W. Conducting Meta-Analyses in R with the metafor Package. *J. Stat. Softw.* **2010**, *36*, 1–48.
39. Rosenberg, M.S.; Rothstein, H.R.; Gurevitch, J. Effect Sizes: Conventional Choices and Calculations. In *Handbook of Meta-Analysis in Ecology and Evolution*; Koricheva, J., Gurevitch, J., Mengersen, K., Eds.; Princeton University Press: Oxford, UK, 2013; pp. 61–85.
40. Bolker, B. Linear and generalized linear mixed models. In *Ecological Statistics: Contemporary Theory and Application*; Fox, G.A., Negrete-Yankelevich, S., Sosa, V.J., Eds.; Oxford University Press: Oxford, UK, 2015; Chapter 13, pp. 309–333.
41. Pinheiro, J.C.; Bates, D.M. *Mixed-Effects Models in S and S-PLUS*; Springer: New York, NY, USA, 2000.
42. Borenstein, M.; Hedges, L.V.; Higgins, J.P.T.; Rothstein, H.R. *Introduction to Meta-Analysis*; John Wiley & Sons: New York, NY, USA, 2009; p. 450.
43. Higgins, J.P.T.; Thompson, S.G.; Deeks, J.J.; Altman, D.G. Measuring inconsistency in meta-analyses. *Br. Med. J.* **2003**, *327*, 557–560.
44. Gurevitch, J.; Nakagawa, S. Research synthesis methods in ecology. In *Ecological Statistics: Contemporary Theory and Application*; Princeton University Press: Princeton, NJ, USA; Oxford, UK, 2015; Chapter 9.
45. Nakagawa, S.; Santos, E.S.A. Methodological issues and advances in biological meta-analysis. *Evolut. Ecol.* **2012**, *26*, 1253–1274.
46. Mupepele, A.C.; Walsh, J.C.; Sutherland, W.J.; Dormann, C.F. An evidence assessment tool for ecosystem services and conservation studies. *Ecol. Appl.* **2016**, *26*, 1295–1301.
47. Rosenthal, R. The file drawer problem and tolerance for null results. *Psychol. Bull.* **1979**, *86*, 638–641.
48. Koricheva, J.; Gurevitch, J.; Mengersen, K. (Eds.). *Handbook of Meta-Analysis in Ecology and Evolution*; Princeton University Press: Princeton, NJ, USA, 2013; p. 498.
49. Egger, M.; Davey Smith, G.; Schneider, M.; Minder, C. Bias in meta-analysis detected by a simple, graphical test. *BMJ Clin. Res.* **1997**, *315*, 629–34.
50. Gerstner, K.; Dormann, C.F.; Stein, A.; Manceur, A.M.; Seppelt, R. Effects of land use on plant diversity—A global meta-analysis. *J. Appl. Ecol.* **2014**, *51*, 1690–1700.
51. Silsbee, D.G.; Larson, G.L. A comparison of streams in logged and unlogged areas of Great Smoky Mountains National Park. *Hydrobiologia* **1983**, *102*, 99–111.
52. Clinton, B.D. Stream water responses to timber harvest: Riparian buffer width effectiveness. *For. Ecol. Manag.* **2011**, *261*, 979–988.
53. Siemion, J.; Burns, D.A.; Murdoch, P.S.; Germain, R.H. The relation of harvesting intensity to changes in soil, soil water, and stream chemistry in a northern hardwood forest, Catskill Mountains, USA. *For. Ecol. Manag.* **2011**, *261*, 1510–1519.
54. Meininger, W.S. The iNfluence of Contemporary Forest Management on Stream Nutrient Concentrations in an Industrialized Forest in the Oregon Cascades. Ph.D. Thesis, Oregon State University, Corvallis, OR, USA, 2011.

55. Neal, C.; Reynolds, B.; Norris, D.; Kirchner, J.W.; Neal, M.; Rowland, P.; Wickham, H.; Harman, S.; Armstrong, L.; Sleep, D.; et al. Three decades of water quality measurements from the Upper Severn experimental catchments at Plynlimon, Wales: An openly accessible data resource for research, modelling, environmental management and education. *Hydrol. Process.* **2011**, *25*, 3818–3830.
56. Graynoth, E. Effects of logging on stream environments and faunas in Nelson. *N. Z. J. Mar. Freshw. Res.* **1979**, *13*, 79–109.
57. Cummins, T.; Farrell, E.P. Biogeochemical impacts of clearfelling and reforestation on blanket-peatland streams II. Major ions and dissolved organic carbon. *For. Ecol. Manag.* **2003**, *180*, 557–570.
58. Wood, S.N. Thin plate regression splines. *J. R. Stat. Soc. Ser. B Stat. Methodol.* **2003**, *65*, 95–114.
59. Rothe, A.; Kölling, C.; Moritz, K. Waldbewirtschaftung und Grundwasserschutz—Der aktuelle Kenntnisstand. *AFZ Der Wald* **1998**, *53*, 291–295.
60. Meesenburg, H.; Horváth, B.; Meiwes, K.J. Stoffhaushalt von Waldökosystemen NW-Deutschlands unter hoher Stickstoffbelastung. *Ber. Freibg. Forstl. Forsch.* **2003**, *49*, 57–69.
61. Puhlmann, H.; Von Wilpert, K. Silvicultural management options to preserve the seepage-water quality under forests - Case study Conventwald. *Hydrol. Wasserbewirsch.* **2009**, *53*, 96.
62. Prescott, C.E. The influence of the forest canopy on nutrient cycling. *Tree Physiol.* **2002**, *22*, 1193–1200.
63. Tilman, D.; Reich, P.B.; Knops, J.M.H. Biodiversity and ecosystem stability in a decade-long grassland experiment. *Nature* **2006**, *441*, 629–632.
64. Lang, A.C.; von Oheimb, G.; Scherer-Lorenzen, M.; Yang, B.; Trogisch, S.; Bruelheide, H.; Ma, K.; Härdtle, W. Mixed afforestation of young subtropical trees promotes nitrogen acquisition and retention. *J. Appl. Ecol.* **2014**, *51*, 224–233.
65. Vitousek, P.M.; Howarth, R.W. Nitrogen limitation on land and in the sea: How can it occurs? *Biogeochemistry* **1991**, *13*, 87–115.
66. Kreutzweiser, D.P.; Hazlett, P.W.; Gunn, J.M. Logging impacts on the biogeochemistry of boreal forest soils and nutrient export to aquatic systems: A review. *Environ. Rev.* **2008**, *16*, 157–179.
67. Göttlein, A.; Baumgarten, M.; Huber, C.; Weis, W.; Papen, H.; Butterbach-Bahl, K.; Gasche, R. Femel- und Kahlhieb im Vergleich, Ökologie der Mischwaldbegründung in einem stickstoffbelasteten Fichtenbestand. *LWF aktuell Magazin für Wald Wissenschaft und Praxis* **2003**, *41*, 6–8.
68. Wenger, W. Bedeutung des Waldes für die Trinkwassergewinnung. *LWF aktuell Magazin für Wald Wissenschaft und Praxis* **2002**, *34*, 3–8.
69. Miller, H.G. The influence of stand development on nutrient demand, growth and allocation. *Plant Soil* **1995**, *168–169*, 225–232.
70. Kölling, C.; Neustifter, H. Stickstoffeintrag in Wälder und Nitratkonzentration im Sickerwasse—Ergebnisse aus dem Messnetz der Bayerischen Waldklimastationen. *AFZ Der Wald* **1997**, *52*, 1107–1110.
71. Weis, W. Beeinflusst der Standort den Nitrataustrag? *LWF aktuell Magazin für Wald Wissenschaft und Praxis* **2002**, *34*, 21–24.
72. Borchert, H. Veränderung des Waldes in Bayern in den letzten hundert Jahren. *LWF Wissen* **2002**, *58*, 2–50.
73. Knoke, T.; Stimm, B.; Ammer, C.; Moog, M. Mixed forests reconsidered: A forest economics contribution on an ecological concept. *For. Ecol. Manag.* **2005**, *213*, 102–116.
74. Shepard, J.P. Water quality protection in bioenergy production: The US system of forestry Best Management Practices. *Biomass Bioenergy* **2006**, *30*, 378–384.
75. Bundesministerium für Ernährung, Landwirtschaft und Verbraucherschutz. *Waldbericht der Bundesregierung*; Technical Report; BMELV: Berlin, Germany, 2009.
76. Kindler, E. Analyse zur erweiterten Nachhaltigkeitsberichterstattung deutscher Länderforstbetriebe. *Appl. Agric. For. Res.* **2015**, *2*, 131–144.
77. BMELV. *German Forests—Nature and Economic Factor*; Technical Report; Bundesministerium für Ernährung, Landwirtschaft und Verbraucherschutz: Berlin, Germany, 2011.
78. Galloway, J.N. Acid deposition: Perspectives in time and space. *Water Air Soil Pollut.* **1995**, *85*, 15–24.
79. Sucker, C.; Wilpert, K.V.; Puhlmann, H. Acidification reversal in low mountain range streams of Germany. *Environ. Monit. Assess.* **2011**, *174*, 65–89.
80. Paré, D.; Thiffault, E. Nutrient budgets in forests under increased biomass harvesting scenarios. *Curr. For. Rep.* **2016**, *2*, 81–91.

81. Campbell, J.L.; Hornbeck, J.W.; Mitchell, M.J.; Adams, M.B.; Castro, M.S.; Driscoll, C.T.; Kahl, J.S.; Kochenderfer, J.N.; Likens, G.E.; Lynch, J.A.; et al. Input-output budgets of inorganic nitrogen for 24 forest watersheds in the Northeastern United States: A review. *Water Air Soil Pollut.* **2004**, *151*, 373–396.
82. Vitousek, P.M. Nitrate losses from disturbed ecosystems. *Science* **1979**, *204*, 469–474.
83. Neary, D.G.; Ice, G.G.; Jackson, C.R. Linkages between forest soils and water quality and quantity. *For. Ecol. Manag.* **2009**, *258*, 2269–2281.
84. Sudduth, E.B.; Perakis, S.S.; Bernhardt, E.S. Nitrate in watersheds: Straight from soils to streams? *J. Geophys. Res. Biogeosci.* **2013**, *118*, 291–302.
85. Gisi, U. *Bodenökologie*; Thieme Verlag: Stuttgart, Germany, 1997; p. 350.
86. Rüetschi, D.; Wülser, R. *Die künstliche Grundwasseranreicherung der Wasserversorgung Basel*; Technical Report; Annual Report of the Lake Constance-Rhine-Waterworks Association: Basel, Switzerland, 1999, pp. 71-84
87. Fenn, M.E.; Poth, M.A.; Aber, J.D.; Baron, J.S.; Bormann, B.T.; Johnson, D.W.; Lemly, A.D.; McNulty, S.G.; Ryan, D.F.; Stottlemyer, R. Nitrogen excess in North American ecosystems: Predisposing factors, ecosystem responses, and management strategies. *Ecol. Appl.* **1998**, *8*, 706–733.
88. Schleppi, P.; Waldner, P.; Hegg, C. Einfluss des Waldes auf Nitrat-Gehalte im Wasser. *Bündner Wald* **2003**, *4*, 27–30.
89. Bredemeier, M.; Cohen, S.; Godbold, D.L.; Lode, E.; Pichler, V.; Schleppi, P. (Eds.). *Forest Management and the Water Cycle—An Ecosytem-Based Approach*; Springer: New York, NY, USA, 2011.
90. Covington, W.; Sackett, S. Soil mineral nitrogen changes following prescribed burning in ponderosa pine. *For. Ecol. Manag.* **1992**, *54*, 175–191.
91. Ledgard, S.; Sprosen, M.; Penno, J.; Rajendram, G. Nitrogen fixation by white clover in pastures grazed by dairy cows: Temporal variation and effects of nitrogen fertilization. *Plant Soil* **2001**, *229*, 177–187.
92. Stednick, J.D. Long-term water quality changes following timber harvesting. In *Hydrological and Biological Responses to Forest Practices*; Stednick, J.D., Ed.; Springer: New York, NY, USA, 2008; Chapter 10, pp. 139–155.
93. Mengersen, K.; Schmid, C.H.; Jennions, M.D.; Gurevitch, J. Statistical Models and Approaches to Inference. In *Handbook of Meta-Analysis in Ecology and Evolution*; Princeton University Press: Princeton, NJ, USA; Oxford, UK, 2013; Chapter 8, pp. 89–107.
94. Binkley, D.; Burnham, H.; Lee Allen, H. Water quality impacts of forest fertilization with nitrogen and phosphorus. *For. Ecol. Manag.* **1999**, *121*, 191–213.
95. Boggs, J.; Sun, G.; Jones, D.; McNulty, S.G. Effect of soils on water quantity and quality in Piedmont forested headwater watersheds of North Carolina. *J. Am. Water Resour. Assoc.* **2013**, *49*, 132–150.

12

Biocontrol of *Fusarium circinatum* Infection of Young *Pinus radiata* Trees

Eugenia Iturritxa [1], Tyler Trask [2], Nebai Mesanza [1,2], Rosa Raposo [3,4], Margarita Elvira-Recuenco [3] and Cheryl L. Patten [2,*]

1 Production and Plant Protection, Neiker Tecnalia, Apartado 46, Vitoria Gasteiz 01080, Spain; eiturritxa@neiker.eus (E.I.); nebaimesanzaiturritza@gmail.com (N.M.)

2 Department of Biology, University of New Brunswick, P.O. Box 4400, Fredericton, NB E3B 5A3, Canada; tyler.tr@unb.ca

3 INIA-CIFOR, Ctra. La Coruña Km 7.5, Madrid 28040, Spain; raposo@inia.es (R.R.); elvira@inia.es (M.E.-R.)

4 Sustainable Forest Management Research Institute, University of Valladolid-INIA, Avda. Madrid 44, Palencia 34004, Spain

* Correspondence: pattenc@unb.ca

Academic Editors: Robert A. Blanchette and Timothy A. Martin

Abstract: Pitch canker, caused by the fungus *Fusarium circinatum*, is a major disease of *Pinus radiata* currently controlled to some extent in nurseries by good hygiene and application of synthetic fungicides. The aim of this study was to evaluate alternative strategies to control fungal infections in nurseries and young pine plantations. The antagonistic effects of biocontrol bacteria and essential oils against *F. circinatum* in vitro and in young *P. radiata* trees were assessed. *Pseudomonas fluorescens*, *Erwinia billingiae*, and *Bacillus simplex* reduced the growth of the fungus in vitro by 17%–29%, and decreased the density of the mycelial mat. In young *P. radiata* trees, the length of *F. circinatum* lesions was reduced by 22%–25% by the same bacterial strains. Direct application of cinnamon and/or clove essential oils to wounds in stems of two-year-old *P. radiata* trees also limited the damage caused by *F. circinatum*. Lesion length was reduced by 51% following treatment with cinnamon oil (10% v/v), and by 45% following treatment with clove oil (15% v/v) or a combination of both oils. However, the oils were toxic to younger trees. The biocontrol bacteria and essential oils show promise as prophylactic treatments to reduce the devastating effects of *F. circinatum* on *P. radiata*.

Keywords: *Fusarium circinatum*; *Pinus radiata*; biocontrol bacteria; essential oils

1. Introduction

The fungus *Fusarium circinatum* Nirenberg & O'Donnell is the causative agent of pitch canker, a major disease of *Pinus radiata* D. Don or Monterey pine [1]. The disease is characterized by resinous lesions at the site of infection on the trunk and branches, and branch dieback (necrosis from the branch tip to the site of infection) as a result of vascular tissue obstruction [2]. Infection reduces growth of the tree and may result in death, often from increased susceptibility to biotic and abiotic stresses [2]. Infected seedlings have a high mortality rate, especially in nurseries, where root infections can be extensive [3,4]. The pathogen spreads through contaminated seeds, and through dissemination of spores via wind, rain, or insect vectors; however, disease results only when wounds in the protective bark are infected.

F. circinatum outbreaks and epidemics are common in nurseries and plantations in many regions of the world and are responsible for high mortality and reduced yields, causing substantial economic losses. The fungus is believed to have originated in Mexico and spread north into the native stands of *P. radiata* in the southeastern US [5]. It is now a major threat in the coastal southwestern US [6],

Chile [7], northern Spain [8], Portugal [9], South Africa [3], and potentially in regions of Australia, New Zealand, China, and Brazil, where *P. radiata* is widely planted and climatic conditions are highly suited for establishment of the pathogen [10]. Although *F. circinatum* is primarily a pathogen of *P. radiata* [11], an important commercial species, other *Pinus* spp. such as *P. muricata* D. Don (bishop pine), *P. taeda* L. (loblolly pine), *P. elliotii* Engelm. *var. elliotii* (slash pine), and *Pseudotsuga menziesii* (Mirb) Franco (Douglas fir) are also susceptible to the pathogen [12,13].

At present, there are few effective options to treat or prevent infections of this fungal pathogen. Good forest management such as planting less susceptible varieties of pine, rapid removal of infected trees, and quarantine measures, including restrictions on importation of wood products, are the only mechanisms that have proven effective in reducing the spread of *F. circinatum*. Measures to prevent seed contamination such as hot water treatment can reduce infections in nurseries [14]. Although effective, widespread application of fungicides such as benomyl may promote resistance and are no longer allowed in some countries (e.g., EU directive 91/414/CEE) [15].

Biological control agents have been used in agriculture over the last century as an alternative to chemical pesticides and have proven effective in protecting crops against many pathogens, including fungi [16–18]. In contrast, less has been done to investigate the potential of biocontrol agents, such as beneficial microorganisms, to protect pine trees from fungal pathogens. Commercial and native strains of the biocontrol fungi *Trichoderma harzianum* Rifai and *T. viride* Pers. were shown to be effective against *F. circinatum* in vitro, however, they did not inhibit growth of the pathogen on *P. radiata* seeds [15] or seedlings [19]. Application of methyl jasmonate to induce systemic resistance also failed to protect pine seedlings from disease caused by *F. circinatum* and other fungi [20]. Essential oils extracted from plants are also known to have fungicidal properties [21–23]. Several essential oils, including cinnamon, oregano, thyme, lavender, tea tree, Japanese mint, clove, rose geranium, and lemongrass were shown to inhibit growth of *F. circinatum* in vitro [15,24], although the high concentrations required to inhibit the fungus may also inhibit seed germination [15]. To address the need for more effective strategies to mitigate the damage caused by *F. circinatum*, we evaluated the efficacy of pine rhizobacterial isolates and essential oils (cinnamon and clove oil) to inhibit fungal growth in vitro and reduce symptoms of fungal infection in young *P. radiata* trees.

2. Materials and Methods

2.1. Microorganisms

The *F. circinatum* strain (CECT20759) used in this study was isolated from a *P. radiata* plantation in Gipuzkoa, Spain, and is associated with vegetative compatibility group (VCG) A and Mating type 2 (Mat-2), the only mating type present in the Basque Country [15]. Isolate CECT20759 does not show significant differences in virulence in relation to Mating type 1 [15]. To assess the specificity of the biocontrol agents in pathogen control, two other fungal pathogens of forest tree species that spread through infection of seeds and seedlings in nurseries were also tested. *Diplodia sapinea* (Fr.) Fuckel 1870 strain DPV24 was isolated from an adult *P. radiata* tree in Bizkaia, Spain [25], and *Neofusicoccum parvum* (Pennycook & Samuels) Crous, Slippers & A.J.L. Phillips 2006, CMW37773, was isolated from an adult *Eucalyptus globulus* Labill tree in Bizkaia, Spain [26]. The fungi were cultured on potato dextrose agar (PDA; Cultimed) at 20 °C in the dark.

Bacterial strains *Pseudomonas fluorescens* S32R2, *Bacillus simplex* S11R41, and *Erwinia billingiae* S31R1 were isolated from the endorhizosphere and *E. billingiae* S23L3 from the ectorhizosphere of a healthy pine tree (*P. radiata*) in a plantation in Abadiano, Bizkaia, Spain, with a high density of diseased trees, and were shown to inhibit the growth of fungal pathogens *Heterobasidion annosum* and *Armillaria mellea* in vitro and pathogenic effects in *P. radiata* [27]. For in vitro and in planta fungal antagonism assays, single bacterial colonies were inoculated into Luria Bertani broth and grown for two days at room temperature (20–25 °C). Cells were washed twice with 0.03 M $MgSO_4$ and the final suspension adjusted to an optical density measured at 600 nm of 0.5. Laboratory strain *Escherichia coli* TOP10

(Invitrogen Life Technologies, Carlsbad, CA, USA) and grass rhizosphere isolate *P. fluorescens* J11 (C. Patten culture collection), used as controls, were prepared in the same manner.

2.2. *In Vitro Bacterial-Fungal Antagonism Assay*

Bacteria and fungi were prepared as described above and inoculated into wells of a six-well plate (Nunc, Sigma-Aldrich Co. LLC., St. Louis, MO, USA) containing International Streptomyces Project 2 (ISP2) agar (4 g/L yeast extract, 10 g/L malt extract, 4 g/L glucose, and 20 g/L agar, pH 7.3; [28]). In preliminary experiments, this medium was determined to be suitable for growth of the bacterial and fungal strains. Bacterial cultures, prepared in triplicate from independent colonies, were applied in a thin line using a sterile inoculation loop. A sterile toothpick was used to transfer a small amount (1 mm^3) of the fungal mycelium to the agar plates. The plates were incubated for one week at 23 °C and 53% humidity. Inhibition of fungal growth by the bacteria was determined by calculating the Area Inhibition Percentage (AIP): AIP (%) = $(A - B)/A \times 100$, where A and B are the surface area covered by the fungus in control (no bacteria) and treated (with bacteria) plates, respectively. Differences in the AIP among the treatments were analyzed by one-way analysis of variance (ANOVA) with Waller-Duncan post-hoc test using SPSS software (SPSS Inc., Chicago, IL, USA). Values of $p < 0.05$ were considered significant.

2.3. *In Vivo Bacterial Biocontrol Assay in* P. radiata

Two-year-old *P. radiata* trees (Deba Seed Orchard, HS-Q-20/002, Basque Country, Spain) were grown at the Explotaciones Forestales Jiménez Araba nursery (Vitoria, Spain) before transfer to a biosafety level 2 greenhouse for cultivation at 18 ± 5 °C, with a relative humidity of 55%–60% and without supplemental light throughout the experimental period (May–August). The trees were contained in 300 mL pots with a mix of peat moss (2/3 volume), perlite (1/3 volume), and NPK fertilizer (N = 200–450 mg/L, P_2O_5 = 200–500 mg/L, K_2O = 300–550 mg/L). After four weeks of acclimatization to greenhouse conditions, approximately 2 cm were removed from the top of the trees to expose the pith, and 5 μL of bacterial culture, prepared as described above, or 0.03 M $MgSO_4$ as a negative control were added to the wound. After one week, trees were reinoculated with the bacterial suspensions. The trees were not watered for three days following each inoculation to allow the bacteria to infect the tree tissue. At least 30 trees were inoculated with each bacterial strain and the pots containing the treated trees were arranged in trays in a randomized block design.

After the second inoculation, the trees were left for one week to allow the bacteria to proliferate and the trees to recover from removal of their tops, and then were infected with an aqueous fungal spore suspension prepared as follows. Sterile water (1 mL) was added to plates containing seven-day-old cultures of *F. circinatum*. A sterile microscope slide was used to suspend the mycelium and spores. The suspension was then filtered through two layers of sterile gauze to remove the mycelia and collect the spores. The density of spores was quantified using a Neubauer chamber and sterile water was added to obtain a final concentration of 25 spores per μL. Small wounds, deep enough to reach the sapwood, were made on the main stem using a drill bit (1.6 mm diameter), and 2 μL of the aqueous spore suspension were deposited into each wound. Control trees were treated in the same manner but were inoculated with 2 μL of sterile water. Six weeks after inoculation, bark and phloem were removed with pruning scissors and the length of the lesions around inoculation points was measured using an electronic caliper [29]. The experiment was repeated once in the same manner.

Data was analyzed by a two-way ANOVA using the statistical package JMP (v7, SAS Institute Inc., Cary, NC, USA) with block (tray) and bacterial treatment as predictor variables and lesion size as the response variable. Significant differences ($p < 0.05$) among the treatments were determined using a post-hoc Tukey honest significant difference (HSD) test.

2.4. *In Vitro Essential Oils Inhibition Assay*

Cinnamon, fennel, and clove oils (Sanoflore Laboratory, Gigors et Lozeron, France) were tested at several concentrations (5%, 10%, 15%, 50%, and 100% *v*/*v*) for the ability to inhibit fungal growth in vitro. Sweet almond oil (Guinama Laboratory, Valencia, Spain) was used as a diluent as it did not affect the growth of the fungus or the development of spores in preliminary tests. Each well of a 12-well plate (Nunc, Sigma-Aldrich Co. LLC., St. Louis, MO, USA) containing PDA was inoculated with *F. circinatum* mycelium as described above, and approximately 100 μL of the prepared oils were applied to cover the surface of the well. To determine if the vapors from the oils impacted fungal growth, six outer wells in each plate were treated with the oils, while six inner wells were left untreated. Six replicates of each oil concentration were applied.

After one week, the coverage of the mycelial mat was measured using electronic calipers and AIP was calculated as described above. The oil concentrations that completely inhibited the growth of the fungus were assessed for fungicidal or fungistatic activity by transferring a small amount of mycelia from the oil-treated plates, in triplicate, to fresh medium without oil. A treatment was deemed fungicidal if no new mycelial growth was detected and fungistatic if growth was initiated on the oil-free medium.

The impact of essential oil application on the biocontrol bacterial strains was analyzed by inoculating a bacterial suspension (prepared as described above) and, after one day, essential oils that showed fungicidal activity into wells of a six-well plate (Nunc, Sigma-Aldrich Co. LLC., St. Louis, MO, USA) containing ISP2 agar. The plates were incubated for one week at 32 °C. Inhibition of bacterial growth was assessed by comparing growth of bacteria treated with the oils to that for the control treatment without oils.

2.5. *In Vivo Essential oils Inhibition Assay in* P. radiata

To assess the phytoxicity of the essential oils, the lowest concentration of cinnamon and/or clove oil (5 μL) determined to have a fungicidal effect in vitro was applied to the excised tops of one-year-old *P. radiata* trees in the same manner as the bacterial treatments, and to small wounds made in the stem of two-year-old *P. radiata* trees using a drill bit as described above (12 trees/treatment). The oils were prepared in almond oil as the diluent and an almond oil treatment was included as a control. Lesions were measured four weeks after application of oil treatments. Data was analyzed by a two-way ANOVA with tissue age and oil treatment as predictor variables and lesion size as the response variable. Significant differences ($p < 0.05$) among the treatments were determined using a post-hoc Tukey HSD test.

To assess the antifungal effect of the essential oils in vivo, 5 μL of a *F. circinatum* spore suspension, prepared as described above, were applied to the stem wounds one day after treatment with each oil or combination of oils (30 trees/treatment). Almond oil, the diluent for the essential oils, was applied as a control treatment. The trees were maintained in the greenhouse as described above. After six weeks the size of the lesions was measured using a digital caliper. Data was analyzed by a one-way ANOVA with oil treatment as the predictor variable and lesion size as the response variable. Significant differences ($p < 0.05$) among the treatments were determined using a post-hoc Tukey HSD test.

3. Results

3.1. *Antagonism of* F. circinatum *by Bacteria In Vitro*

The rhizobacterial strains significantly inhibited the growth of *F. circinatum* on ISP2 agar plates ($F_{5,12} = 27.46, p < 0.001$). The AIP was greater for *P. fluorescens* S32R2, *E. billingiae* S31R1, and *E. billingiae* S23L3 treatments than for *B. simplex* S11R41 (Figure 1). Fungal growth was arrested as it approached the boundary of bacteria and microscopic examination of the interface between the mycelia and the bacteria revealed a decrease in the density of the mycelial mat. *E. coli* also inhibited fungal growth with an AIP value similar to that of four biocontrol strains. Treatment with the grass root isolate

P. fluorescens J11 resulted in a low AIP value and the fungal mycelium grew over most of the surface of the agar, therefore this strain was not tested further. Neither *D. pinea* nor *N. parvum* growth were inhibited by any of the bacterial strains (AIP = 0).

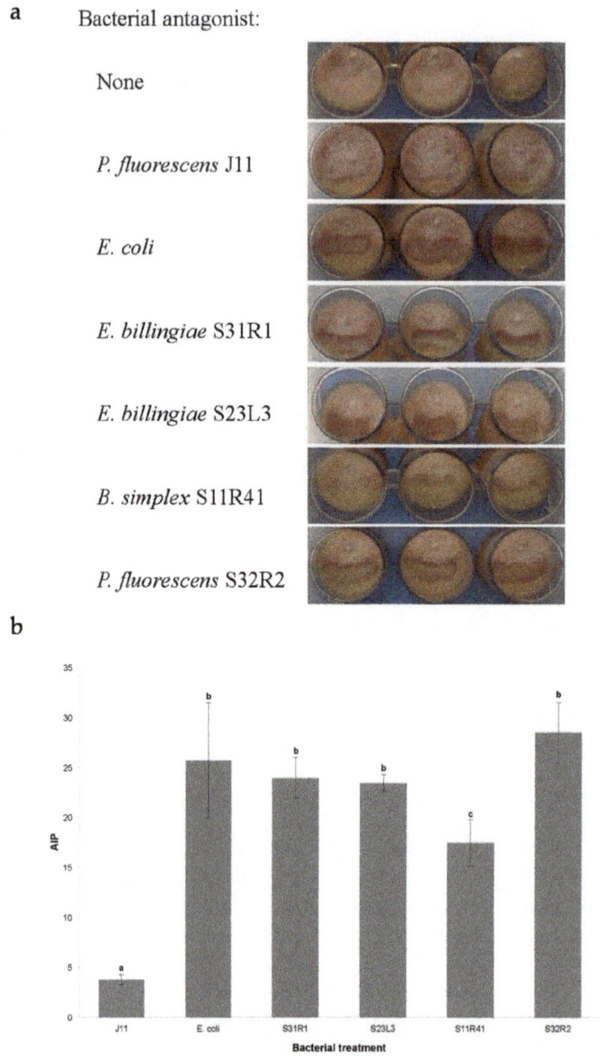

Figure 1. Antagonism of *F. circinatum* by bacteria in dual culture plate assays on International Streptomyces Project 2 (ISP2) agar (**a**); Area inhibition percent (AIP) of the tested bacterial strains against *F. circinatum* (**b**). Error bars show the standard deviation of the three independent replicates. Statistically significant differences of $p < 0.05$ between treatments are indicated with different lowercase letters. S31R1: *E. billingiae* S31R1; S23L3: *E. billingiae* S23L3; S11R41: *B. simplex* S11R41; S32R2: *P. fluorescens* S32R2.

3.2. *Inhibition of* F. circinatum *Lesions by Bacteria in Young* P. radiata *Trees*

Treatment of *P. radiata* with the pine rhizobacterial isolates significantly reduced the size of the lesions caused by *F. circinatum* infection (Experiment (Exp.) 1, $F_{6,211} = 13.52$, $p < 0.0001$; Exp. 2, $F_{6,189} = 17.13$, $p < 0.0001$). The reduction in the length of the lesions was similar among the bacterial strains, about 78% of the length of untreated *F. circinatum* lesions six weeks after introduction of the fungus, although in the first experiment the effect of *B. simplex* S11R41 was not significantly different from the control treatment without bacteria (Figure 2). Neither treatment with *E. coli* nor $MgSO_4$ reduced the size of the fungal lesions. While tray position (block) influenced lesion size (Exp. 2, $F_{7,189} = 4.20$, $p = 0.0002$) due to one of the eight trays of trees having a smaller mean lesion length compared to the others, there was no significant interaction between tray position and bacterial treatment (Exp. 2, $F_{42,189} = 0.99$, $p = 0.49$).

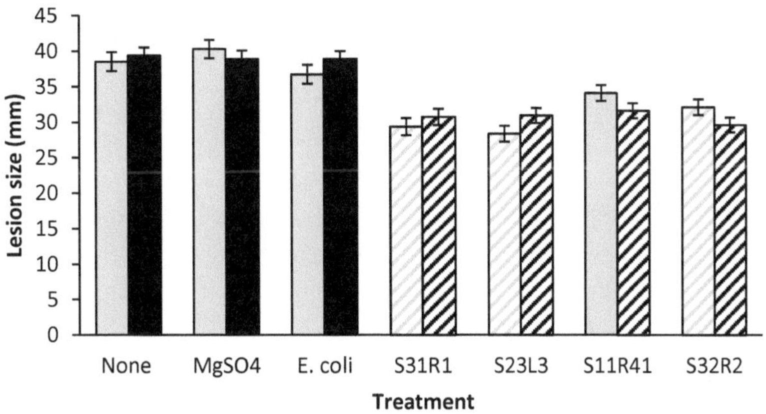

Figure 2. Size of *F. circinatum* lesions on two-year-old *P. radiata* trees treated with pine rhizobacteria. Values shown are the means of at least 30 lesions ± standard error in two separate experiments (Exp. 1, solid and hatched grey bars; Exp. 2, solid and hatched black bars); hatched bars indicate lesions that are significantly different from those on untreated (*F. circinatum* only) trees ($p < 0.05$).

3.3. *Effect of Essential Oils In Vitro*

Application of undiluted fennel, cinnamon, or clove oil directly to the wells containing *F. circinatum* completely inhibited the growth of the fungus (Table 1), and the vapors emitted by the undiluted oils were sufficient to limit the growth of the untreated fungus in the central wells of the plate. The 50% dilution of each of the oils produced similar inhibitory results. Cinnamon oil diluted to 10% and clove oil to 15% also completely inhibited fungal growth and were fungicidal. However, the inhibitory effect was reduced at lower oil concentrations to 82.7 AIP in the case of 5% cinnamon oil, and 87.7 and 69.0 AIP for 10% and 5% clove oil, respectively. Fennel oil, even when undiluted, had a fungistatic activity and there was no inhibition of growth at concentrations below 50%. Mycelia grew throughout the area covered by 5%–15% fennel oil, and therefore it was excluded from further testing. All of the essential oils at all of the concentrations tested inhibited the growth of the bacterial strains.

Table 1. Antifungal activities of essential cinnamon, clove, and fennel oils diluted with sweet almond oil. Area inhibition percent (AIP) was assessed after one week of exposure to the oil. The oils were determined to be fungicidal if no growth occurred after an inoculum was transferred to fresh media without oil or fungistatic if growth resumed after transfer. ND, not determined.

Oil	Concentration (% *v*/*v*)	AIP	Effect on *F. circinatum*
Cinnamon	5	82.7 ± 1.0	ND
	10	100	Fungicidal
	15	100	Fungicidal
	50	100	Fungicidal
	100	100	Fungicidal
Clove	5	69.0 ± 2.1	ND
	10	87.7 ± 1.8	ND
	15	100	Fungicidal
	50	100	Fungicidal
	100	100	Fungicidal
Fennel	5	0	ND
	10	0	ND
	15	0	ND
	50	100	Fungistatic
	100	100	Fungistatic

3.4. *Inhibition of* F. circinatum *Lesions by Essential Oils in Young* P. radiata *Trees*

The essential oils had a significant toxic effect on *P. radiata* trees ($F_{3,88} = 31.76$, $p = 0.0001$) although phytotoxicity was dependent on the age of the tree tissue ($F_{3,88} = 30.32$, $p = 0.0001$). One-year-old seedlings exhibited signs of phytotoxicity within three days of treatment with the cinnamon and clove oils. Symptoms were similar to those produced by herbicide treatments, including epinasty or curling and twisting of needles and stems followed by necrosis [30]. Chlorotic tissue extended to 14.18 and 18.21 mm from the site of cinnamon and clove oil application, respectively, compared to 1.74 mm for the control almond oil treatment, and was especially pronounced for the combined cinnamon and clove oil treatment (24.16 mm) (Table 2). In contrast, the essential oils were not toxic to older *P. radiata* tissues. The average lesion size was 3.31 mm and none of the oil treatments damaged two-year-old pine trees to a greater extent than the control treatment (Table 2). Due to the toxicity of the oils to one-year-old pine seedlings, their anti-fungal effect was tested only in older trees.

Table 2. Phytotoxicity of essential oils in one- and two-year-old *P. radiata* tissue. Cinnamon and clove oils were diluted to the indicated concentration in sweet almond oil. Values shown are the means of 12 lesions ± standard error. Statistically significant ($p < 0.05$) differences in lesion sizes among the treatments are indicated with a different lowercase letter.

Essential Oil Treatment	Concentration (% *v*/*v*)	Lesion Size (mm)	
		1-Year Old Tissue	2-Year-Old Tissue
Almond (diluent)	100	1.74 ± 0.18 [c]	3.27 ± 0.11 [c]
Cinnamon	10	18.22 ± 1.38 [b]	3.19 ± 0.14 [c]
Clove	15	14.18 ± 0.97 [b]	3.16 ± 0.12 [c]
Cinnamon + clove	10, 15	24.16 ± 2.94 [a]	3.62 ± 0.13 [c]

Treatment of wounds in stems of two-year-old *P. radiata* trees with clove and/or cinnamon oils prior to *F. circinatum* infection reduced the size of the lesions caused by the fungus ($F_{3,116} = 33.74$, $p < 0.0001$). Six weeks after fungal infection, the length of the lesions treated with 10% cinnamon oil was reduced by 51% compared to those treated with the diluent sweet almond oil (Table 3). Clove oil (15%) or a combination of the two oils reduced lesion length by 45%.

Table 3. Efficacy of essential oils in reducing the size of *F. circinatum* lesions in *P. radiata*. Essential oils were applied one day prior to infection with *F. circinatum*, and lesions were measured six weeks after the infection. Cinnamon and clove oils were diluted to the indicated concentration with sweet almond oil. Values shown are the means of 30 lesions ± standard error. Statistically significant ($p < 0.05$) differences in lesion sizes among the treatments are indicated with a different lowercase letter.

Essential Oil Treatment	Concentration (% *v*/*v*)	Lesion Size (mm)
Almond (diluent)	100	38.08 ± 1.98 [a]
Cinnamon	10	18.60 ± 1.38 [b]
Clove	15	20.83 ± 1.29 [b]
Cinnamon + clove	10, 15	20.79 ± 1.50 [b]

4. Discussion

4.1. *Biocontrol Bacteria*

The biocontrol bacteria *P. fluorescens* S32R2, *B. simplex* S11R41, *E. billingiae* S31R1, and *E. billingiae* S23L3, isolated from the roots of a healthy pine tree, inhibited *F. circinatum* growth in vitro and reduced the symptoms of pitch canker disease on young *P. radiata* trees. In a previous study, Mesanza et al. [27] reported the antagonist effect of these bacterial strains against the pine pathogens *H. annosum* and *A. mellea*. In that study, *P. fluorescens* S32R2 and *B. simplex* S11R41 had the greatest antagonistic effect in vitro and in vivo. In the present study, *P. fluorescens* S32R2 and the two strains of *E. billingiae* reduced *F. circinatum* lesion length to the greatest extent in young *P. radiata* trees. In contrast, the pine pathogens *D. sapinea* and *N. parvum* were not inhibited by any of the bacterial strains. This may suggest that the bacteria can outcompete some fungi, but not others, for nutrients and/or that they produce anti-fungal compounds that are effective against some fungi.

Pseudomonads are well known as biocontrol agents that are effective against several plant pathogens including pathogenic fungi such as *Rhizoctonia solani, R. oryzae, Pythium ultimum,* and *P. irregulare* [31]. *P. fluorescens* reduced the severity of *Fusarium* (mainly *F. oxysporum*)-induced wilt on a variety of crops [32–36]. Pseudomonads produce several antifungal metabolites including cyclic lipopeptides, phloroglucinols, phenazines, pyoluteorin, and pyrrolnitrin [37] or may induce local or systemic resistance in the plant against the pathogen [38]. Fungal inhibition is not characteristic of all pseudomonads, as indicated by the inability of *P. fluorescens* J11 to inhibit growth of *F. circinatum* in vitro.

Erwinia spp. are mainly plant pathogens, however, *E. billingiae* is part of the normal commensal microbiota of pome fruit trees and is well adapted to plant colonization including resistance to environmental stresses [39–41]. It is an antagonist of the closely related species *E. amylovora* that causes fire blight on apples and pears [42]. Currently, the mechanism by which *E. billingiae* inhibits fungal growth is not known.

B. simplex is a plant growth-promoting bacterium that can enhance the biomass of tomato and wheat [43,44], and induce rooting and root growth of kiwifruit stem cuttings [45] In addition, this bacterium showed antifungal effects against *F. oxysporum* in vitro [46] and *Phytophthora erythroseptica*, a pathogen of potato [47].

Although the biocontrol bacteria were isolated from the pine rhizosphere, they were able to inhibit *F. circinatum* growth on aerial tissues, suggesting that the bacterial strains are well adapted to grow in both plant environments. Plant epiphytic strains of *P. fluorescens* are common (e.g., [48]) and *E. billingae* strains have been isolated from the phyllosphere of apple, hawthorn, and pear [39]. The ability to colonize both the rhizo- and phyllosphere of a plant is a desirable characteristic of a bacterial treatment to control *F. circinatum*, which can also infect pine roots [49,50].

The bacteria were introduced through the excised tops of the trees to ensure that they and/or their antifungal products were able to reach the vascular tissue for possible systemic dissemination. This was done to optimize the potential for interactions between the fungus and bacteria/bacterial metabolites. While this is an aggressive treatment, control plants that were wounded in a similar manner (but not inoculated with fungus or bacteria), did not exhibit any signs of additional damage and recovered well. Cutting the tops of pine trees (typically 4–6 cm) is used in practice to obtain cuttings and to reinforce the root system before planting [51].

While *E. coli* inhibited *F. circinatum* in vitro, it was ineffective against the fungus in vivo. The difference in efficacy may reflect the mechanism by which *E. coli* inhibits *F. circinatum*, perhaps by outcompetition for nutrients. The *E. coli* strain used as a control in this study is a common laboratory strain selected for rapid growth under laboratory conditions. While the ISP2 agar used to test antifungal activity in vitro was determined to be a suitable growth substrate for both the fungal and bacterial isolates, in vivo growth factors such as nutrients, pH, and osmolarity are determined by the tree and may not be suitable for proliferation of *E. coli*, which is not a plant-associated bacterium. On the other hand, the biocontrol strains are native to pine trees and are adapted to growth under those conditions.

4.2. Essential Oils

Essential oils extracted from a variety of plants have been proposed as safe, natural alternatives to synthetic fungicides. Many in vitro studies have reported the efficacy of essential oils against fungi including sapstain fungi (*Ophiostoma piceae, Aureobasidum pullulans, Alternaria alternata, Gliocladium viride*) [52], *Fusarium* spp. [15,24,53], and *D. pinea* [15]. However, plant conditions may change the efficiency of essential oils, therefore fungal inhibition must be assessed in situ. Soylu et al. [54] reported the efficacy of oregano essential oil on tomato plants against *Botrytis cinerea*. Clove oil applied on guava seedlings roots completely inhibited *F. oxysporum*, *Botryodiplodia theobromae*, and *R. solani* [55], and cumin, geranium, and basil essential oils were effective against six species of *Fusarium* both in vitro and on cumin plants when applied directly to soil during planting in concentrations ranging from 2% to 8% (v/v) [56]. In this study, the direct application of cinnamon and/or clove oils to wounds in the stems of two-year-old *P. radiata* trees limited the damage to the trees by *F. circinatum*. The size of the lesions was reduced to the same extent by the individual and combined oils treatments. However, essential oils may be phytotoxic depending on their composition and concentration, the age of the plant, and the method of application [57,58]. In our study, cinnamon (10%) and clove (15%) oils had a toxic effect when applied to one-year-old *P. radiata* seedlings, especially when the two oils were combined, but were not toxic to older trees.

The antifungal activity of essential oils may be attributed to reactive aldehydes, flavonoids, or terpenes that disrupt cell membranes [59,60]. Cinnamaldehyde, a major component of cinnamon and some other essential oils, is a potent antifungal agent. When applied to tobacco, cinnamaldehyde reduced symptoms of tobacco black shank disease caused by *Phytophthora parasitica* var. *nicotianae* [61]. It also inhibited the growth of the tree pathogenic fungi *R. solani*, *Ganoderma australe*, *Collectotrichum gloeosporioides*, and to a lesser extent, *Fusarium solani* [62], and was active against *Phytophthora cactorum* and *F. circinatum* [23]. Eugenol, a phenylpropanoid, is a major component of clove oil and is known to have antifungal properties [63], including toxicity to several species of *Fusarium* [64]. Cinnamaldehyde and eugenol may be responsible for the inhibitory effect of cinnamon and clove oils, respectively, against *F. circinatum* demonstrated in this study.

5. Conclusions

The results presented herein demonstrate the ability of four pine rhizobacterial isolates, and cinnamon and clove essential oils, to inhibit the growth of *F. circinatum* in vitro and in young *P. radiata* trees. These biocontrol agents show promise as early, prophylactic treatments in nurseries to supplement existing control measures such as seed sanitization and/or cultivation of more resistant tree species to reduce fungal infections. However, the biocontrol treatments cannot be combined as the essential oils are toxic to the antagonistic bacteria. The oils are also toxic to *P. radiata* seedlings and therefore their application is restricted to trees older than one year. The use of natural compounds and indigenous antagonistic organisms may represent an alternative to the use of synthetic fungicides to control *F. circinatum* infections in nurseries and young pine plantations; however, additional studies are needed to optimize the method of application, including treatments to protect against root infections. The future of control strategies in forest ecosystems should be directed to an integrated system in which biocontrol agents and natural products play a crucial role.

Acknowledgments: This work was supported by Spanish Project RTA 2013-00048-C03-03 INIA and the European Union Healthy Forest Project LIFE14 ENV/ES/000179 to E.I., a grant from the New Brunswick Innovation Foundation to C.L.P., and a grant from the European Union 7th Framework Programme (Marie Curie Action) to E.I. and C.L.P.

Author Contributions: E.I., M.E.-R., and R.R. conceived and designed the experiments; E.I., M.E.-R., R.R., and T.T. performed the experiments; E.I., M.N., T.T., and C.L.P. analyzed the data; E.I., M.N., and C.L.P. contributed reagents/materials/analysis tools; C.L.P. wrote the paper.

Conflicts of Interest: The authors declare no conflict of interest. The funding sponsors had no role in the design of the study; in the collection, analyses, or interpretation of data; in the writing of the manuscript, and in the decision to publish the results.

References

1. Nirenberg, H.I.; O'Donnell, K. New *Fusarium* species and combinations within the *Giberrella fujikuroi* species complex. *Mycologia* **1998**, *90*, 434–458. [CrossRef]
2. Wingfield, M.J.; Hammerbacher, A.; Ganley, R.J.; Steenkamp, E.T.; Gordon, T.R.; Wingfield, B.D.; Coutinho, T.A. Pitch canker caused by *Fusarium circinatum*—A growing threat to pine plantations and forests worldwide. *Australas. Plant Pathol.* **2008**, *37*, 319–334. [CrossRef]

3. Viljoen, A.; Wingfield, M.J.; Marasas, W.F.O. First report of *Fusarium subglutinans* f. sp. *Pini* on seedlings in South Africa. *Plant Dis.* **1994**, *78*, 309–312.
4. Aegerter, B.J.; Gordon, T.R. Rates of pitch canker induced seedling mortality among *Pinus radiata* families varying in levels of genetic resistance to *Gibberella circinata* (anamorph *Fusarium circinatum*). *For. Ecol. Manag.* **2006**, *235*, 14–17. [CrossRef]
5. Wikler, K.; Gordon, T.R. An initial assessment of genetic relationships among populations of *Fusarium circinatum* in different parts of the world. *Can. J. Bot.* **2000**, *78*, 709–717. [CrossRef]
6. Gordon, T.R.; Storer, A.J.; Wood, D.L. The pitch canker epidemics in California. *Plant Dis.* **2001**, *85*, 1128–1139. [CrossRef]
7. Wingfield, M.J.; Jacobs, A.; Coutinho, T.A.; Ahumada, R.; Wingfield, B.D. First report of the pitch canker fungus, *Fusarium circinatum*, on pines in Chile. *Plant Pathol.* **2002**, *51*, 397. [CrossRef]
8. Landeras, E.; Garcia, P.; Fernandez, Y.; Brana, M. Outbreak of pitch canker caused by *Fusarium circinatum* on *Pinus* spp. in Norther Spain. *Plant Dis.* **2005**, *89*, 1015. [CrossRef]
9. Braganca, H.; Diogo, E.; Moniz, F.; Amaro, P. First report of pitch canker on pines caused by *Fusarium circinatum* in Portugal. *Plant Dis.* **2009**, *93*, 1079. [CrossRef]
10. Ganley, R.J.; Watt, M.S.; Manning, L.; Iturritxa, E. A global climatic risk assessment of pitch canker disease. *Can. J. For. Res.* **2009**, *39*, 2246–2256. [CrossRef]
11. Martínez-Álvarez, P.; Pando, V.; Diez, J.J. Alternative species to replace Monterey pine plantations affected by pitch canker caused by *Fusarium circinatum* in northern Spain. *Plant Pathol.* **2014**, *63*, 1086–1094. [CrossRef]
12. Gordon, T.R.; Kirkpatrick, S.C.; Aegerter, B.J.; Wood, D.L.; Storer, A.J. Susceptibility of Douglas fir (*Pseudotsuga menziesii*) to pitch canker, caused by *Gibberella circinata* (anamorph = *Fusarium circinatum*). *Plant Pathol.* **2006**, *55*, 231–237. [CrossRef]
13. European and Mediterranean Plant Protection Organization. Data sheets on quarantine pests: *Gibberella circinata*. *EPPO Bull.* **2005**, *35*, 383–386.
14. Agusti-Brisach, C.; Perez-Sierra, A.; Armengol, J.; Garcia-Jimenez, J.; Berbegal, M. Efficacy of hot water treatment to reduce the incidence of *Fusarium circinatum* on *Pinus radiata* seeds. *Forestry* **2012**, *85*, 629–635. [CrossRef]
15. Iturritxa, E.; Desprez-Loustau, M.L.; Garcia-Serna, I.; Quintana, E.; Mesanza, N.; Aitken, J. Effect of alternative disinfection treatments against fungal canker in seeds of *Pinus radiata*. *Seed Technol.* **2011**, *33*, 88–110.
16. Howell, C.R. Mechanisms employed by *Trichoderma* species in the biological control of plant diseases: The history and evolution of current concepts. *Plant Dis.* **2003**, *87*, 4–10. [CrossRef]
17. Haas, D.; Defago, G. Biological control of soil-borne pathogens by fluorescent pseudomonads. *Nat. Rev. Microbiol.* **2005**, *3*, 307–319. [CrossRef] [PubMed]
18. Pliego, C.; Ramos, C.; de Vicente, A.; Cazorla, F. Screening for candidate bacterial biocontrol agents against soilborne fungal plant pathogens. *Plant Soil* **2011**, *340*, 505–520. [CrossRef]
19. Martínez-Álvarez, P.; Alves-Santos, F.M.; Diez, J.J. *In vitro* and *in vivo* interactions between *Trichoderma viride* and *Fusarium circinatum*. *Silva Fenn.* **2012**, *46*, 303–316. [CrossRef]
20. Vivas, M.; Martín, J.A.; Gil, L.; Solla, A. Evaluating methyl jasmonate for induction of resistance to *Fusarium oxysporum*, *F. circinatum* and *Ophiostoma novo-ulmi*. *For. Syst.* **2012**, *21*, 289–299. [CrossRef]
21. Mimica-Duvic, N.; Bozin, B.; Sokovic, M.; Mihajlovic, B.; Matavulj, M. Antimicrobial and antioxidant activities of three *Mentha* species essential oils. *Planta Med.* **2003**, *69*, 413–419.
22. Bozin, B.; Mimica-Dukic, N.; Simin, N.; Anackov, G. Characterization of the volatile composition of essential oil of some Lamiaceae species and the antimicrobial and antioxidant activities of the entire oils. *J. Agric. Food Chem.* **2006**, *54*, 1822–1828. [CrossRef] [PubMed]
23. Lee, Y.S.; Kim, J.; Lee, S.G.; Oh, E.; Shin, S.C.; Park, I.K. Effects of plant essential oils and components from Oriental sweetgum (*Liquidambar orientalis*) on growth and morphogenesis of three phytopathogenic fungi. *Pest. Biochem. Physiol.* **2009**, *93*, 138–143. [CrossRef]
24. Seseni, L.; Regnier, T.; Roux-van der Merwe, M.P.; Mogale, E.; Badenhorst, J. Control of *Fusarium* spp. causing damping-off of pine seedlings by means of essential oils. *Ind. Crops Prod.* **2015**, *76*, 329–332. [CrossRef]
25. Iturritxa, E.; Ganley, R.J.; Raposo, R.; Garcia-Serna, I.; Mesanza, N.; Kirkpatrick, S.C.; Gordon, T.R. Resistance levels of Spanish conifers against *Fusarium circinatum* and *Diplodia pinea*. *For. Pathol.* **2013**, *43*, 488–495.
26. Iturritxa, E.; Slippers, B.; Mesanza, N.; Wingfield, M.J. First report of *Neofusicoccum parvum* causing canker and die-back of *Eucalyptus* in Spain. *Australas. Plant Dis. Notes* **2011**, *6*, 57–59. [CrossRef]

27. Mesanza, N.; Iturritxa, E.; Patten, C.L. Native rhizobacteria as biocontrol agents of *Heterobasidion annosum* s.s. and *Armillaria mellea* infection of *Pinus radiata*. *Biol. Control* **2016**, *101*, 8–16. [CrossRef]
28. Shirling, E.B.; Gottlieb, D. Methods for characterization of *Streptomyces* species. *Int. J. Syst. Bacteriol.* **1966**, *16*, 313–340. [CrossRef]
29. Schmale, D.G.; Gordon, T.R. Variation in susceptibility to pitch canker disease, caused by *Fusarium circinatum*, in native stands of *Pinus muricata*. *Plant Pathol.* **2003**, *52*, 720–725. [CrossRef]
30. Tran, H.; Harrington, K.C.; Robertson, A.W.; Watt, S.M. Assessment of herbicides for selectively controlling broom (*Cytisus scoparius*) growing with radiata pine (*Pinus radiata*) in New Zealand. *N. Z. J. For. Sci.* **2016**, *46*, 1–13. [CrossRef]
31. Mavrodi, O.V.; Walter, N.; Elateek, S.; Taylor, C.G.; Okubara, P.A. Suppression of *Rhizoctonia* and *Pythium* root rot of wheat by new strains of *Pseudomonas*. *Biol. Control* **2012**, *62*, 93–102. [CrossRef]
32. Corrêa, B.O.; Schafer, J.T.; Moura, A.B. Spectrum of biocontrol bacteria to control leaf, root and vascular diseases of dry bean. *Biol. Control* **2014**, *72*, 71–75. [CrossRef]
33. Al-Mughrabi, K.I. Biological control of *Fusarium* dry rot and other potato tuber diseases using *Pseudomonas fluorescens* and *Enterobacter cloacae*. *Biol. Control* **2010**, *53*, 280–284. [CrossRef]
34. Manikandan, R.; Saravanakumar, D.; Rajendran, L.; Raguchander, T.; Samiyappan, R. Standardization of liquid formulation of *Pseudomonas fluorescens* Pf1 for its efficacy against *Fusarium* wilt of tomato. *Biol. Control* **2010**, *54*, 83–89. [CrossRef]
35. Srivastava, R.; Khalid, A.; Singh, U.S.; Sharma, A.K. Evaluation of arbuscular mycorrhizal fungus, fluorescent *Pseudomonas* and *Trichoderma harzianum* formulation against *Fusarium oxysporum* f. sp. *lycopersici* for the management of tomato wilt. *Biol. Control* **2010**, *53*, 24–31.
36. Dhingra, O.D.; Coelho-Netto, R.A.; Rodrigues, F.Á.; Silva, G.J.; Maia, C.B. Selection of endemic nonpathogenic endophytic *Fusarium oxysporum* from bean roots and rhizosphere competent fluorescent *Pseudomonas* species to suppress *Fusarium*-yellow of beans. *Biol. Control* **2006**, *39*, 75–86. [CrossRef]
37. Beneduzi, A.; Ambrosini, A.; Passaglia, L.M.P. Plant growth-promoting rhizobacteria (PGPR): Their potential as antagonists and biocontrol agents. *Gen. Mol. Biol.* **2012**, *35*, 1044–1051. [CrossRef]
38. Van Loon, L.C. Plant responses to plant growth-promoting rhizobacteria. *Eur. J. Plant Pathol.* **2007**, *119*, 243–254. [CrossRef]
39. Mergaert, J.; Hauben, L.; Cnockaert, M.C.; Swings, J. Reclassification of non-pigmented *Erwinia herbicola* strains from trees as *Ennrinia billingiae* sp. nov. *Int. J. Syst. Bacteriol.* **1999**, *49*, 377–383. [CrossRef] [PubMed]
40. Jakovljevic, V.; Jock, S.; Du, Z.; Geider, K. Hypersensitive response and acyl-homoserine lactone production of the fire blight antagonists *Erwinia tasmaniensis* and *Erwinia billingiae*. *Microb. Biotechnol.* **2008**, *1*, 416–424. [CrossRef] [PubMed]
41. Palacio-Bielsa, A.; Roselló, M.; Llop, P.; López, M.M. *Erwinia* spp. from pome fruit trees: Similarities and differences among pathogenic and non-pathogenic species. *Trees* **2012**, *26*, 13–29. [CrossRef]
42. Jakovljevic, V.; Du, Z.; Jock, S.; Geider, K. Epiphytic *Erwinias* to control fire blight. In Proceedings of the 10th International Workshop on Fire Blight, Bologna, Italy, 5–9 July 2004; Bazzi, C., Mazzucchi, U., Eds.; Acta Horticulturae, No. 704. International Society for Horticultural Science: Leuven, Belgium, 2006; pp. 289–292.
43. Hassen, A.I.; Labuschagne, N. Root colonization and growth enhancement in wheat and tomato by rhizobacteria isolated from the rhizoplane of grasses. *World J. Microbiol. Biotechnol.* **2010**, *26*, 1837–1846. [CrossRef]
44. Barneix, A.J.; Saubidet, M.I.; Fatta, N.; Kade, M. Effect of rhizobacteria on growth and grain protein in wheat. *Agron. Sustain. Dev.* **2005**, *25*, 505–511. [CrossRef]
45. Erturk, Y.; Ercisli, A.; Haznedar, A.; Cakmakci, R. Effects of plant growth promoting rhizobacteria (PGPR) on rooting and root growth of kiwifruit (*Actinidia deliciosa*) stem cuttings. *Biol. Res.* **2010**, *43*, 91–98. [CrossRef] [PubMed]
46. Schwartz, A.; Ortiz, I.; Maymon, M.; Herbold, C.; Fujishige, N.; Vijanderan, J.; Villella, W.; Hanamoto, K.; Diener, A.; Sanders, E.; et al. *Bacillus simplex*—little known PGPB with anti-fungal activity—Alters pea legume root architecture and nodule morphology when coinoculated with *Rhizobium leguminosarum* bv. viciae. *Agronomy* **2013**, *3*, 595–620. [CrossRef]
47. Adiyaman, T.; Schisler, D.A.; Slininger, P.J.; Sloan, J.M.; Jackson, M.A.; Rooney, A.P. Selection of biocontrol agents of pink rot based on efficacy and growth kinetics index rankings. *Plant Dis.* **2011**, *95*, 24–30. [CrossRef]

48. Lindow, S.E.; Desurmont, C.; Elkins, R.; McGourty, G.; Clark, E.; Brandl, M.T. Occurrence of indole-3-acetic acid-producing bacteria on pear trees and their association with fruit russet. *Phytopathology* **1998**, *88*, 1149–1157. [CrossRef] [PubMed]
49. Garbelotto, M.; Schweigkofler, W.; Shaw, D. First report of *Fusarium circinatum*, causal agent of pitch canker disease, from the roots of mature Aleppo pines in California. *Plant Health Prog.* **2007**. [CrossRef]
50. Mitchell, R.G.; Steenkamps, E.T.; Coutinho, T.A.; Wingfield, M.J. The pitch canker fungus, *Fusarium circinatum*: Implications for South African forestry. *South. For.* **2011**, *73*, 1–13. [CrossRef]
51. Mead, D.J. *Sustainable Management of Pinus radiata Plantations*; FAO Forestry Paper No. 170; FAO: Rome, Italy, 2013.
52. Dawson-Andoh, B.E.; Lovell, R.; Kamdem, D.P. Inhibitory and compatibility effects of essential oils on sapstain and biological control fungi. *J. Essent. Oil Res.* **2000**, *12*, 509–515. [CrossRef]
53. Amri, I.; Gargouri, S.; Hamrouni, L.; Hanana, M.; Fezzani, T.; Jamoussi, B. Chemical composition, phytotoxic and antifungal activities of *Pinus pinea* essential oil. *J. Pest Sci.* **2012**, *85*, 199–207. [CrossRef]
54. Soylu, M.; Kurt, S.; Soylu, S. *In vitro* and *in vivo* antifungal activities of the essential oils of various plants against tomato grey mould disease agent *Botrytis cinerea*. *Int. J. Food Microbiol.* **2010**, *143*, 183–189. [CrossRef] [PubMed]
55. Hamad, Y.K.; Fahmi, M.M.; Zaitoun, F.M.; Ziyada, S.M. Role of essential oils in controlling fungi that cause decline disease of guava. *Int. J. Pure Appl. Biosci.* **2015**, *3*, 143–151. [CrossRef]
56. Hassem, M.; Moharam, A.M.; Zaied, A.A.; Saleh, F.E.M. Efficacy of essential oils in the control of cumin root rot disease caused by *Fusarium* spp. *Crop Prot.* **2010**, *29*, 1111–1117. [CrossRef]
57. Cloyd, R.A.; Galle, C.L.; Keith, S.R.; Kalscheur, N.A.; Kemp, K.E. Effect of commercially available plant-derived essential oil products on arthropod pests. *J. Econ. Entomol.* **2009**, *102*, 1567–1579. [CrossRef] [PubMed]
58. Morcia, C.; Tumino, G.; Salhi, N.; Terzi, V. What's the role of essential oils in cereal protection? In *Microbial Pathogens and Strategies for Combating Them: Science, Technology and Education*; Méndez-Vilas, A., Ed.; Formatex Research Center: Badajoz, Spain, 2013; Volume 2, pp. 1412–1418.
59. Shao, X.; Cheng, S.; Wang, H.; Yu, D.; Mungai, C. The possible mechanism of antifungal action of tea tree oil on *Botrytis cinerea*. *J. Appl. Microbiol.* **2013**, *114*, 1642–1649. [CrossRef] [PubMed]
60. Zuzarte, M.; Gonçalves, M.J.; Cavaleiro, C.; Canhoto, J.; Vale-Silva, L.; Silva, M.J.; Pinto, E.; Salgueiro, L. Chemical composition and antifungal activity of the essential oils of *Lavandula viridis* L'Hér. *J. Med. Microbiol.* **2011**, *60*, 612–618. [CrossRef] [PubMed]
61. Lu, M.; Han, Z.; Yao, L. *In vitro* and *in vivo* antimicrobial efficacy of essential oils and individual compounds against *Phytophthora parasitica* var. *nicotianae*. *J. Appl. Microbiol.* **2013**, *115*, 187–198. [CrossRef] [PubMed]
62. Lee, H.C.; Cheng, S.S.; Chang, S.T. Antifungal property of the essential oils and their constituents from *Cinnamomum osmophloeum* leaf against tree pathogenic fungi. *J. Sci. Food Agric.* **2005**, *85*, 2047–2053. [CrossRef]
63. Núñez, L.; Aquino, M.D.; Chirife, J. Antifungal properties of clove oil (*Eugenia caryophylata*) in sugar solution. *Braz. J. Microbiol.* **2001**, *32*, 123–126. [CrossRef]
64. Morcia, C.; Mainati, M.; Terzi, V. *In vitro* antifungal activity of terpinen-4-ol, eugenol, carvone, 1,8-cineole (eucalyptol) and thymol against mycotoxigenic plant pathogens. *Food Addit. Contam.* **2012**, *29*, 415–422.

13

Using Fractal Analysis in Modeling the Dynamics of Forest Areas and Economic Impact Assessment: Maramureș County, Romania

Radu-Daniel Pintilii [1], Ion Andronache [1], Daniel Constantin Diaconu [1,*], Răzvan Cătălin Dobrea [2], Martina Zeleňáková [3], Rasmus Fensholt [4], Daniel Peptenatu [1], Cristian-Constantin Drăghici [1] and Ana-Maria Ciobotaru [1]

1 University of Bucharest—Research Center for Integrated Analysis and Territorial Management; 4-12, Regina Elisabeta Avenue; 030018 Bucharest, Romania; pinty_ro@yahoo.com (R.-D.P.); andronacheion@email.su (I.A.); peptenatu@yahoo.fr (D.P.); cristian.draghici@geo.unibuc.ro (C.C.D.); ciobotaruanamaria@inbox.lv (A.-M.C.)

2 Bucharest University of Economic Studies—Faculty of Management; 6, Piața Romană Square; 010374 Bucharest, Romania; rdobrea@yahoo.com

3 Institute of Environmental Engineering, Technical University of Kosice, Vysokoskolska 4, Kosice 04200, Slovakia; martina.zelenakova@tuke.sk

4 University of Copenhagen, Department of Geosciences and Natural Resource Management, Oster Voldgade 10, DK-1350 Copenhagen, Denmark; rf@ign.ku.dk

* Correspondence: ddcwater@yahoo.com

Academic Editors: Rodney J. Keenan and Timothy A. Martin

Abstract: This study uses fractal analysis to quantify the spatial changes of forest resources caused by an increase of deforested areas. The method introduced contributes to the evaluation of forest resources being under significant pressure from anthropogenic activities. The pressure on the forest resources has been analyzed for Maramureș County, one of the most deforested counties in Romania. In order to evaluate this, the deforested areas were calculated for the period of 2001–2014, by using the Global Forest Change 2000–2014 database. The Fractal Fragmentation Index (*FFI*) and Fixed Grid 2D Lacunarity (*FG2DL*) were used to quantify the degree of fragmentation and dispersion of the forested areas, and thereby the extent to which a forest area is affected by deforestation. The process of quantifying the pressure on forested areas included the creation of a database for the period of 2000–2014 containing economic activities (turnover) related to woody recourses, important indicators of forest exploitation. Taken together, the results obtained indicate a dramatic increase in deforested areas (over 19,122 ha in total for the period of analysis), in Maramureș County.

Keywords: deforested areas; forest resources; economic pressure; territorial management; fractal analysis; lacunarity

---

1. Introduction

Forest areas represent one of the most important and complex terrestrial ecosystems. In 2010, forested areas covered over 4 billion hectares globally, representing 31% of the total land area [1]. Management of forest areas is one of the greatest challenges for policy-makers, faced with significant pressure exerted on forest resources by various sectors of the economy, which require a growing wood volume [2]. The complexity of the effects of deforestation renders territorial management strategies more expensive. The forest, by its products, has several functions with implications for the economy, the environment, and society. This multi-functionality defines the key role forested areas play in the fight against pollution, poverty, and environmental protection. This role is recognized by a number of official documents both globally [3] and in the European Union [4]. According to these documents, the increase in deforested areas is a matter of general interest, and vigorous action is required to reduce their effects. In this context, reducing deforested areas is a priority for decision-makers, which may bring a number of important benefits to local communities, including reducing carbon emissions, watershed protection, biodiversity, and soil quality conservation [5–11]. The most important causes that determine the expansion of the deforested areas globally identified by several studies are the conversion of forested areas into farmland and the price of timber [12–16]. This causes overexploitation of forests, with negative consequences with respect to ecosystem functionality and economic revenue [17]. The analysis of the deforested areas' evolution and the causal drivers becomes particularly important, especially from the economic point of view, by its implications on the sustainable development of local communities [18–22]. Expanding deforested surfaces causes this important resource to be increasingly reduced for local communities with direct consequences for livelihoods. Romania's forested area is estimated at about 29% of the total area of the country, much below the average EU level of 40% [4]. Currently, it is subjected to increasing pressure from socio-economic factors, reflected by legal or illegal increases in logging of the forested areas. In this context, the study of the forest area dynamics plays a key role in developing and implementing strategies to ensure better forest monitoring [23–25].

After 1990, following the fall of communism, an increasing pressure was exerted on forest areas in Romania. The development of methods to analyze the dynamics of forested areas constitutes an immediate necessity, given that Romania is facing a reduction in forested areas each year. Additionally, mapping and quantification of illegal logging, which has escalated in recent years, requires the development of analysis methods based on the textural imaging analysis of the forest. Thus, fractal analysis can be used complementary to traditional Geographic Information Systems (GIS)-based analyses.

An innovative method in shaping the economic pressure on the forest's resources is represented by the fractal analysis. The fractal dimension is a measure of complexity; the measure to which the fractal "fills" the space, quantifying the degree of irregularity and fragmentation of a geometrical structure or of an object of nature often having superior value as compared to the topological dimension [21]. The fractal theory was adopted in geography and has been applied in shoreline analysis, land relief appearance, the appearance of clouds, in river basin studies, and the distribution curves of the climatic and hydrological parameters [25]. Fractal analysis has also been applied for studying forest characteristics based on fractal analysis and remote sensing imagery covering different forest regions of the world [26–28].

Within this study, we aimed to determine if fractal analysis can be useful and whether it can provide additional information in the spatial analysis of the effects incurred by deforestation, imposed by the economic pressure (increasingly more intense) on the forest's resources in a case study area (Maramureș County) located in the northernmost part of Romania.

2. Materials and Methods

2.1. Preprocessing of Satellite Imagery

To assess the deforested area, the Global Forest Change 2000–2014 database was used, provided by the Department of Geographical Sciences, Maryland University. The database is the result of the analysis of 654,178 Landsat 7 ETM + images characterizing global forest extent and change from 2000 through 2014 [29]. Remotely sensed images and GIS analyses can provide relevant forest change estimates at some scale and level of accuracy [29,30]. Based on this, the deforested area and the wooded land areas for the period of 2000–2014 was calculated for Maramureș County and five of the most deforested territorial administrative units (Borșa, Poienile, Repedea, Săpânța, and Vișeu de Sus). Initially, the Hansen_GFC2015 [29] maps of the forested and deforested areas were extracted in tiff format using the ArcGIS platform and the resolution chosen was 3509 × 2481 (170.72 km × 120.71 km). We selected this resolution because it provides sufficient spatial detail for a good fractal analysis. The used method started with a conversion of the initial projection of Hansen_GFC, into the Stereographic 70 coordinate system (EPSG 31700), specific for Romania.

Initially, the extent of the forested and the deforested areas were analyzed, over the 15 year period, both for the entire Maramureș County, and for the five territorial administrative units (TAUs) characterized by the highest forest cover of the county (Figure 1).

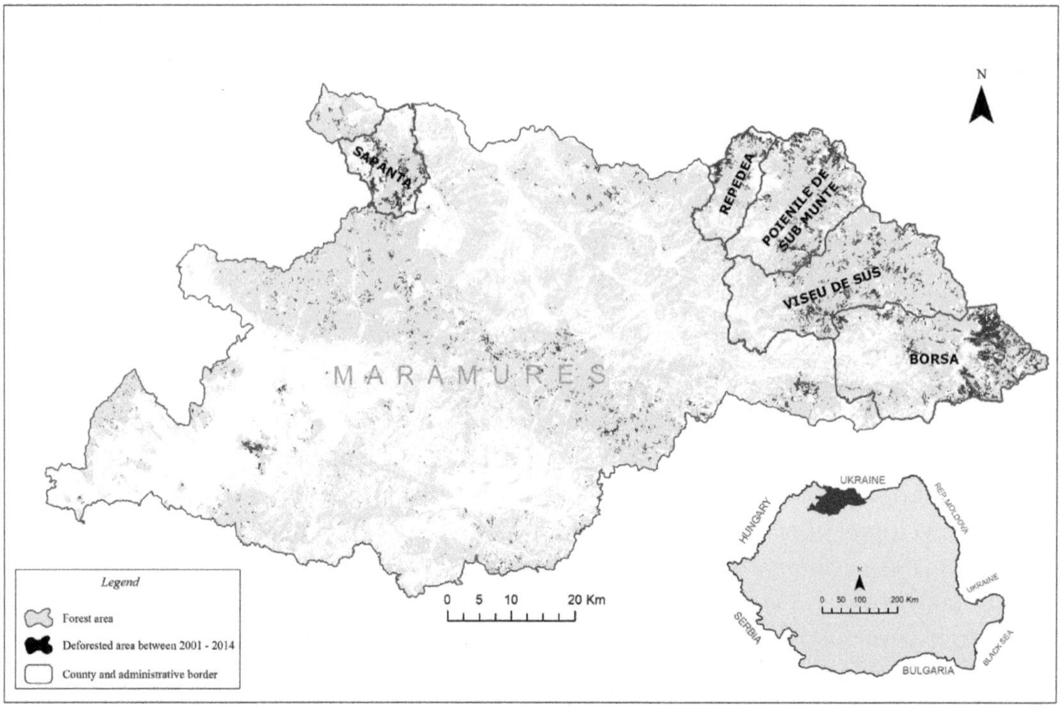

Figure 1. Geographical study area of Maramureș County and selected territorial administrative units (TAU), located in Northern Romania. Data Source: [29].

2.2. Data Analysis

All pixels were transformed into points, with the dimension of the pixels determining the area covered with vegetation. The change in number of forest pixels from year to year helped us to determine and to quantify the deforestation process, and its economic exploitation. The wood-related economic activities were analyzed in detail, by using a complex database of turnover. For this purpose, the NACE codes (statistical classification of economic activities in the European Community) related to (a) sylviculture and other forestry activities; (b) logging; (c) sawing and plantation of woods; (d) joinery installations; and (e) wholesale of woods, construction materials, and sanitary equipment were selected because these turnover values reflect the economic activity related to timber harvesting. Fractal analysis was used to derive other evidence, to corroborate our findings. The images, first obtained from the ArcGIS platform [31], were manually binarized, subsequently calculating the km^2 areas using ImageJ 1.51 g software [32]. The resulting binarized images were subject to fractal analysis using IQM 3.3 software [33]. The method used to determine the fractal dimension was the Fractal Fragmentation Index (to determine the fractal fragmentation of forest areas as a result of deforestation) [34] and Fixed Grid 2D Lacunarity (to determine how heterogeneous the area taken up by deforestation is) [35].

The Fractal Fragmentation Index (*FFI*) quantifies, in a single value, the information obtained from the fractal analysis on mass concentration, but also on the tortuosity of the perimeters describing the fractal fragmentation which can be interpreted as a compaction index, as well [34]:

$$FFI = D_A - D_P \quad (1)$$

where *FFI* is the fractal fragmentation index, D_A is the fractal dimension of the summed-up areas, and D_P is the fractal dimension of the summed up perimeters.

According to [34] $FFI = 0$, when $D_P = D_A$, which means that the forested areas are represented only in very small areas, characterized by punctual shapes. As the *FFI* is close to 0, the forested areas are highly fragmented, dispersed, smaller, and fewer, or of a tentacular and sprawling pattern. As the *FFI* tends to increase towards 1, the forested areas become larger and more compact, being arranged in clusters. $FFI = 1$ is recorded only when the forested areas are geometrically perfect and 100% compact, without any discontinuity ($D_P = 1$ and $D_A = 2$).

D_P and D_A were obtained using the Pyramid Dimension, implemented in IQM [36]. Fixed Grid 2D Lacunarity (*FG2DL*) was used to calculate the degree of heterogeneity by the variation of deforested areas distribution and it is determined using the following equation:

$$FG2DL = (CV_{FG2DL})^2 \tag{2}$$

where CV_{FG2DL} is the coefficient of variation [37,38].

As the deforestation process is more heterogenous and chaotic, the value *FG2DL* increases, and vice-versa.

The ArcGIS platform was used to graphically illustrate the geographical position and the distribution of the economic indicator. Finally, correlations between deforestation and the economic exploitation of wood were used to determine the impact of economic pressure on forested areas.

3. Results

3.1. Spatio-Temporal Deforestation Rates

The forest-covered areas have generally decreased during the period of analysis, but at a variable rate, at the Maramureș County level (total deforested area is 19,122.8 ha), with maximum rates in 2004, 2007, and 2012 and minimum rates in 2001–2003 (Figure 2).

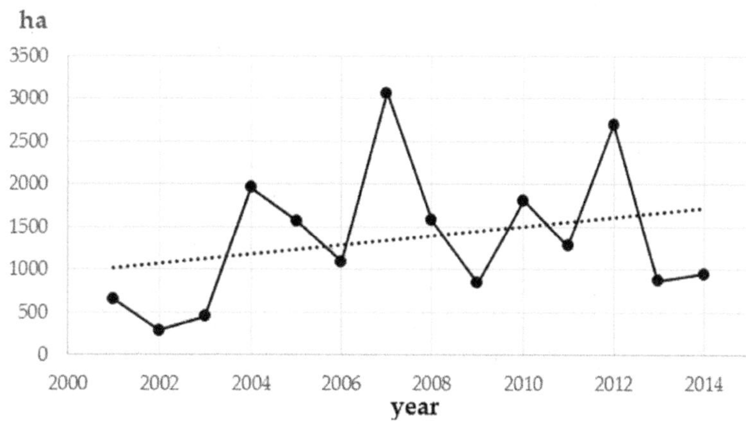

Figure 2. Deforestation rates in Maramureș County (2001–2014).

In four out of five TAUs, the analysis of deforestation indicates high levels in 2007 and 2012, except for the Săpânța TAU, where deforestation was at its maximum level in 2004. In all cases at county and territorial levels, deforestation was lowest in 2002–2003, except for the Săpânța TAU showing the lowest deforestation in 2014.

During the period from 2000 to 2014, 13,177.3 ha have been deforested in the five TAUs, representing 69% of the entire deforested area of Maramureș County. Borșa was the most deforested TAU (5,848.8 ha, corresponding to 41.32% of the entire deforested area of Maramureș County), whereas

Repedea represented the TAU of lowest deforestation, with a decrease of 1,085.7 ha, corresponding to 8.23% of the deforested area of Maramureș County.

3.2. Spatio-Temporal Patterns of Forest Fragmentation

The low annual *FFI* values of the Maramureș County (all numbers below 0.12; Figure 3) indicate that the county is characterized by large fragmentation of forested area. Along with the expansion of deforestation in the period analyzed, the *FFI* of the forested areas of the Maramureș County decreased by 0.0211 (Figure 3), indicating a continued increase in the fragmentation of forested areas especially in the northwest (NW) of the county. The largest decreases of the *FFI* values were registered in 2004, 2007, and 2012 and the smallest reduction of the *FFI* occurred in 2002 (0.0005), corresponding to the years of highest and lowest deforestation rates. Out of the five most forested TAUs from the analyzed Maramureș County, Borșa had the lowest *FFI* in 2000, while Repedea (the TAU with the most compact forest of Maramureș County) had the highest *FFI*. Deforestation during the 2000–2014 period caused an increased fragmentation of the forested areas, with *FFI* declines ranging between 0.0704 (Poienile) and 0.0806 (Săpânța), which were the two TAUs showing the strongest fragmentation of the forested areas during the period of analysis. The largest fragmentation of forested areas was recorded in 2004 in Săpânța in the NW of the county, and in 2007 in the other four TAUs from the eastern part of the county. At the lower end, the smallest *FFI* declines were recorded in 2001 (Borșa and Poienile), 2002 (Repedea), 2009 (Vișeu de Sus), and 2014 (Săpânța, when the difference between *FFI* 2013 and *FFI* 2014 was 0).

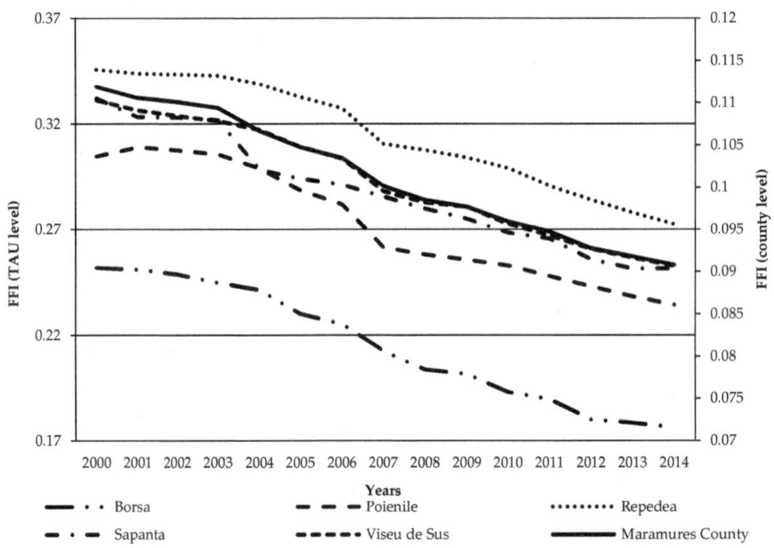

Figure 3. The temporal evolution of the *FFI* (Fractal Fragmentation Index) (2000–2014) at the county and TAU levels.

3.3. Relationship between Deforestation and Fragmentation/Heterogeneity

We noted that the *FFI* achieved a very good quantification of the spatial effects of deforestation on forested areas. Thus, the evolution of *FFI* closely follows the evolution of the reduction in the forested areas with correlation coefficients (R^2) ranging between 0.99 for Maramureș County and 0.98 for the Săpânța territorial administrative unit (Figure 4).

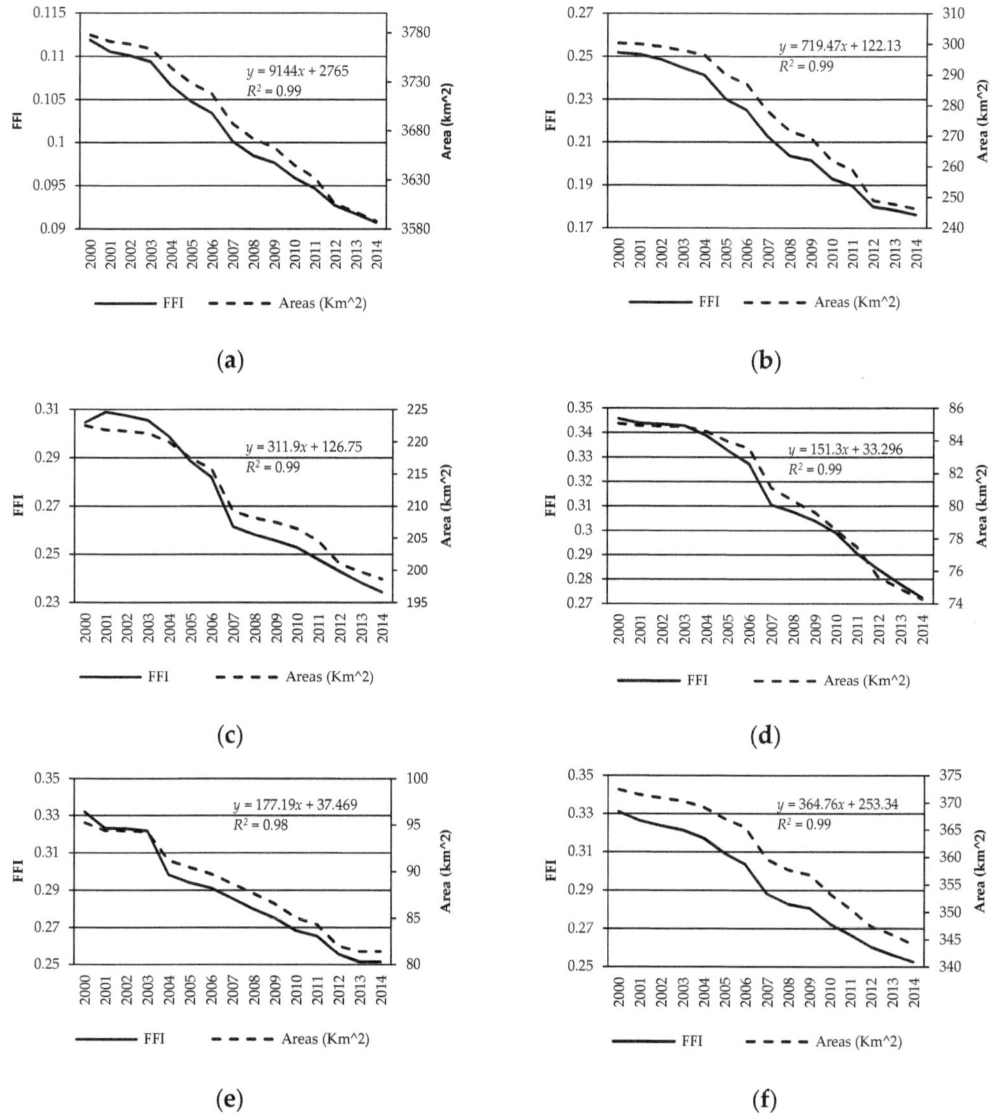

Figure 4. Temporal evolution of *FFI* (Fractal Fragmentation Index) and forested areas: (**a**) Maramureș County; (**b**) Borșa; (**c**) Poienile; (**d**) Repedea; (**e**) Săpânța; (**f**) Vișeu de Sus.

The *FG2DL* value (lacunarity) (Figure 5) allows for a quantification of the degree of the forested area's heterogeneity, indicating whether deforestation occurred in a homogeneous or heterogeneous/dispersed manner. For Maramureș County, the highest *FG2DL* value was registered in 2002, when rather limited areas were deforested. In 2007, with larger areas of deforestation, the *FG2DL* had the lowest value. Thus, when deforestation was much higher in this area, it was more organized, and more compact.

At the TAU level, the same pattern is observed: when the deforested areas are more consistent, the lacunarity is lower, and vice versa. Thus, the largest deforestation dispersion was recorded in 2001 (Borșa) and 2002 (Poienile, Repedea, Săpânța, Vișeu de Sus). Deforestation occurred in the most compact manner in 2004 (Săpânța), in 2007 (Repedea, Poienile and Vișeu de Sus), and in 2012 (Borșa). The highest *FG2DL* value from the data analyzed was registered in 2002 in Poienile when only small areas were deforested. The lowest *FG2DL* value was recorded in Săpânța, coincides with relatively high rates of deforestation in 2004. At the time of maximum deforestation at the TAU level in Borșa 2012, (998.5 ha), an *FG2DL* value of 0.6327 was registered, indicating that deforestation in Borșa was made in compact and homogeneous clusters, but dispersed as compared to Săpânța.

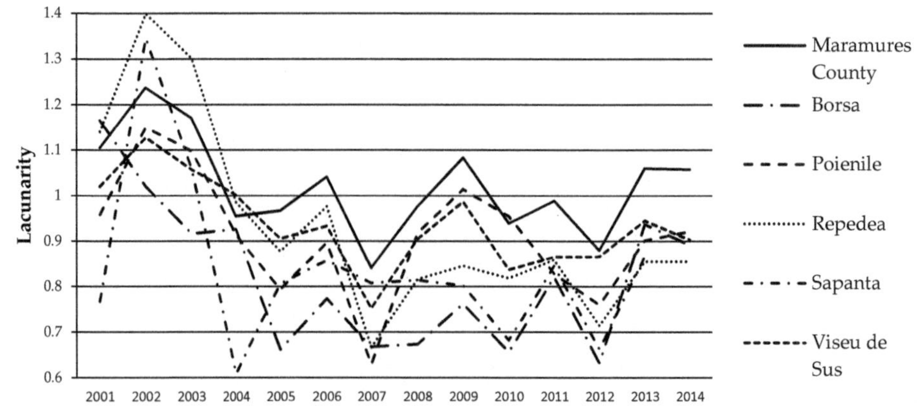

Figure 5. Temporal evolution of *FG2DL* (Fixed Grid 2D Lacunarity) (2000–2014) at the county and TAU levels.

At the county level, the correlation coefficient between *FG2DL* and the deforested areas is high (R^2 value of 0.91), but it is lower at the TAU level: 0.61 in Săpânța and 0.83 in Vișeu de Sus, being the lowest and highest value. Hence, as R^2 increases, there is a closer correlation between the deforested areas and *FG2DL* and, thus, the higher *FG2DL* values indicate that deforestations are in more compact and homogeneous clusters, but dispersed between each other.

The correlation coefficient (R^2) between *FFI* and *FG2DL* has the highest values in Repedea (0.51), where the decrease of the *FG2DL* was more significant due to a lower *FFI* (more compact deforestation), and the lowest value in Poienile (0.20), where there are years when the lowering of the *FFI* involved higher values of the *FG2DL* (dispersed deforestation), like in 2009 (0.26 for *FFI*) and 1.01 for *FG2DL*) (Figure 6).

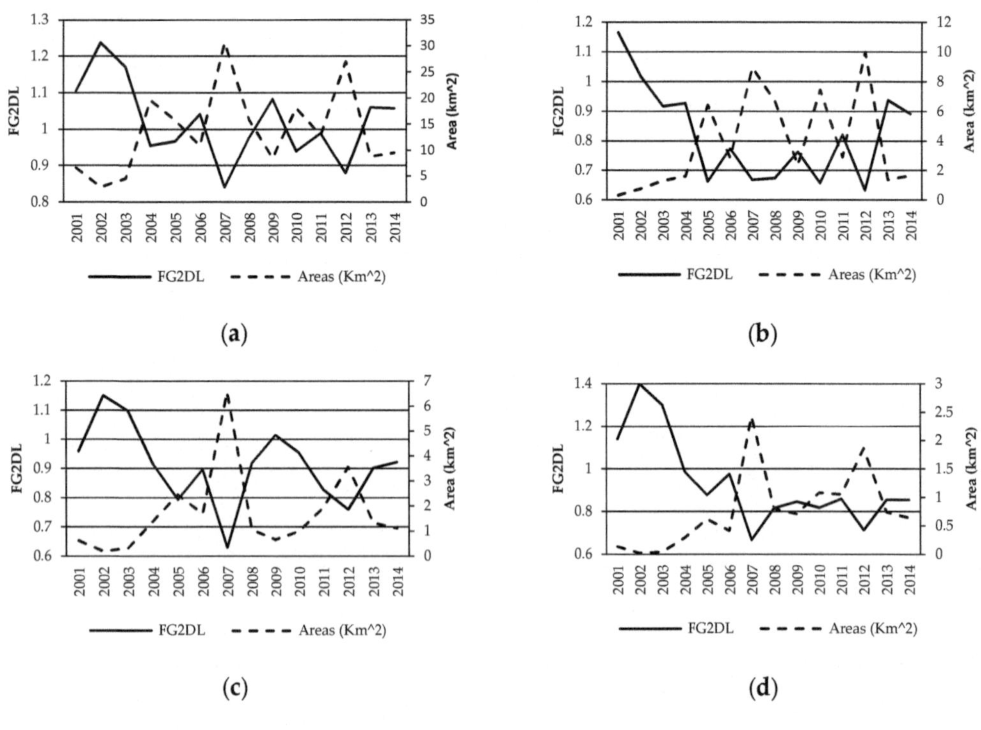

Figure 6. *Cont.*

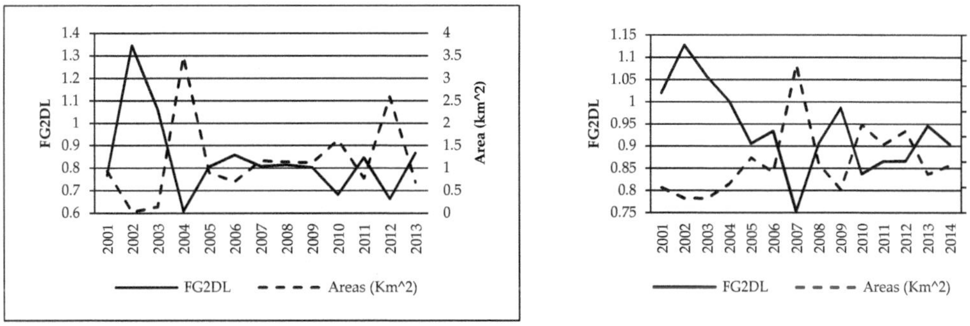

(e) (f)

Figure 6. Temporal evolution of *FG2DL* (Fixed Grid 2D Lacunarity) and deforested areas: (**a**) Maramureș County; (**b**) Borșa; (**c**) Poienile; (**d**) Repedea; (**e**) Săpânța; (**f**) Vișeu de Sus.

3.4. Spatio-Temporal Patterns of Economic Activity Related to Timber Harvesting

Figure 7a–e represents the distribution of the turnover of (a) sylviculture and other forestry activities; (b) logging; (c) sawing and plantation of woods; (d) joinery installations; and (e) wholesale of woods, construction materials, and sanitary equipment for all administrative units in the county of Maramureș. The spatial representation of these economic activities, for the county of Maramureș, is provided at the level of the TAU as given by the NACE codes (statistical classification of economic activities in the European Community). What can be seen is that an important part of the turnover reflecting the economic activity related to timber harvesting can be found in some municipalities that do not have large forested areas, but are rather centered nearby Baia Mare, Borșa and Vișeu de Sus. This pattern can be explained by the development of subsidiaries of multinational companies, which was conducted mainly in the county capital (Baia Mare) and other cities in the county of Maramureș (Vișeu de Sus, Baia Sprie, Sighetu Marmației, etc.).

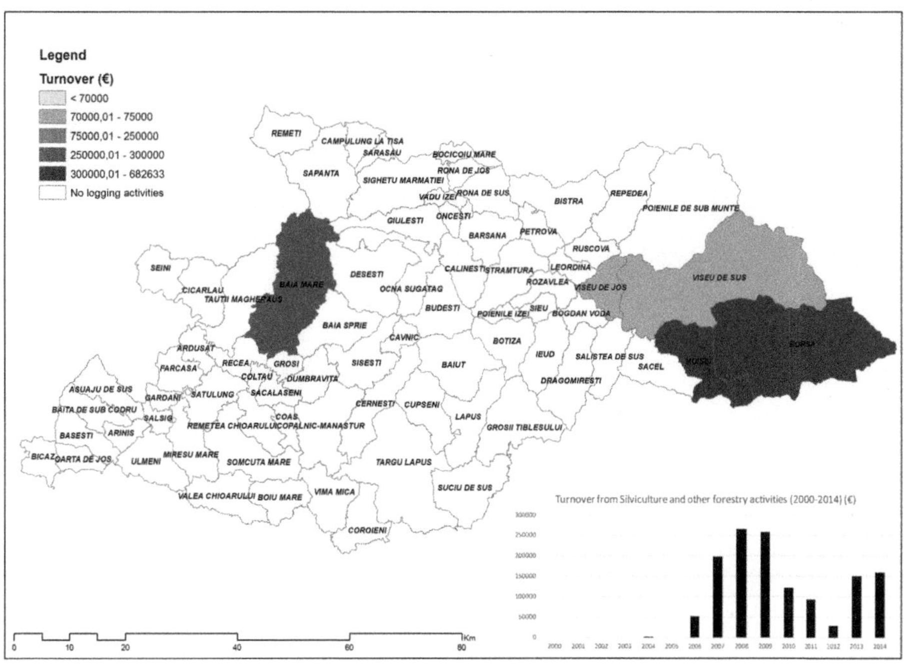

(a)

Figure 7. *Cont.*

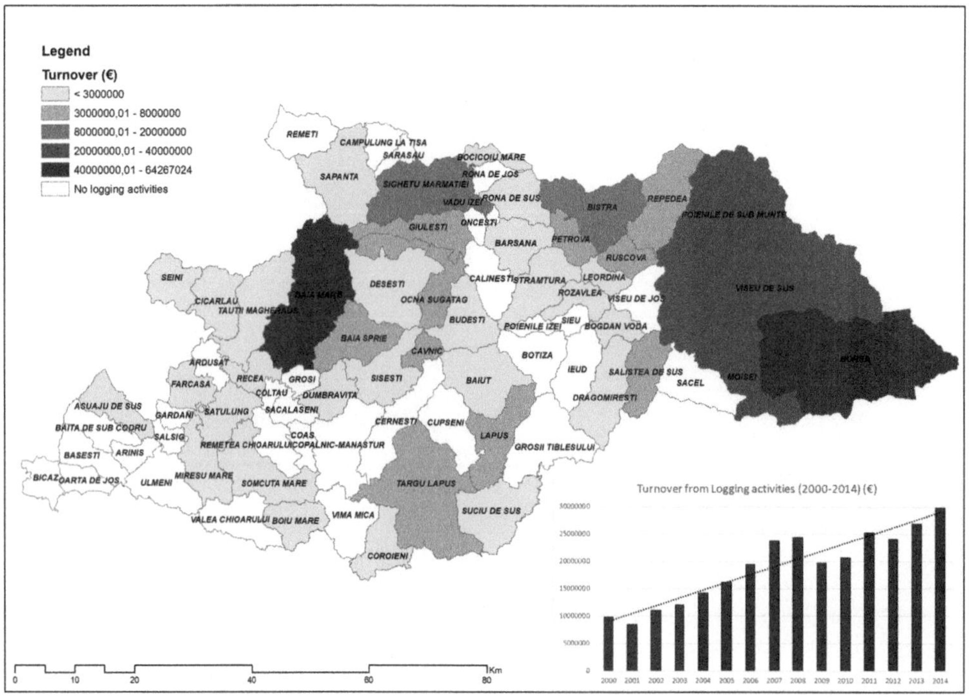

(**b**)

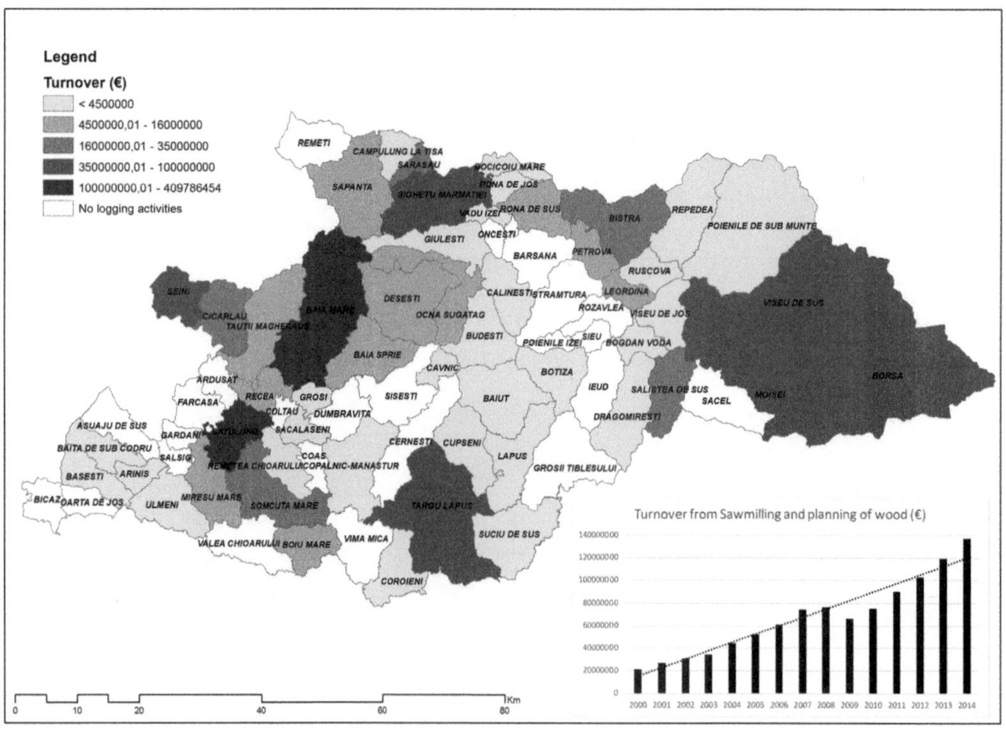

(**c**)

Figure 7. *Cont.*

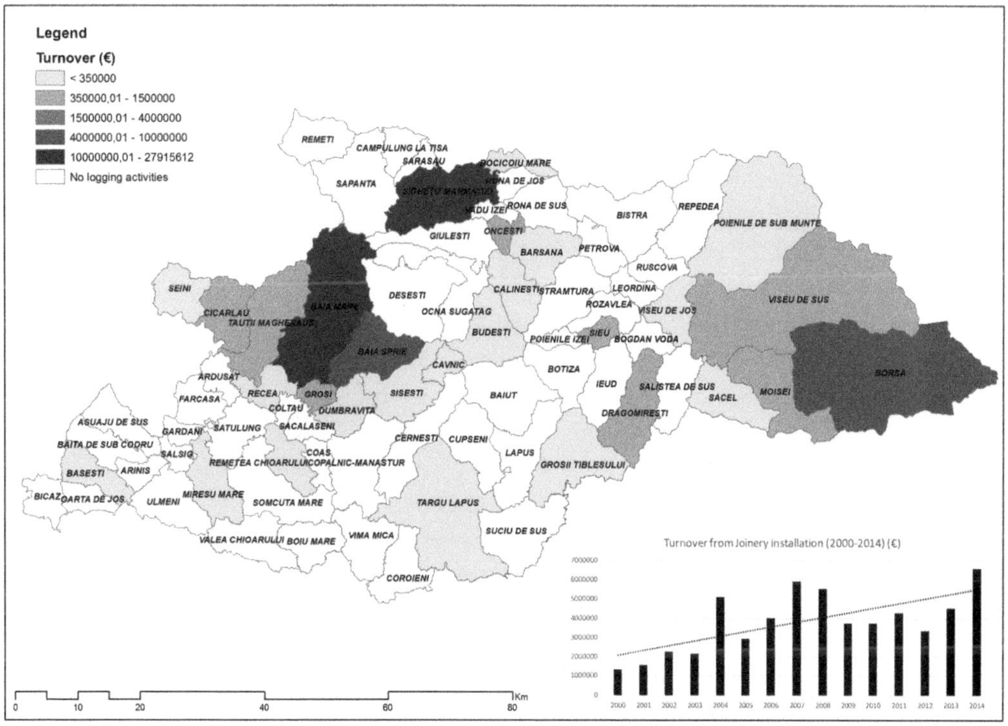

(d)

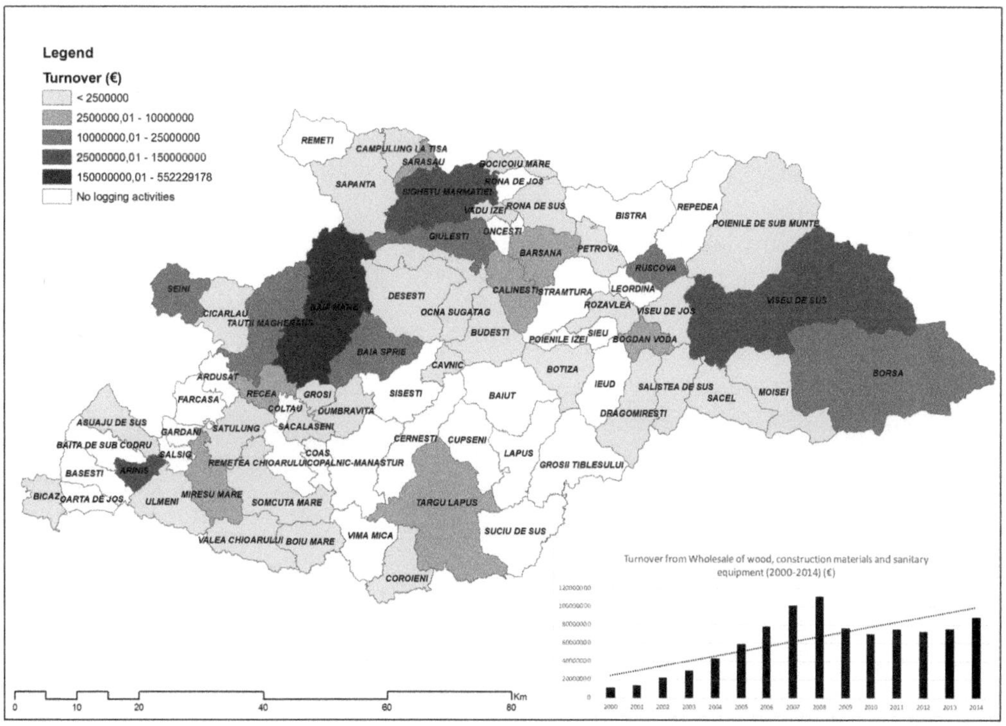

(e)

Figure 7. (**a**) Turnover from sylviculture and other forestry activities (NACE Code 0210). (**b**) Turnover from logging activities (NACE Code 0220). (**c**) Turnover from Sawmilling and planting of wood activities (NACE Code 1610). (**d**) Turnover from Joinery installation (NACE Code 4332). (**e**) Turnover from wholesale of wood, construction materials, and sanitary equipment (NACE Code 4673).

4. Discussion

The evolution of the forested areas in the Maramureș County of Northern Romania in the period of 2000–2014 reveals a general downward trend. This is caused by an increasing exploitation through legal and illegal logging, reflected by the economic changes over this period (Figure 7a–e) to a large extent governed by changes in the legislation [39]. The box-counting method is generally used for the fractal analysis of the forested areas, to describe spatial patterns [26,34,40] and their dynamics in the deforestation process [27], or to describe root morphology [28]. Here we used fractal analyses to study the evolution of fragmentation related to the space occupied by forested areas and affected by deforestation, by means of the Fractal Fragmentation Index [34], using the Pyramid Dimension method, as well as the degree of heterogeneity and the dispersion of the deforested areas by Fixed Grid 2D Lacunarity [35]. In doing this, the Global Forest Change 2000–2014 database [29] was used, which is based on satellite remote sensing Landsat imagery. Our findings confirm our hypothesis, that the use of fractal analyses on Landsat imagery in a 30 m spatial resolution is very useful in providing quantitative information on the spatio-temporal patterns of deforestation.

In a previous study [34], the *FFI* was used to determine the fragmentation or compaction of forests. However, the deforested and regenerated areas were analyzed only at the county level, not at TAU's level, without including the temporal dimension. In this paper, we determined changes in the *FFI* over a 15 year period and at a detailed spatial level of territorial administrative units (TAUs). By doing so, it can be noticed that the *FFI* can play an important role in quantifying the evolution of the fragmentation degree of the forested areas, indicating the mode and the extent to which deforestation spatially fragments forested areas. The correlation between the *FFI* and the absolute values of the forested areas provided us with valuable information. Overall, very high R^2 were obtained, suggesting that *FFI* can accurately quantify deforestation. However, in some instances of massive deforestation, the R^2 is reduced, as the *FFI* does not decrease correspondingly to deforestation rates, indicating that deforestation activities were compact in shape, and forested areas are not likely affected by fragmentation.

In the previous studies by Sengupta and Vinoy [34,40], lacunarity was used to underline the degree of heterogeneity of the deforested areas. In this study, we present another method to determine lacunarity, namely *FG2DL*, which is fast and efficient in processing time series of high-resolution image data from the Landsat archive and is expected to be a highly useful method in relation to the emerging data archives of Senitnel-2 data. The study demonstrates that the *FG2DL* can be valuable in quantifying the degree of homogeneity or spatial heterogeneity of the deforested areas, providing new quantitative information on the dispersion of deforestation.

The combined use of the *FFI* and the *FG2DL* (correlation analysis) provided us with complementary information regarding the impact of the dispersed or compact deforestation upon the fragmentation of the forested areas. The correlation between the absolute values of the deforested areas and the *FG2DL* provided useful information, as in most situations an increase in the deforested areas is characterized by a decrease in *FG2DL*. However, some important exceptions occur when the deforested areas are more compact and homogeneous than the county average. The *FFI* offers information about the fragmentation/compactation of the forest areas and, in addition, *FG2DL* offers information about the heterogeneity/homogeneity of deforestation in the forest areas. So, in some cases an increase in fragmentation is not associated with an increase in heterogeneity causing a rather low correlation coefficient; between 0.19 (Poienile) and 0.51 (Repedea). The present research is applied only at the Maramureș County level and at its five most deforested territorial administrative units (TAUs), only by using the Global Forest Change 2000–2014 database.

The observed spatial displacement between the highest turnovers from timber harvesting occurring in urbanized municipalities and the origin of the woody resources primarily in rural municipalities is likely to be driven by the development of subsidiaries of multinational companies. This situation occurred due to the general precariousness of the legislation in Romania (forest legislation) [41], being rather permissive, which stimulated and encouraged logging, which eventually stimulated economic growth in the region. The two main problems that arise from this are that the turnover is obtained from municipalities, which do not have forested areas. This can be termed areal migration of the turnover, which comes from municipalities with small budgets (but rich in forest mass) and is transferred to other urbanized municipalities. The other important aspect is that the forest activities that have developed in forested municipalities do not provide added value to the woody resources (e.g., logging activities in contrast to furniture production, which produces more value and implies more qualified employees) [42–44].

According to our findings, the analysis of the economic pressure on the forest resources in Maramureș County indicated that the forested areas are ranked fifth in Romania, out of 41 counties. However, when considering the fractal dimension, Maramureș County holds a ninth place due to the structuring/patterns of the forested areas into relatively compact areas (described by *FG2DL*), and also due to other strongly fragmented forest areas (described by *FFI*). As for deforested areas Maramureș County is ranked third, after Suceava and Harghita (located east and south east of Maramureș) in the period 2000–2014. In total, 19,122.8 hectares were deforested in the Maramureș County; a fact also reflected by the small value of the *FFI*: 0.09 in 2014.

5. Conclusions

Deforestation is an important phenomenon which may create major imbalances between territorial administrative units. These imbalances may cause development difficulties by multiplying their negative effects, which may affect the environment (floods, pollution, landslides) and the socio-economic components (labor force). The identification and research of the causes which determine the expansion of the deforested areas is a major concern both for researchers and for decision-makers, due to the complex relations established between forests and the other components of the territorial systems.

Fractal analysis is a tool (method) to assess the geographic space-time phenomena and therefore can be valuable improving our knowledge of the spatial organization of deforestation in forested areas. The *FFI* provides important information on the textural effect of deforestation on forested areas, quantifying the degree of fragmentation whereas lacunarity is a method for the *FFI*, deemed useful to quantify how deforestation is made in relation to the degree of heterogeneity/homogeneity. We show here that the individual, as well as combined, use of the two methods *FFI* and Lacunarity (*FG2DL*) provide new and complementary information to traditional deforestation rates, that can be relevant in developing strategies in forest management.

Among the causes having a primary impact on deforestation in Maramureș County, northern Romania, we specify the following: low income of the local communities, excessive development of the economic sectors based on wood exploitation and processing against the local capacity of support, and legislative gaps. Thus, the need becomes obvious for effective territorial management strategies to mitigate the imbalances generated by mass forest cutting, especially illegal cutting. This is especially important for those local systems that have their functional profile based on the resources offered by the forest.

The specific territorial management strategies must aim primarily at the decision-making process of institutional reorganization, so that the decisions taken may be effective.

Acknowledgments: This work was supported by a grant of the Romanian National Authority for Scientific Research and Innovation, CNCS-UEFISCDI, project number PN-II-RU-TE-2014-4-0835. The authors would like to thank Simion Adrian-Gabriel of the University of Bucharest, Romania, and George N. Zaimes, from the Eastern Macedonia & Thrace Institute, Greece, for their help.

Author Contributions: All authors have equal contribution to the preparation of this scientific paper.

Conflicts of Interest: The authors declare no conflict of interest.

References

1. Food and Agriculture Organization. Global Forest Resources Assessment 2010, Main Report, FAO Forestry Paper. Available online: http://www.fao.org/docrep/013/i1757e/i1757e.pdf (accessed on 25 September 2016).
2. Pravalie, R.; Sirodoev, I.; Peptenatu, D. Detecting climate change effects on forest ecosystems in Southwestern Romania using Landsat TM NDVI data. *J. Geogr. Sci.* **2014**, *24*, 815–832. [CrossRef]
3. UNCED-Agenda 21 1992, United Nations Conference on Environment & Development Rio de Janerio, Brazil, 3–14 June 1992, Chapter 11. Available online: https://sustainabledevelopment.un.org/content/documents/Agenda21.pdf (accessed on 14 August 2016).
4. The UE Forest Action Plan 2007–2011. Available online: http://ec.europa.eu/agriculture/fore/publi/2007_2011/brochure_en.pdf (accessed on 14 August 2016).
5. Houghton, R.A. *Tropical Deforestation as a Source of Green House Gas Emissions*; Mountinho, P., Schwartzman, S., Eds.; Tropical Deforestation and Climate Change, IPAM: Belem, Brazil; Environmental Defense: Washington, DC, USA, 2005; pp. 13–21.
6. Peptenatu, D.; Sirodoev, I.; Pravalie, R. Quantification of the aridity process in south-western Romania. *J. Environ. Health Sci. Eng.* **2013**, *11*, 1–5. [CrossRef] [PubMed]
7. Stickler, C.M.; Nepstad, D.C.; Coe, M.T.; McGrath, D.G.; Rodrigues, H.O.; Walker, W.S.; Soares-Filho, B.S.; Davidson, E.A. The potential ecological costs and cobenefits of REDD: A critical review and case study from the Amazon region. *Glob. Chang. Biol.* **2009**, *15*, 2803–2824. [CrossRef]
8. Murărescu, O.; Murătoreanu, G.; Frînculeasa, M. Agrometeorological drought in the Romanian plain within the sector delimited by the valleys of the Olt and Buzău Rivers. *J. Environ. Health Sci. Eng.* **2014**, *12*, 152. [CrossRef] [PubMed]
9. Thu-Ha Dang, P.; Brouwer, R.; Davidson, M. The economic costs of avoided deforestation in the developing world: A meta-analysis. *J. For. Econ.* **2014**, *20*, 1–16.
10. Lawrence, D.; Vandecar, K. Effects of tropical deforestation on climate and agriculture. *Nat. Clim. Chang.* **2015**, *5*, 27–36. [CrossRef]
11. Parks, N. Deforestation-related climate impacts may vary by soil. *Front. Ecol. Environ.* **2014**, *12*, 204.
12. Brown, S.; Ariel, E.; Lugo, A.E. Tropical secondary forests. *J. Trop. Ecol.* **1990**, *6*, 1–32. [CrossRef]
13. Salam, M.A.; Noguchi, T. Factors influencing the loss of forest cover in Bangladesh: An analysis from socioeconomic and demographic perspectives. *J. For. Res.* **1998**, *3*, 145–150. [CrossRef]
14. Ickowitz, A. Shifting cultivation and forest pressure in Cameroon. *Agric. Econ.* **2011**, *42*, 207–220. [CrossRef]
15. Roy, P.; Murthy, M.; Roy, A.; Kushwaha, S.; Singh, S.; Jha, C.; Behera, M.; Joshi, P.; Jagannathan, C.; Karnatak, H.; et al. Forest fragmentation in India. *Curr. Sci.* **2013**, *105*, 774–780.
16. Sasaki, N. Carbon emissions due to land-use change and logging in Cambodia: A modeling approach. *J. For. Res.* **2006**, *11*, 397–403. [CrossRef]
17. Turner, M.G. Landscape Ecology: The Effect of Pattern on Process. *Annu. Rev. Ecol. Syst.* **1989**, *20*, 171–197. [CrossRef]
18. Gios, G. Multifunctionality and the management of Alpine forest. In *The Multifunctional Role of Forest—Policies. Methods and Case Studies*; EFI Proceedings 55; Cesaro, L., Gatto, P., Pettenella, D., Eds.; European Forest Institute: Joensuu, Finland, 2008; pp. 47–54.
19. Zhang, J.; Alavalapati, J.R.R.; Shrestha, R.K.; Hodges, A.W. Economic impacts of closing national forests for commercial timber production in Florida and Liberty County. *J. For. Econ.* **2005**, *10*, 207–223. [CrossRef]
20. Goio, I.; Gios, G.; Pollini, C. The development of forest accounting in the province of Trento (Italy). *J. For. Econ.* **2008**, *14*, 177–196. [CrossRef]
21. Paletto, A.; Ferretti, F.; Cantiani, P.; De Meo, I. Multi-functional approach in forest landscape management planning: An application in Southern Italy. *For. Syst.* **2012**, *21*, 68–80. [CrossRef]
22. Daniels, S.E.; Hyde, W.F.; Wear, D.N. Distributive effects of forest service attempt to maintain community stability. *For. Sci.* **1991**, *37*, 245–260.
23. Petrişor, A.I. Assessment of the green infrastructure of Bucharest using Corine and urban atlas data. *Urban. Archit. Constr.* **2015**, *6*, 19–24.
24. Juutinen, A.; Kosenius, A.K.; Ovaskainen, V. Estimating the benefits of recreation-oriented management in state-owned commercial forests in Finland: A choice experiment. *J. For. Econ.* **2014**, *20*, 396–412. [CrossRef]

25. Andronache, I.C. Integrated fractal analysis of the morpology and dynamics of the floodplain of the river Danube. In *Fractal Analysis in the Fluvial Geomorphology*; Ivanov, A.V., Pozdnyakov, A.V., Eds.; University Book: Moscow, Russia, 2013; pp. 114–132. (In Russian)
26. Virkkala, R. Ranges of northern forest passerines—A fractal analysis. *Oikos* **1993**, *67*, 218–226. [CrossRef]
27. Sun, J.; Southworth, J. Remote sensing-based fractal analysis and scale dependence associated with forest fragmentation in an Amazon Tri-National Frontier. *Remote Sens.* **2013**, *5*, 454–472. [CrossRef]
28. Eamus, D.; Chen, X.; Kelly, G.; Hutley, L.B. Root biomass and root fractal analyses of an open Eucalyptus forest in a savannah of north Australia. *Aust. J. Bot.* **2002**, *50*, 31–41. [CrossRef]
29. Hansen, M.C.; Potapov, P.V.; Moore, R.; Hancher, M.; Turubanova, S.A.; Tyukavina, A.; Thau, D.; Stehman, S.V.; Goetz, S.J.; Loveland, T.R.; et al. High-resolution global maps of 21st-century forest cover change. *Science* **2013**, *342*, 850–853. [CrossRef] [PubMed]
30. Roesch, F.A. A simulation of image-assisted forest monitoring for national inventories. *Forests* **2016**, *7*, 1–23. [CrossRef]
31. Malek, Ž.; Scolobig, A.; Schröter, D. Understanding land cover changes in the Italian Alps and Romanian Carpathians combining remote sensing and stakeholder interviews. *Land* **2014**, *3*, 52–73. [CrossRef]
32. Sousa, P.Q. Decreasing deforestation in the Southern Brazilian Amazon—The role of administrative sanctions in Mato Grosso State. *Forests* **2016**, *7*, 66. [CrossRef]
33. Schneider, C.A.; Rasband, W.S.; Eliceiri, K.W. NIH Image to ImageJ: 25 years of image analysis. *Nat. Methods* **2012**, *9*, 671–675. [CrossRef] [PubMed]
34. Kainz, P.; Mayrhofer-Reinhartshuber, M.; Ahammer, H. IQM: An Extensible and Portable Open Source Application for Image and Signal Analysis. *PLoS ONE* **2015**, *10*, e0116329. [CrossRef] [PubMed]
35. Andronache, I.C.; Ahammer, H.; Jelinek, H.F.; Peptenatu, D.; Ciobotaru, A.M.; Draghici, C.C.; Pintilii, R.D.; Simion, A.G.; Teodorescu, C. Fractal analysis for studying the evolution of forests. *Chaos Solitons Fractals* **2016**, *91*, 310–318. [CrossRef]
36. Reiss, M. IQM Plugin frac2D. 2016. Available online: https://sourceforge.net/projects/iqm-plugin-frac2d/ (accessed on 17 September 2016).
37. Ahammer, H.; Mayrhofer-Reinhartshuber, M. Image pyramids for calculation of the box counting dimension. *Fractals* **2012**, *20*, 281–293. [CrossRef]
38. Karperien, A.; Ahammer, H.; Jelinek, H.F. Quantitating the subtleties of microglial morphology with fractal analysis. *Front. Cell. Neurosci.* **2013**, *7*. [CrossRef] [PubMed]
39. Sedjo, R.A.; Wisniewski, J.; Sample, A.V.; Kinsman, J.D. The economics of managing carbon via forestry: Assessment of existing studies. *Environ. Resour. Econ.* **1995**, *6*, 139–165. [CrossRef]
40. Pintilii, R.D.; Andronache, I.C.; Simion, A.G.; Draghici, C.C.; Peptenatu, D.; Ciobotaru, A.M.; Dobrea, R.C.; Papuc, R.M. Determining forest resources evolution by fractal analysis (Suceava-Romania). *Urban. Archit. Constr.* **2016**, *7*, 31–42.
41. Romanian Government Decision 46/2008, Forestry Code. Available online: http://apepaduri.gov.ro/wp-content/uploads/2014/08/Legea-nr.-46-din-2008-Codul-Silvic-.pdf (accessed on 15 September 2016).
42. Gounaridis, D.; Zaimes, G.N.; Koukoulas, S. Quantifying spatio-temporal patterns of forest fragmentation in Hymettus Mountain, Greece. *Comput. Environ. Urban Syst.* **2014**, *46*, 35–44. [CrossRef]
43. Peptenatu, D.; Pintilii, R.D.; Draghici, C.; Merciu, C.; Mateescu, R.D. Management of environment risk within emergency territorial systems. Case study—The influence area of the Bucharest City. *J. Environ. Prot. Ecol.* **2012**, *13*, 2360–2370.
44. Pravalie, R.; Sirodoev, I.; Peptenatu, D. Changes in the forest ecosystems in areas impacted by aridization in south-western Romania. *J. Environ. Health Sci. Eng.* **2014**, *12*. [CrossRef] [PubMed]

14

Evaluating the Effects of Carbon Prices on Trade-Offs between Carbon and Timber Management Objectives in Forest Spatial Harvest Scheduling Problems

Huiyan Qin [1,†], Lingbo Dong [2,†] and Yingli Huang [1,*]

1. Department of Forestry Economics, College of Economics & Management, Northeast Forestry University, Harbin 150040, China; huiyanqin@hotmail.com
2. Department of Forest Management, College of Forestry, Northeast Forestry University, Harbin 150040, China; lingbodong@hotmail.com

* Correspondence: huangyinglinefu@163.com

† These authors contributed equally to this work.

Academic Editors: Thomas J. Straka and Timothy A. Martin

Abstract: To mitigate global climatic changes, long-term carbon trading and carbon taxes have been implemented in many countries. However, carbon prices have varied in many of these regions. Therefore, the goal of this paper was to evaluate the effects of carbon prices on trade-offs between forest carbon and timber management objectives in spatial harvest scheduling problems. The objective function of the planning problem was designed to maximize the discounted net present value of harvested timber and the differences of carbon stocks in living tree biomass between the beginning and end of the planning horizon (*DoC*) within a 30-year time frame for a large forest region in northeast China. The constraints primarily related to maintaining an even flow of harvested timber and guaranteeing the maximum opening size. Forest developments were simulated using a set of standard stand-level growth and yield models, and the solutions of the planning problem were generated using the standard version of a simulated annealing algorithm. The effects of a wide range of carbon prices on the harvested timber and *DoC* levels were examined. The results showed that the trade-offs between forest harvested timber and the *DoC* displayed a typical nonlinear tendency as carbon prices increased. The current carbon prices (i.e., 25, 50 and 75 ¥/ton) in China had no significant effects on the optimal forest management plans compared with a scenario that used a carbon price of zero. The minimum carbon price that can provide the financial incentive for the forests to function as a significant carbon sink was estimated to be somewhat over 800 ¥/ton. This result could be useful in determining the appropriate carbon offset standard in this region.

Keywords: forest planning; multi-function forest; ecosystem services; combinatorial optimization; adjacency constraints; simulated annealing

---

1. Introduction

Forest ecosystems provide a number of economic (e.g., timber and non-timber products), ecological (e.g., carbon sequestration, water resources, and biodiversity) and social values (e.g., recreation and employment opportunities for local residents) [1]. However, in many countries, forest management policies have generally focused on timber production over the past decades. The carbon sequestration of forest ecosystems is now receiving substantially more attention as a result of global climate change. For example, China is an extensively forested country with productive forests that in 2013 covered approximately 22% of the land area, provided a volume of 15,137 million

m^3, and 8427 million stored tons of carbon [2]. The annual volume of forest growth was estimated to be 283 million m^3, the annual harvest was 84 million m^3, and the annual carbon sequestration in forests was estimated to be 115 million tons in the 8th Forest Resources Inventory (2009–2013) [2]. Therefore, China's forests play important roles as carbon sinks in mitigating global climatic changes [3]. However, the carbon emissions from fossil fuels in China reached 2260 million tons in 2010 [4]. It is estimated that approximately 5% of these emissions can be sequestered in forest biomass. Obviously, China faces serious pressure to reduce carbon emissions from fossil fuels and increase carbon stocks in forest biomass. Thus, understanding the carbon dynamics of the forest ecosystem and its response to forest management efforts and climate change has become one of the most important research topics in forestry.

Previous studies have demonstrated that planting trees in suitable areas and increasing forest biomass carbon stocks through adaptive forest management are among the most effective alternative strategies for reducing atmospheric CO_2 concentrations [5]. The first strategy has been widely implemented in many countries over past decades. For example, Fang et al. [3] estimated that planted forests (e.g., forestation and reforestation) in China have sequestered 0.45 Pg of carbon since the mid-1970s. The second strategy has been long overlooked. However, in the last few decades, it has drawn more attention worldwide [6–8]. The literature indicates that adaptive forest management strategies could significantly affect carbon dynamics in forest ecosystems and thus are effective for mitigating global climate change. The carbon benefits of forest management prescriptions (e.g., selective cutting) typically vary according to the methods used to utilize trees (e.g., part or whole tree), the methods used to harvest trees (e.g., thinning from below vs. from above), the rotation length, the management intensity, and the length of time after a thinning treatment [9–11]. In addition, species composition, development stage, and characteristic variables (e.g., stand density and site quality) within a stand influence the carbon benefits of different management prescriptions. In general terms of short-term monitoring, live biomass carbon pools have significantly decreased relative to the no-harvest forest because trees have been directly removed from stands. However, because the soil carbon pool is more resistant to forest management changes, it may not distinctly change [10,12,13]. For long-term monitoring, the differences between the total and all sub-carbon pools of forest ecosystems may be not significant [7]. In addition, forest management prescriptions produce certain end-use products that can store significant amounts of carbon for much longer periods than dead wood and residue materials (e.g., branches and barks) in forest ecosystems [14]. From this perspective, most forest management prescriptions should be effective in improving the amount of carbon in forest ecosystems.

However, an experiment to evaluate the effects of these forest management prescriptions on improving the level of carbon stocks across a broad forest landscape would be costly and time-consuming. Therefore, the forest management planning process, which can provide details of where, when and how management prescriptions should be scheduled across a landscape, has been widely used to support the assessment of economic and ecological (e.g., wildlife habitat, water quality and carbon sequestration) goals during the last few decades [1,15–17]. Several studies have incorporated the carbon objective into the forest management planning model [1,16,18,19]. These studies evaluate trade-offs between harvested timber and carbon sequestration and examine the effects of different constraints or policies on forest carbon management objectives using traditional and exact mathematical techniques, such as goal programming and linear programming. However, harvest adjacency and green-up constraints have become some of the most important components of laws, regulations and forest certification programs worldwide [20–22]. These constraints may have significant effects on the fragmentation of the forest landscape and other ecological processes. However, few studies have incorporated the necessary spatial requirements (i.e., the adjacency and green-up constraints) into forest carbon objective management planning models in the manner used by Boston and Bettinger [21] and Borges et al. [23] for forest economic and commodity production goals. An exception is the research of Dong et al. [24], who incorporated an area restriction model [25] and forest spatial value [26] in their planning problems but without investigating trade-offs between harvested timber and carbon sequestration.

Based on a broad survey of the forestry literature, one can state that numerous variables affect forest management decision making (e.g., forest characteristics, management policies and economic factors). Of these variables, economic factors might be one of the clearest and most credible bases for forest management practices. Today, long-term carbon trading or carbon taxes have been implemented in many countries. However, carbon prices vary from region to region and country to country [27]. For example, the carbon prices in countries that have adopted carbon trading are typically less than 100 ¥/ton (1¥≈0.16$; e.g., Europe: 31 ¥/ton, China: 50 ¥/ton, New Zealand: 81 ¥/ton). In contrast, the carbon prices in countries that employ carbon taxes are approximately 1000 ¥/ton (e.g., Sweden: 818 ¥/ton). Therefore, the quantitative assessment of the effects of carbon prices on decision making is an important prerequisite when a forest planning model is applied in practice.

Thus, the aim of this paper is to evaluate the effects of carbon prices on trade-offs between forest carbon and timber management objectives in spatial harvest scheduling problems for a large forest region in northeast China. The objective function of the planning problem is designed to maximize the discounted net present value (NPV) of harvested timber and the differences of carbon stocks in living tree biomass between the beginning and end of the planning horizon (*DoC*) within a 30-year time frame, i.e., three 10-year time periods. The constraints primarily consider maintaining an even flow of harvested timber and guaranteeing the maximum opening size. A wide set of carbon prices was employed to evaluate trade-offs between forest harvest levels and the *DoC*. The results could be useful in determining the appropriate carbon offset standard for this region.

2. Materials and Methods

2.1. Forest Planning Data Sets

The study area was located at the Pangu forest farm (52°41′57.1″ N, 123°51′56.5″ E), Heilongjiang Province, northeast China (Figure 1). The farm is characterized by rolling mountainous terrain with elevations that range from 230 to1397 m. The average slope of the land is approximately 8.63°, with a standard deviation of 6.95°. The climate is a typical cold-temperate continental monsoon climate with a mean annual precipitation of 428 mm, which is primarily concentrated in July and August. The mean annual temperature is −2.4 °C, with the lowest temperature of −46 °C typically occurring in January and the highest temperature of 37 °C typically occurring in July. The period of snow cover can be as long as five months, and the thickness of the snow cover usually varies from 30 to 50 cm. The soil is dominated by dark-brown coniferous forest soil, with a small amount of meadow soil and boggy soil. The main canopy species consist of *Larix gmelinii* (Rupr.) Kuzen., *Pinus sylvestris* var. mongolica, *Picea koraiensis* Nakai, *Betula platyphylla* Suk., and *Populus davidiana* Dode.

The total forest area is 123,434 ha, which was divided into 6421 management units in the forest inventory that was performed by the Forestry Center of Great Xing'an Mountain in 2009. The vegetation in this area could be classified into five different forest types, in which natural *Larix gmelinii* forest (NLG), natural *Betula platyphylla* forest (BPF), mixed coniferous forest (CF), mixed broadleaf–conifer forest (BCF), and mixed broad-leaved forest (BF) account for 27%, 18%, 30%, 19%, and 3%, respectively, of the total area of the forestland. In addition, approximately 3% non-forest lands coexist within the landscape. The initial age distribution is as follows: younger than 20 years: 6%; 20–39 years: 10%; 40–59 years: 35%, 60–79 years: 35%, 80–99 years: 11%, and more than 100 years: 3%.

The planning horizon of this analysis was 30 years, which was divided into three 10-year periods. The potential management prescriptions for each management unit included three selective cutting intensities of total stand volume (i.e., 10%, 20% and 30%) or no harvest. These management prescriptions were designed to satisfy the requirements of the forest management policy of Heilongjiang Province, northeast China [28]. The forest development forecasts were accomplished with a standard stand-level growth and yield model system created by Wang [29]. This simulation system consists of five components: (1) a maximum density-size line model; (2) a recruitment model; (3) a survival model; (4) a basal area prediction model; and (5) a volume prediction model. All the models were established based on commonly applied empirical and theoretical functions using a large national forest resource inventory dataset. From the perspective of harvested timber, the upper-thinning prescription would be implemented for each management unit once the stand age reached a user-specific minimum thinning age. The minimum harvest ages primarily depended on the dominant tree species, which was determined based on previous studies that address optimal stand management treatments [30,31]. The minimum harvest ages for NLG, CF and BCF were assumed to be 40 years; for BPF and BF, 30 years. The carbon stocks in the living tree biomass continually accumulate in a logged forest as a result of the growth of residual and recruitment trees. Overall, twelve alternative uneven-age management prescriptions (i.e., three time periods × three intensities of selective cutting + no harvest) were simulated for each stand within the planning horizon.

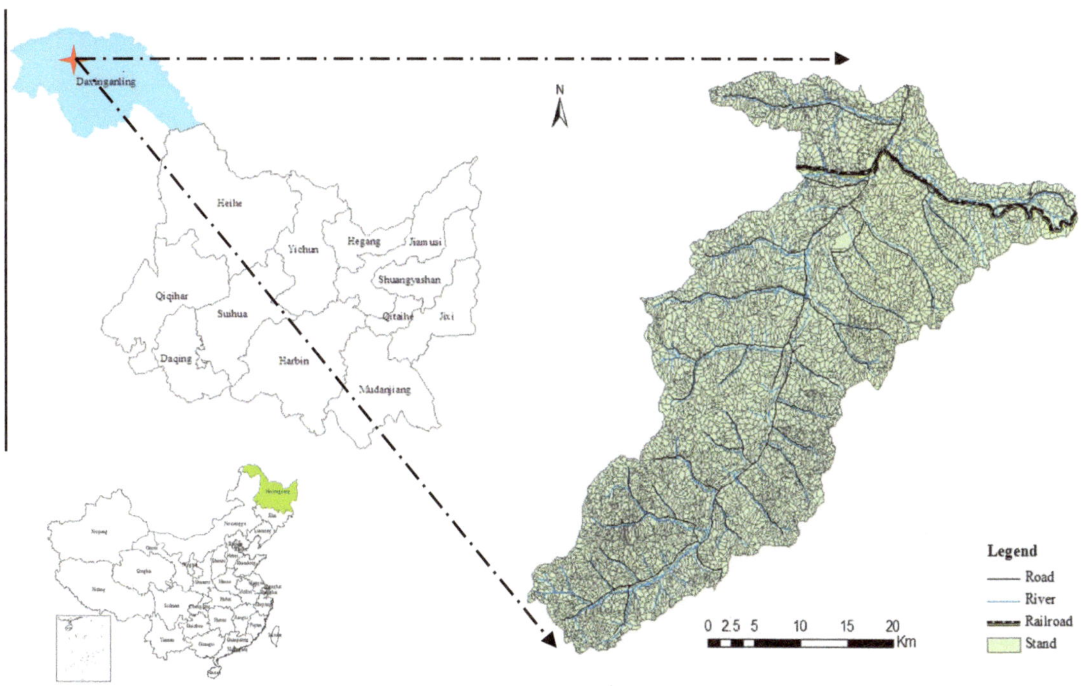

Figure 1. Study area of Pangu forest farm in Heilongjiang Province, northeast China.

The timber prices for different forest types were retrieved from the 2012 price list of the Forestry Department of Heilongjiang Province, northeast China: 980¥/m^3 for NLP,930¥/m^3 for BPF, 1020¥/m^3 for CF, 1100¥/m^3 for BCF, and 800¥/m^3 for BF. The mean costs of management prescriptions included selection harvesting and silvicultural and transportation activities, which were only associated with the area of management unit: 800¥/ha, 1600¥/ha and 2400¥/ha for low intensity (10%), medium intensity (20%), and high intensity (30%) selective cutting prescriptions, respectively. However, the mean carbon-trading price in China was only approximately 50¥/ton (with a lowest price of approximately 20 ¥/ton and a highest price of 75 ¥/ton) during the last two years (2014–2015) [32]. All these variables were tested in our planning problem.

2.2. Forest Planning Problems

To evaluate the effects of carbon prices on trade-offs between forest carbon and timber management objectives in spatial harvest scheduling problems, a reasonable multi-objective forest management planning model should be initially formulated. The objective function of the planning problem discussed in this paper is designed to maximize the NPV of harvested timber and the *DoC* when subjected to several constraints. Therefore, the forest planning problem can be formulated as follows:

$$\text{Maximize Z} = \text{NPV}^{\text{timber}} + \text{NPV}^{\text{carbon}} - \text{NPV}^{\text{harvest}} \tag{1}$$

subject to:

$$HV_t = \sum_{i}^{M}\sum_{j}^{N} A_i \cdot V_{ijt} \cdot X_{ijt} \forall \text{t} \tag{2}$$

$$PV_t^{timber} = \sum_{i}^{M}\sum_{j}^{N} P_s \cdot (A_i \cdot V_{ijt} \cdot X_{ijt}) \forall \text{ t} \tag{3}$$

$$NPV^{timber} = \sum_{t=1}^{T} \frac{PV_t^{timber}}{(1+p)^{(t-0.5)\cdot TPL}} \tag{4}$$

$$PV_{\text{t}}^{harvest} = \sum_{i}^{M}\sum_{j}^{N} A_i \cdot P_{\text{j}} \cdot X_{ijt} \forall \text{t} \tag{5}$$

$$NPV^{harvest} = \sum_{t=1}^{T} \frac{PV_t^{harvest}}{(1+p)^{(t-0.5)\cdot TPL}} \tag{6}$$

$$CS_t = \sum_{i=1}^{M}\sum_{j=1}^{N} A_i \cdot C_{ijt} \cdot X_{ijt} \forall \text{t} \tag{7}$$

$$\Delta_{\text{t}} = CS_t - CS_{t-1} \forall \text{t} \tag{8}$$

$$PV_t^{carbon} = P_c \cdot \Delta_t \forall \text{t} \tag{9}$$

$$NPV^{carbon} = \sum_{t=1}^{T} \frac{PV_t^{carbon}}{(1+p)^{(t-0.5)\cdot TPL}} \tag{10}$$

$$B_{lt} HV_t - HV_{t+1} \leq 0 \quad \text{t} = 1,\ 2,\ \cdots, \text{T} - 1 \tag{11}$$

$$-B_{ht} HV_t + HV_{t+1} \leq 0 \quad \text{t} = 1,\ 2,\ \cdots,\ \text{T} - 1 \tag{12}$$

$$A_i \cdot X_{ijt} + \sum_{k \in N_i \cup S_i} A_k \cdot X_{kjt} \leq U_{\max} \tag{13}$$

$$\sum_{\text{i}}^{M} Age_{it} \geq Age_{\text{s}}^{\text{min}} \forall \text{t} \tag{14}$$

$$\sum_{t=1}^{T} X_{ijt} \leq 1 \forall \text{i} \tag{15}$$

$$X_{ijt} \in \{0,1\} \tag{16}$$

where $\text{NPV}^{\text{timber}}$, $\text{NPV}^{\text{carbon}}$ and $\text{NPV}^{\text{harvest}}$ are the discounted net present value of harvested timber, the *DoC* and harvest activities, respectively; ${PV_t}^{timber}$, ${PV_t}^{carbon}$ and ${PV_t}^{harvest}$ are the present value of the harvested timber, the *DoC* and harvest activities in period *t*, respectively; HV_t and CS_t are the amounts of actual scheduled harvested timber and carbon stock residues in living tree biomass in period *t*, respectively (whereby CS_0 represents the amount of carbon stocks in living tree biomass at the beginning of the planning horizon); *M*, *N*, and *T* are the total number of management units, potential management activities for each unit, and planning periods, respectively; A_i and A_k are the areas of management units *i* and *k*, respectively,; V_{ijt} is the amount of harvested timber during period *t* from management unit *i* when managed under prescription *j*; X_{ijt} is a binary variable that indicates whether (1) or not (0) management unit *i* would be harvested with prescription *j* during time period *t*; P_{s} represents the mean stumpage price for forest type *s* (which represents one of the five forest types, i.e., NLG, CF, BCF, BPF or BF); P_j is the harvest costs for management prescription *j*; TPL is the length of the time period; C_{ijt} is the amount of carbon stocks for management unit *i* after harvest with management prescription *j* in time period *t*; Δ_t denotes the differences of carbon stocks in living tree biomass from time period *t* to *t*-1; B_{lt} represents the lower bound on decreases in the harvest level between period *t* and *t*+1; B_{ht} represents the upper bound on increases in the harvest level between period *t* and *t*+1; *k* is a management unit adjacent to unit *i* or to a management unit that is adjacent to unit *i*; N_i is the entire set of all management units adjacent to management unit *i*; S_i is the entire set of all management units adjacent to the set of management units (N_i) adjacent to management unit *i* (which were formulated in the form of a recursive function, as shown in Murray [25]); U_{max} is the assumed maximum concurrent harvest area; Age_{it} is the stand age of management unit *i* in time period *t*; Age_s^{min} is the minimum harvest age assumed for forest type *s*; and *p* is the discount rate expressed as a percentage.

Equation (1) specifies the objective function of the planning problem: to maximize the discounted net present value from harvested timber and the *DoC* during the planning horizon minus the costs of harvest activities. As implemented here, all revenues (or costs) associated with harvested timber, the *DoC* and management prescriptions would be discounted using a 3% discount rate at the middle of each time period. Equation (2) consists of harvest accounting constraints that aggregate the harvest volume for each time period to harvest accounting variables (HV_t). Equation (3) similarly aggregates the present value of harvested timber from each time period to timber revenue accounting variables (PV_t^{timber}). Equation (4) is used to aggregate the total discounted net present value of timber production (NPV^{timber}) during the entire time horizon.

Equation (5) assigns the present value of harvest activities from each time period to harvest costs accounting variables ($PV_t^{harvest}$). Equation (6) accumulates the total discounted net present value of harvest costs ($NPV^{harvest}$) during the entire time horizon. Equation (7) consists of the residual carbon stocks accounting constraints that assign the carbon stocks after harvest under the assigned management prescription for each time period to carbon accounting variables (CS_t). Equation (8) indicates the differences of the amount of carbon stocks in living tree biomass between two adjacent time periods. Equation (9) assigns the present value of the carbon objective from time period *t* to the carbon revenue accounting variables (PV_t^{carbon}). Equation (10) represents the total NPV of the *DoC* (NPV^{carbon}) during the entire time frame. Since an even flow constraint of harvested timber typically plays an important role in ensuring full utilization of felling equipment and labor, constraint sets (11) and (12) are employed to restrict the deviation proportion of harvested timber from one period to the next within 10% in this analysis.

Constraint set (13) assists with maximum harvest opening size constraints, which was termed an area restriction model (ARM) by Murray [25]. The concepts of the area restriction model are typically complex and complicated when integrated into a specific planning problem. Briefly, the constraints allow the concurrent harvest of adjacent management units if the combined area of the contiguous harvests area does not exceed the maximum harvest opening size (U_{max}). As implemented here, adjacency only refers to the management units that share a common boundary (i.e., not a single point). Details regarding maximum harvest opening size constraints can be found in Murray [25], McDill et al [33], and Boston and Bettinger [21].

Constraint (14) ensures that the minimum harvest age for assigning a potential management prescription should not be violated. Constraint (15) indicates that each management unit can only be harvested no more than once during the entire time horizon. Finally, constraint (16) requires that decision variables should be binary, which means that a management prescription should be applied to an entire management unit rather than to part of it when it is assigned.

2.3. Forest Planning Process

Simulated annealing was developed and used to schedule the potential prescription for each management unit for the forest planning problem. This technique was initially inspired by an algorithm developed by Metropolis et al. [34] to simulate the cooling process of heated materials. Subsequently, it was introduced as a technique for solving optimization problems in the early 1980s [35]. It is a typical neighborhood search technique that only examines a single solution to a forest plan and then proposes a change to the status of a single management unit (i.e., randomly change harvest time or prescription), i.e., by generating a new candidate solution. Then, the following three logical processes are gradually conducted. (**1**) If any constraints are violated, the process reverts to the previous feasible solution (S), and then a new change of management unit and its status (i.e., harvest time period and prescription) are randomly selected, i.e., generating a new candidate solution (S'); (**2**) Changes to the solution are evaluated with respect to Constraints (11)–(15). If the new candidate solution is feasible (i.e., does not violate any constraints) and the results from this proposed solution are better than those of the best previous solution, then the proposed solution becomes the best solution (S^*) and the current solution (S). However, (**3**) if the results of the proposed solution are worse than those of the best previous solution, then the proposed solution can only be accepted with a probability of $Prob = \text{Exp}((f(S') - f(S^*))/Temperature)$, where $f()$ is the objective function formulation. The necessary parameters for simulated annealing are initial temperature (T_0), final temperature (ft), cooling rate (cr) and the number of iterations at each temperature ($nrep$). Obviously, if the initial temperature is very high, the algorithm will accept all proposed solutions (i.e., $Prob = 1$), which resembles a random search process. The probability for accepting a poor solution will decrease continuously (i.e., $0 < Prob < 1$) with the decrease in current temperature T. However, if the initial temperature is too low, the search process will reject all poor solutions (i.e., $Prob = 0$) and become a greedy hill-climbing method.

The performance of simulated annealing typically depends on the parameters used [36]. Therefore, forest planners should carefully evaluate the effects of parameters prior to applying them in forest management planning practices. To identify the set of parameters that will result in optimal solutions, wide ranges of T_0 (i.e., 10^4, 5×10^4, 10^5, 5×10^5 and 10^6), cr (i.e., 0.95, 0.975, 0.99, 0.995 and 0.999), and $nrep$ (i.e., from 100 to 500 in step of 100) were considered using trial-and-error tests, as in Boston and Bettinger [37]. Finally, the values of parameters T_0, ft, cr and $nrep$ were set as 10^4, 10, 0.995 and 100, respectively, which resulted in 137,900 iterations per independent run. To minimize the effects of potential parameter combinations with respect to the objective function value, the parameter values were held constant for all planning scenarios. The forest planning process was developed using the Visual Basic 6.0 programming language. All solutions (i.e., forest plans) were generated on a 2.6 GHz Core i5 processor and a Windows 7 operating system.

3. Results

The carbon-trading prices in China between 2014 and 2015 varied from 20 to 75 ¥/ton. When these prices were entered into the planning model, nearly all management units were harvested using a high intensity (30%) selective cutting prescription. Therefore, a wide range of carbon prices that varied from zero to 5000 ¥/ton was evaluated.

The levels of NPV of assigned harvest timber and its corresponding *DoC* both presented typical nonlinear tendencies with increases in carbon prices (Figure 2) primarily because of the necessity of maintaining an even flow of harvested volume and the harvest adjacency constraint. The NPV of harvested timber increased moderately when the carbon price was less than 500 ¥/ton and then decreased more quickly with continuous increases in carbon price. The mean NPV of assigned harvest timber given a carbon price of zero was approximately ¥2120.94 million during the entire horizon, and a zero NPV of assigned harvest timber could be reached at a carbon price somewhat over 5000 ¥/ton, which was substantially higher than the current carbon trading prices (i.e., 20–75 ¥/ton) in China. The forest might function as a significant source of carbon because the amount of *DoC* within the entire planning horizon was less than zero (i.e., from −0.59 to −0.01 million tons) when employed at carbon prices below 800 ¥/ton. However, with increases in the carbon price, the carbon-sink function of the forest might predominate, which varied from 0.19 to 1.82 million tons within the entire time frame. These values correspond to an average annual net carbon change from −0.16 to 0.52 ton/ha.

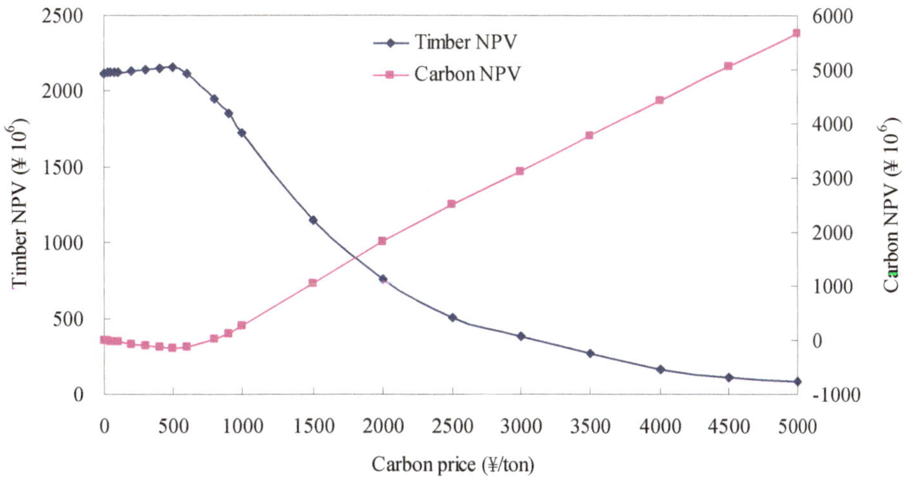

Figure 2. Developments of the discounted net present values of harvested timber and the DoC together with carbon prices that varied from zero to 5000 ¥/ton, where *DoC* represents the differences of carbon stocks in living tree biomass between the beginning and the end of the planning horizon.

The levels of carbon stocks in living tree biomass at the end of the planning horizon (*CsE*) as a function of assigned harvest timber within the entire time frame also exhibited a significant nonlinear tendency (Figure 3). The relevancies between harvested timber and the *CsE* could be fitted as a typical quadratic polynomial: $CsE = -0.184HT^2 + 0.017HT + 6.1783$ (R^2=0.998), where *HT* represents the level of harvested timber and R^2 represents the coefficient of determination between the predicted and observed values. For the minimum harvested timber level (1.91 million m^3), the level of the *CsE* was approximately 6.23 million tons, which was increased by 45% compared with the levels of carbon stocks at the beginning of the planning horizon. However, for the maximum harvested timber level (3.71 million m^3), the *CsE* level decreased by 13% (i.e., 3.73 million tons) compared with the levels of carbon stocks at the beginning of the planning horizon. As previously mentioned, the differences of carbon stocks between the beginning and end of the planning horizon were not significant when a carbon price of 800 ¥/ton was employed.

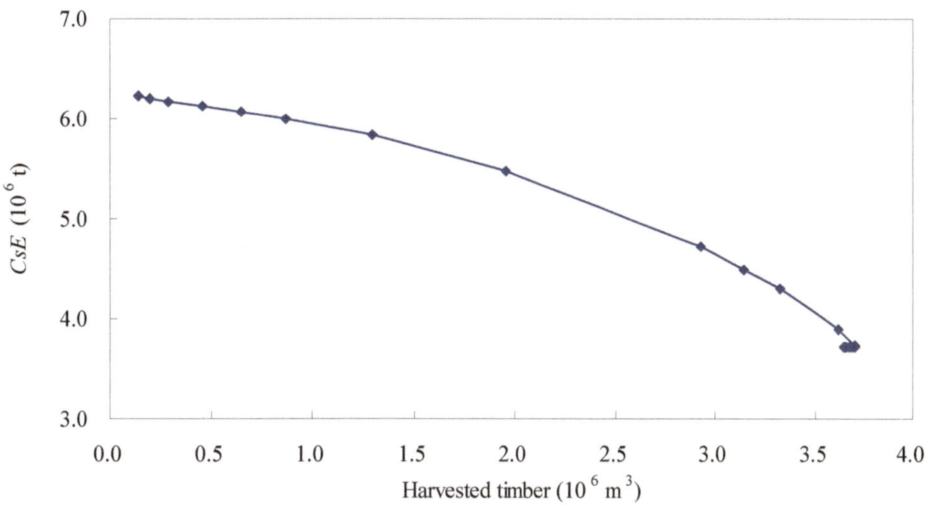

Figure 3. Trade-offs between harvested timber and the *CsE* for a 30-year time frame when carbon prices varied from zero to 5000 ¥/ton, where *CsE* represents the levels of carbon stocks in living tree biomass at the end of the planning horizon.

The levels of harvested timber were nearly consistent when the carbon price was less than 500 ¥/ton and then quickly decreased with increases in the carbon price (Figure 4a). The allocation patterns of harvested timber for each time period reflected the constraints of even flow of harvest volume (Equations (11) and (12)) when evaluated for different carbon prices. However, the pattern of *DoC* changes was reversed (Figure 4b), i.e., the *DoC* increased with carbon prices. The minimum carbon price that could increase the levels of total carbon stocks at the end of the planning horizon was also somewhat over 800 ¥/ton in our analysis. The mean percentages of the *DoC* for the first and second time periods decreased significantly (i.e., from 39% to 17% and from 37% to 15%), whereas the mean percentages of the *DoC* for the third time period increased from 24% to 68% when carbon prices were less than 500 ¥/ton. However, small differences in the percentages among different carbon prices that varied from 1000 to 5000 ¥/ton were observed, i.e., 40%, 34% and 26%, for the three time periods. Obviously, the *DoC* levels decreased significantly with the increasing time period. A possible explanation is that the growth rates of forest biomass for young- and middle-age stands are typically significantly larger than that of mature and over-mature forests.

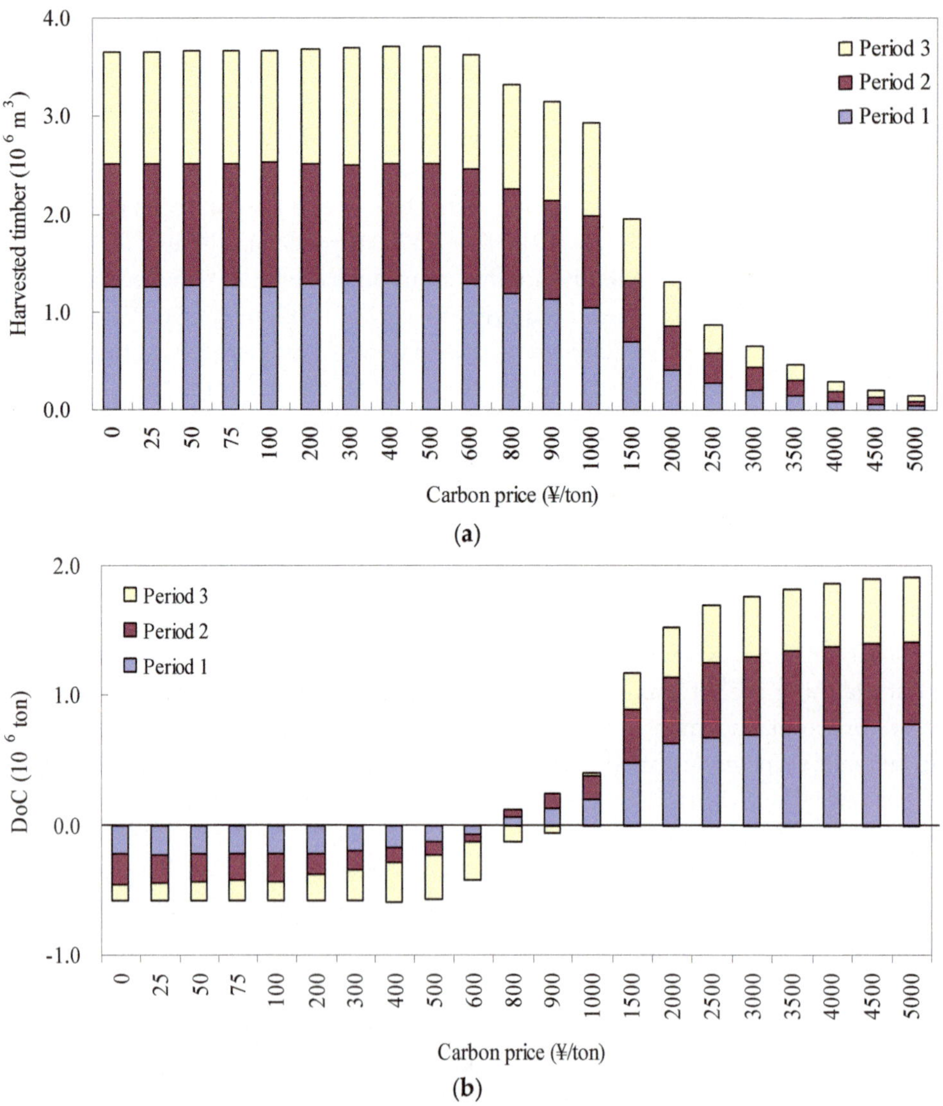

Figure 4. Levels of harvested timber (**a**) and the *DoC* (**b**) for each time period when carbon prices varied from zero to 5000 ¥/ton, where *DoC* represents the differences of carbon stocks in living tree biomass between the beginning and end of the planning horizon.

The optimal forest management plans indicated that carbon prices had significant effects on the optimal spatial and temporal assignments of alternative management prescriptions (Table 1). The percentages of no harvest units increased significantly from 0.2% (0 ¥/ton) to 88% (5000 ¥/ton) as the carbon price increased. However, the percentages of harvested units with alternative management prescriptions decreased dramatically from 95% (0 ¥/ton) to 7% (5000 ¥/ton). Because of the constraint of the even flow of harvested timber (i.e., Equations (11) and (12)), the total percentages of harvested units in each period were approximately identical. Within each planning period for all planning scenarios, the percentages of harvested units with a high-intensity selective cutting prescription presented a typical downtrend as carbon prices increased. However, the percentages of harvested units with a low-intensity selective cutting prescription increased moderately.

Table 1. Percentage of harvested units with alternative management prescriptions when a carbon price of 0, 50, 500, 800, 1000, and 5000 ¥/ton was used.

Period [(a)]	Prescription [(b)]	Carbon Price (¥/ton)					
		0	50	500	800	1000	5000
−1	−1	4.36	4.36	4.36	4.36	4.36	4.36
0	0	0.22	0.23	0.40	5.44	9.61	88.40
1	1	0.00	0.00	0.00	0.00	0.02	0.47
	2	0.51	0.34	0.30	1.76	3.82	0.44
	3	42.19	41.68	39.90	32.08	26.15	0.19
2	1	0.00	0.00	0.00	0.25	0.70	2.27
	2	0.02	0.03	0.03	3.36	7.09	0.16
	3	29.81	30.09	27.91	22.92	16.57	0.03
3	1	0.00	0.03	0.09	5.26	8.91	3.69
	2	0.37	0.45	0.93	8.47	13.88	0.00
	3	22.52	22.78	26.07	16.09	8.91	0.00

[(a)] represents the three different time periods; [(b)] represents the three intensities of the selective cutting prescriptions: low intensity (10%), medium intensity (20%), and high intensity (30%). Period = −1 and prescription = −1 represents non-forest land, and period = 0 and prescription = 0 indicate that no harvest was assigned.

4. Discussion

To evaluate the effects of carbon prices on trade-offs between forest carbon and timber management objectives in spatial harvest scheduling problems, a persuasive multi-objective forest management planning model was formulated. Similar to Backéus et al. [1] and Charles et al. [18], we integrated carbon management objectives in a wood supply model. However, in addition to using the traditional forest management model, which only includes the typical even flow of harvested timber and minimum harvest age constraints, we considered the adjacency constraints (i.e., ARM) of assigned management prescriptions in temporal and spatial scales. We also employed a well-known heuristic technique (i.e., simulated annealing) capable of greatly accelerating the search process for complex combinatorial optimization problems. The results of the case study indicated that the assigned harvest timber and its corresponding *DoC* both presented typical nonlinear responses as the carbon price increased. This result agreed precisely with Backéus et al. [1] and Raymer et al. [19]. Given the harvest quota management constraints of China's State Forestry Bureau, the realistic annual harvest level was set at approximately 15 thousand m^3 per year in this region for the long term, which resembles the simulated forest management scenario in which a carbon price of 3500 ¥/ton was used. The results of this scenario corresponded to an average annual *DoC* of approximately 0.49 ton/ha, which was consistent with the values (i.e., 0.45 ton/(year·ha)) reported by Fu et al. [38]. Therefore, the planning model presented here can be immediately applied in forest management practices in northeast China. However, the planning model ignored several important factors, which require additional discussion.

The first and perhaps most important factor is that not all the components of carbon sequestration related to forest and forestry were included, such as carbon in the soil, litter, shrub and herbal layers in forest ecosystems. Forest soil carbon stocks typically account for 29%–56% of the total carbon stocks for different development stages of the forests in this region [39]. However, the amount of soil carbon stocks may change dramatically over a short period [12], particularly in forests that grow in high latitudes, such as our study area. Additionally, because of the significant spatial heterogeneity of other vegetation layers (i.e., shrub and herbal), the carbon stocks in these components were not considered. Moreover, the harvested timber within the entire time horizon can also store a significant amount of carbon. Based on an average biomass expansion factor of 0.56 t/m^3 for all forest types within this region [40] and a carbon content of 0.45 (figures that have been widely used in scientific research [1,41]), the carbon stocks of the harvested timber within the entire time horizon for different management planning scenarios were estimated to be 0.36–0.92 million tons. Therefore, the carbon emissions process of various forest products (i.e., sawlogs) also should be integrated into the planning process [1,16].

A second factor that requires additional discussion is the effect of climate change on the forest management planning process. Under boreal conditions, climate changes are expected to increase the annual temperature and precipitation and thus may increase forest growth and productivity [42]. However, the frequency and severity of unproductive events (i.e., fire, wind and insect damage) may also be increased significantly [43]. To evaluate the effects of climate change, a process-based growth model or a modified empirical model can be integrated into the forest planning process [42]. Among the various counterproductive events, forest fires represent a key natural disturbance factor in this region. Hu et al. [44] reported as many as 1614 forest fires with a burned area of approximately 35.23 million hectares from 1965 to 2010. The total carbon emissions of these fires were estimated to be 29.32 million tons, and the mean annual carbon emissions were approximately 0.64 tons per year, which accounted for 6% of the total carbon emissions from forest fires in China. Therefore, these unproductive events should also be integrated into traditional forest planning models, as in Peter and Nelson [45], Couture and Reynaud [46], and Garcia-Gonzalo et al. [47].

The final factor that requires discussion is that forest characteristics, economic factors, and management policies have significant effects on the forest planning process. For example, Borges et al. [48] examined the effects of site productivity on forest harvest scheduling problems that involved green-up and maximum area restrictions. Prices not only significantly affect the search process of heuristics but also reflect the priorities of multi-objective forest management. Therefore, in this study, we only focused on the effects of carbon prices, as in Backéus et al. [1]. However, the prices of various forest products and management activities represent a complex system from the economic perspective. A change to one economic parameter may affect other parameters. Thus, the interactions among such parameters might be important in forest management practices. Finally, the management policies of adjacency and green-up constraints also have significant effects on the forest planning process. Boston and Bettinger [21], Borges et al. [48], and Nalle et al. [49] have confirmed that the economic profitability of an optimal management plan typically increased as the period of green-up constraints and the size of the maximum clear-cut area increased. However, the fragmentation of forest landscapes or habitat patches might be aggravated significantly [32].

5. Conclusions

The case study presented here confirmed that integrating a carbon management objective into the traditional forest harvest scheduling model significantly reduced the levels of assigned harvest timber. However, trade-offs between the carbon management objective and traditional timber harvest presented a typical nonlinear tendency primarily due to the constraints of the even flow of harvested timber and harvest adjacency. The differences of the optimal solutions (i.e., forest management plans) between the current carbon prices (25, 50 and 75 ¥/ton) in China and the assumed minimum carbon price (0 ¥/ton) were not significant. The minimum carbon price that can increase the total carbon stocks at the end of the planning horizon was estimated to be somewhat over 800 ¥/ton. This outcome can be used in determining the appropriate carbon offset standard in this region. However, we recognize that this value might be overestimated because not all products (e.g., berry and mushroom yields) and services (e.g., water conservation, habitat maintenance) from forest ecosystems were considered. The percentage of harvest units also decreased as carbon prices increased.

Acknowledgments: This study was financially supported by Fundamental Research Funds of the Central Universities of China (2572015EC03, 2572015AC03) and the National Social Science Foundation of China (13BJY032). The authors would like to thank the reviewers for their helpful comments and suggestions.

Author Contributions: Y. Huang conceived and designed the experiments. L. Dong contributed analysis tools for the simulated annealing algorithm. H. Qin, L. Dong and Y. Huang analyzed the data and wrote the paper.

Conflicts of Interest: The authors declare no conflict of interest.

References

1. Backéus, S.; Wikström, P.; Lämås, T. Modelling carbon sequestration and timber production in a regional case study. *Silva Fenn.* **2006**, *40*, 615–629. [CrossRef]
2. State Forestry Bureau. *Forest Resources Statistics of China for 2009–2013*; Chinese Forestry Press: Beijing, China, 2014. (In Chinese)
3. Fang, J.; Chen, A.; Peng, C.; Zhao, S.; Ci, L. Changes in forest biomass carbon storage in China between 1949 and 1998. *Science* **2001**, *292*, 2320–2322. [CrossRef] [PubMed]
4. Carbon Dioxide Information Analysis Centre. World's countries Ranked by 2011 Total Fossil-Fuel CO_2 Emissions from Fossil-Fuel Burning, Cement Production, and Gas Flaring. Available online: http://cdiac.ornl.gov/ (accessed on 17 November 2016).
5. Krcmar, E.; Kooten, G.C.V.; Vertinsky, I. Managing forest and marginal agricultural land for multiple tradeoffs: Compromising on economic, carbon and structural diversity objectives. *Ecol. Model.* **2005**, *185*, 451–468. [CrossRef]
6. Saunders, M.; Tobin, B.; Black, K.; Gioria, M.; Nieuwenhuis, M.; Osborne, B.A. Thinning effects on the net ecosystem carbon exchange of a Sitka spruce forest are temperature-dependent. *Agric. For. Meteorol.* **2012**, *157*, 1–10. [CrossRef]
7. Liu, Q.; Cai, H.; Jin, G. Effects of selective cutting on the carbon density and net primary productivity of a mixed broadleaved-Korean pine forest in northeast China. *Chin. J. Appl. Ecol.* **2013**, *24*, 2709–2716.
8. Bravo-Oviedo, A.; Ruiz-Peinado, R.; Modrego, P.; Alonso, R.; Montero, G. Forest thinning impact on carbon stock and soil condition in Southern European populations of *P. sylvestris* L. *For. Ecol. Manag.* **2015**, *357*, 259–267. [CrossRef]
9. Jandl, R.; Lindner, M.; Vesterdal, L.; Bauwens, B.; Baritz, R.; Hagedorn, F.; Johnson, D.W.; Minkkinen, K.; Byme, K.A. How strongly can forest management influence soil carbon sequestration? *Geoderma* **2007**, *137*, 253–268. [CrossRef]
10. Ruiz-Peinado, R.; Bravo-Oviedo, A.; López-Senespleda, E.; Montero, G.; Rio, M. Do thinning influence biomass and soil carbon stocks in Mediterranean maritime pinewoods? *Eur. J. For. Res.* **2013**, *132*, 253–262. [CrossRef]
11. Primicia, I.; Artázcoz, R.; Imbert, J.B.; Puertas, F.; Traver, M.D.C.; Castillo, F.J. Influence of thinning intensity and canopy type on Scots pine stand and growth dynamics in a mixed managed forest. *For. Syst.* **2016**, *25*, e057. [CrossRef]
12. Johnson, D.W.; Curtis, P.S. Effects of forest management on soil C and N storage: Metaanalysis. *For. Ecol.Manag.* **2001**, *140*, 227–238. [CrossRef]
13. Campbell, J.; Alberti, G.; Martin, J.; Law, B.E. Carbon dynamics of a ponderosa pine plantation following a thinning treatment in the northern Sierra Nevada. *For. Ecol. Manag.* **2009**, *257*, 453–463. [CrossRef]
14. Eriksson, E.; Gillespie, A.R.; Gustavsson, L.; Langvall, O.; Olsson, M.; Sathre, R.; Stendahl, J. Integrated carbon analysis of forest management practices and wood substitution. *Can. J. For. Res.* **2007**, *37*, 671–681. [CrossRef]
15. Kurttila, M.; Pukkala, T.; Loikkanen, J. The performance of alternative spatial objective types in forest planning calculations: A case for flying squirrel and moose. *For. Ecol. Manag.* **2002**, *166*, 245–260. [CrossRef]

16. Baskent, E.Z.; Keles, S. Developing alternative forest management planning strategies incorporating timber, water and carbon values: An examination of their interactions. *Environ. Model. Assess.* **2009**, *14*, 467–480. [CrossRef]
17. Başkent, E.Z.; Küçüker, D.M. Incorporating water production and carbon sequestration into forest management planning: A case study in Yalnızçam planning unit. *For. Syst.* **2010**, *19*, 98–111. [CrossRef]
18. Bourque, C.P.A.; Neilson, E.T.; Gruenwald, C.; Perrin, S.F.; Hiltz, J.C.; Blin, Y.A.; Horsman, G.V.; Parker, M.S.; Thorburn, C.B.; Corey, M.M.; et al. Optimizing carbon sequestration in commercial forests by integrating carbon management objectives in wood supply modeling. *Mitig. Adapt. Strateg. Glob. Chang.* **2007**, *12*, 1253–1275. [CrossRef]
19. Annkristin, R.; Terje, G.; Birger, S.; Hans-Fredrik, H.; Even, B. A forest optimization model including carbon flows: Application to a forest in Norway. *For. Ecol. Manag.* **2009**, *258*, 579–589.
20. Dahlin, B.; Sallnas, O. Harvest scheduling under adjacency constraints-A case study from the Swedish sub-alpine region. *Scand. J. For. Res.* **1993**, *8*, 281–290. [CrossRef]
21. Boston, K.; Bettinger, P. An economic and landscape evaluation of the green-up rules for California, Oregon, and Washington (USA). *For. Policy Econ.* **2006**, *8*, 251–266. [CrossRef]
22. Sustainable Forestry Initiative. *SFI 2015–2019 Forest Management Standard*; Sustainable Forestry Initiative Inc.: Washington, DC, USA, 2015.
23. Borges, P.; Bergseng, E.; Eid, T.; Gobakken, T. Impact of maximum opening area constraints on profitability and biomass availability in forest-a large, real world case. *Silva Fenn.* **2015**, *49*, 1347. [CrossRef]
24. Dong, L.; Bettinger, P.; Liu, Z.; Qin, H. Spatial forest harvest scheduling for areas involving carbon and timber management goals. *Forests* **2015**, *6*, 1362–1379. [CrossRef]
25. Murray, A.T. Spatial restrictions in harvest scheduling. *For. Sci.* **1999**, *45*, 45–52.
26. Chen, B.; Gadow, K.V. Timber harvest planning with spatial objectives, using the method of simulated annealing. *Eur. J. Oper. Res.* **2002**, *121*, 25–34.
27. World Bank, Ecofys. *State and Trends of Carbon Pricing*; World Bank, Ecofys: Washington, DC, USA, 2015.
28. State Development Planning Commission; State Forestry Bureau. *The Ecological Protection and Economic Transformation Plans of Great Xing'an Mountain in 2010–2020*; Chinese Forestry Press: Beijing, China, 2010. (In Chinese)
29. Wang, H. Dynamic Simulating System for Stand Growth of Forests in Northeast China. Ph.D. Thesis, Northeast Forestry University, Harbin, China, 2012. (In Chinese)
30. State Forestry Bureau. *The Technical Regulations of National Forest Investigation*; Chinese Forestry Press: Beijing, China, 2003. (In Chinese)
31. Lei, X.; Lu, Y.; Peng, C.; Zhang, X.; Chang, J.; Hong, L. Growth and structure development of semi-natural larch-spruce-fir (*Larix olgensis-Picea jezoensis-Abies nephrolepis*) forests in northeast China: 12-year results after thinning. *For. Ecol. Manag.* **2007**, *240*, 165–177. [CrossRef]
32. Website of China Carbon Emission and Trading. Available online: http://www.tanpaifang.com/ (accessed on 17 November 2016).
33. McDill, M.E.; Rebain, S.A.; Braze, J. Harvest scheduling with area-based adjacency constraints. *For. Sci.* **2002**, *48*, 631–642.
34. Metropolis, N.; Rosenbluth, A.; Rosenbluth, M.; Teller, A.H.; Teller, E. Equation of state calculations by fast computing machines. *J. Chem. Phys.* **1953**, *21*, 1087–1101. [CrossRef]
35. Kirkpatrick, S.; Gelatt, C.D.; Vecchi, M.P. Optimization by simulated annealing. *Science* **1983**, *220*, 671–680. [CrossRef] [PubMed]
36. Pukkala, T.; Heinonen, T. Optimizing heuristic search in forest planning. *Nonlinear Anal.: Real World Appl.* **2006**, *7*, 1284–1297. [CrossRef]
37. Boston, K.; Bettinger, P. An analysis of Monte-Carlo integer programming simulated annealing and tabu search heuristics for solving spatial harvest scheduling problems. *For. Sci.* **1999**, *45*, 292–301.
38. Fu, H.; Yan, W.; Chen, J. Forest carbon storage and its dynamics in Daxing'an Mountains of Inner Mongolia. *J. Arid Land Resour. Environ.* **2013**, *27*, 166–170. (In Chinese)
39. Wei, Y.; Zhou, W.; Zhou, L.; Qin, S.; Dai, L. Carbon storage and its distribution pattern in the natural *larix gmelinii* forest on Daxing'an Mountains. *Acta Ecol. Sin.* **2015**, *35*, 189–195. (In Chinese)
40. Dong, L. Developing Individual and Stand-Level Biomass Equations in NortheastChina Forest Area. Ph.D. Thesis, Northeast Forestry University, Harbin, China, 2015. (In Chinese)

41. Pan, Y.; Birdsey, R.A.; Fang, J.; Houghton, R.; Kauppi, P.E.; Kurz, W.A.; Phillips, O.L.; Shvidenko, A.; Lewis, S.L.; Canadell, J.G.; et al. A large and persistent carbon sink in the world's forests. *Science* **2011**, *333*, 988–993. [CrossRef] [PubMed]
42. Garcia-Gonzalo, J.; Pukkala, T.; Borges, J.G. Integrating fire risk in stand management scheduling: An application to Maritime pine stands in Portugal. *Ann. Oper. Res.* **2014**, *219*, 379–395. [CrossRef]
43. Bettinger, P.; Siry, J.; Merry, K. Forest management planning technology issues posed by climate change. *For. Sci. Technol.* **2013**, *9*, 9–19. [CrossRef]
44. Hu, H.; Wei, S.; Sun, L. Estimation of carbon emissions due to forest fire in Daxing'an Mountains from 1965 to 2010. *Chin. J. Plant Ecol.* **2012**, *36*, 629–644. (In Chinese) [CrossRef]
45. Peter, B.; Nelson, J. Estimating harvest schedules and profitability under the risk of fire disturbance. *Can. J. For. Res.* **2005**, *35*, 1378–1388. [CrossRef]
46. Couture, S.; Reynaud, A. Forest management under fire risk when forest carbon sequestration has value. *Ecol. Econ.* **2011**, *70*, 2002–2011. [CrossRef]
47. Garcia-Gonzalo, J.; Borges, J.G.; Palma, J.H.N.; Zubizarreta-Gerendiain, A. A decision support system for management planning of Eucalyptus plantations facing climate change. *Ann. For. Sci.* **2014**, *71*, 187–199. [CrossRef]
48. Borges, P.; Martins, I.; Berseng, E.; Eid, T.; Gobakken, T. Effects of site productivity on forest harvest scheduling subject to green-up and maximum area restriction. *Scand. J. For. Res.* **2015**, *31*, 507–516. [CrossRef]
49. Nalle, D.J.; Arthur, J.L.; Montgomery, C.A. Economic impacts of adjacency and green-up constraints on timber production at a landscape scale. *J. For. Econ.* **2005**, *10*, 189–205. [CrossRef]

Use of Real-Time GNSS-RF Data to Characterize the Swing Movements of Forestry Equipment

Ryer M. Becker [1,*], Robert F. Keefe [1] and Nathaniel M. Anderson [2]

1 Forest Operations Research Lab, College of Natural Resources, University of Idaho, 875 Perimeter Drive, Moscow, ID 83844-3322, USA; robk@uidaho.edu
2 Rocky Mountain Research Station, U.S. Forest Service, Missoula, MN 59801, USA; nathanielmanderson@fs.fed.us
* Correspondence: beck0667@vandals.uidaho.edu

Academic Editor: Timothy A. Martin

Abstract: The western United States faces significant forest management challenges after severe bark beetle infestations have led to substantial mortality. Minimizing costs is vital for increasing the feasibility of management operations in affected forests. Multi-transmitter Global Navigation Satellite System (GNSS)-radio frequencies (RF) technology has applications in the quantification and analysis of harvest system production efficiency and provision of real-time operational machine position, navigation, and timing. The aim of this study was to determine the accuracy with which multi-transmitter GNSS-RF captures the swinging and forwarding motions of ground based harvesting machines at varying transmission intervals. Assessing the accuracy of GNSS in capturing intricate machine movements is a first step toward development of a real-time production model to assist timber harvesting of beetle-killed lodgepole pine stands. In a complete randomized block experiment with four replicates, a log loader rotated to 18 predetermined angles with GNSS-RF transponders collecting and sending data at two points along the machine boom (grapple and heel rack) and at three transmission intervals (2.5, 5.0 and 10.0 s). The 2.5 and 5.0 s intervals correctly identified 94% and 92% of cycles at the grapple and 92% and 89% of cycles at the heel, respectively. The 2.5 s interval successfully classified over 90% of individual cycle elements, while the 5.0 s interval returned statistically similar results. Predicted swing angles obtained the highest level of similarity to observed angles at the 2.5 s interval. Our results show that GNSS-RF is useful for real-time, model-based analysis of forest operations, including woody biomass production logistics.

Keywords: forest operations; logistics; multi-transmitter; accuracy; precision forestry

---

1. Introduction

Real-time data analysis using Global Navigation Satellite System (GNSS) positioning coupled with data communication over radio frequencies (RF), or GNSS-RF, is an area of increasing interest in harvest operations as this new technology creates opportunities for innovation in operational forestry. Real-time data analysis and decision support may also be useful in the context of woody biomass logistics as new markets for forest residues and salvage wood develop. Increased interest in emerging and existing bioenergy markets is closely linked with reducing the risks that forested lands face from fire, insects, and pathogens [1]. It has been estimated that 15 western states in the US contain over 11 million hectares of forested land that could benefit from treatments to improve resilience and health by initiating active forest management on the landscape [2]. In order to return stands to historic stand dynamics and improve forest health, silvicultural treatments need to be performed, even in stands where such treatments generate large amounts of residues and primarily yield low value products [3].

Despite increased interest and developing markets, and because biomass products from forest residues are generally low in value, landowners and contractors must develop efficient, effective, and sustainable methods for harvesting and gathering forest residues and beetle-killed timber or the long-term feasibility of supporting new wood-based bioenergy markets will be limited [4]. By 2022, it is estimated that 12.4 million dry tons of forest residues will be available annually to be utilized for bio-energy [5]. Higher production rates and lower costs associated with logging systems used in silvicultural systems for salvage and forest restoration operations, including the gathering and processing of other forest residues, will allow forest managers to treat more forest land at lower costs and access a higher proportion of this available biomass feedstock. Not only could this result in healthier forests, but the increased feedstock supply for bioenergy and bio-based product manufacturing could help bolster the economies in communities that rely heavily on the forest products industry. Historically in the US, forest residues have been the primary fuel for bioenergy production, with industrial process heat and power production from forest biomass representing a large portion of overall renewable energy production. Epidemic outbreaks of the mountain pine beetle (*Dendroctonus ponderosae* Hopkins) and subsequent mortality of millions of hectares of lodgepole pine (*Pinus contorta* var. *latifolia* Engelm.) in the northern Rocky Mountains has led to an increased interest in utilizing associated biomass as a bioenergy feedstock option [6]. Identifying harvesting and processing methods that increase production efficiency and reduce costs through real-time positional analysis of operational equipment and workers may increase the feasibility of using forest residues from treatments in beetle-killed timber, as well as improve overall operational efficiency. Harvesting these degraded beetle-killed stands helps promote the reintroduction of healthy, sustainable forests. The ability to accurately define specific machine elements from spatial data acquired from GNSS-RF transponders is a necessary and first step for the development of real-time production analysis and operator decision-support models that would help achieve these goals.

In the western US, a major potential supply of biomass is thinning residues and other materials removed from fuel treatment operations that occur on US national forests [7]. Incorporating GNSS-RF technologies in innovative ways in the forest industry may assist operations foresters and contractors in determining the best processes and methods for timber harvesting through real-time positioning and production logistic modeling, especially in beetle-kill. For example, real-time analysis can help to optimize on-the-fly skid trail layout, the order of harvesting, skidding, and processing, and placement of intermediate feedstock preprocessing depot locations.

The on-board data loggers that are standard equipment in many modern forest machines record several types of data that may be useful for real-time analysis of production, such as log piece size and numbers of stems processed [8,9]. The addition of discrete equipment position, navigation, and timing (PNT) information to the standard data stream would make it possible for operators to account for and adapt to unforeseen delays in machine cycles or site conditions. Experienced operators do this naturally over short time steps on the fly, but computer-aided analysis can help expose patterns in operation that are not always apparent, even to the most experienced personnel. Utilization of real-time position data for individual pieces of equipment and among multiple pieces of equipment in a system could make it possible for higher resolution and higher order complexity operational models that monitor individual equipment cycle elements and suggest efficiency improvements based on variables like terrain, timber quality, and other site characteristics [8,9].

GNSS technology has been studied and in some cases employed in operational forestry contexts including thinning stands, tracking movements of site preparation machinery, aligning logging roads, and positioning and dispatching log trucks along haulage routes in real-time, as well as characterizing soil disturbances related to harvest operations [10–13]. Recently, the use of real-time positioning systems in forestry that employ integrated Global Navigation Satellite System (GNSS) and radio frequency (RF) devices is growing, and is important for the advancement and development of operational precision forestry [14,15]. Precision forestry is an intensive management technique that emphasizes development of operational practices that incorporate technologies and processes to increase productivity, reduce costs, and reduce negative site impacts, especially those on vegetation, soil, and water resources. Advancements in GIS, GNSS, light detection and ranging (LiDAR), real-time analysis, and most recently GNSS-RF has led to increased interest and application of precision forestry techniques. The innovative application of these new technologies opens the door for advancements in operational forestry by increasing efficiencies of forest management activities [11,13]. GNSS uses satellites from the U.S. Global Positioning System (GPS), the Russian global navigation satellite system (GLONASS), and possibly other satellite systems to offer spatial reference data to GNSS receivers around the world. Positioning relies on GNSS receiver ability to communicate with satellite systems to provide location data and is used widely across the globe in consumer, military, and industrial applications. RF are electromagnetic wave frequencies in the range commonly used for communication and radar signals. When GNSS is paired with RF (GNSS-RF) the resultant system has the capability to utilize multiple GNSS-RF transponders that receive positional data through GNSS signals and then instantaneously relay information through RF devices to a receiver. Additionally, communication between transponders and receivers is not reliant on cellular networks nor internet connectivity [14]. This is especially important for real-time positioning applications in forest operations, which often occur in remote areas with limited cellular network coverage.

Earlier work has primarily focused on GNSS positioning technologies that relay the location of equipment to a distant computer, as with dispatch systems deployed in transportation [10] and service monitoring applications included on modern machinery. Real-time systems that communicate equipment and ground-worker locations among one another locally at remote logging sites are an important advancement because the technology opens the door to integrated, simultaneous analysis of data from multiple machines and ground workers interacting with one another within the framework of computer-augmented decision processes, with the objective of improving production logistics and safety [14,15].

Past research has returned promising results related to the ability of GNSS receivers to monitor forestry equipment movements, though dense forest canopies are known to increase GNSS error and reduce the ability to collect precise and accurate measurements in dense stands [10,16–20]. Even when accuracy is reduced due to canopy closure, GNSS can be successful when high accuracy in monitoring object movements is not required [10]. For example, position accuracy to 10 m may be insufficient for safety applications, but acceptable for transportation routing. Landscape topography has also been an obstacle in the use of GNSS in forestry applications due to line of sight obstructions [21]. McDonald and Fulton used GNSS for elemental analysis of skidder cycles and found that GNSS locations could successfully be used to predict cycle durations and distinguish between different elements within those cycles (e.g., grapple, positioning, and travel), though the system was subject to large errors when compared to clock studies (i.e., manual timing by an observer) with regard to specific element durations [22]. Operational cycles measured with GNSS agreed with direct observation times 90% of the overall time studied [22]. Similar accuracies were found when using GNSS transponders to analyze the cycle times of forwarders and when using vibration sensors to assist in the determination of cycle times [23].

While GNSS data have been effective for capturing the location, movements, and overall cycle times of some forest machinery, researchers have encountered problems when attempting to acquire high levels of accuracy in the analysis of specific productive cycle elements [22–25]. Additionally, there has been limited prior research evaluating the quality of productive cycle element characterization using positional information derived from GNSS-RF data. Understanding the accuracies and capabilities of commercially available, consumer-grade GNSS-RF transponders will facilitate development of their application in beetle-killed harvest logistics and in production analysis in forest operations more broadly.

The objective of this study was to determine the effectiveness of multi-transmitter GNSS-RF units for characterizing the cycle elements of a log loader at three intervals of GNSS-RF signal transmission and two locations along the machine boom, using time and positional data sent by the transponders.

GNSS-RF positional data were compared statistically to manually recorded time intervals for the same cycle elements. Our goals in determining the accuracy of the technology were two-fold: (1) to foster innovation in operations research, especially in the safe, efficient characterization of cycle elements and continued development of accurate, real-time model-assisted decision support and analysis; and (2) to facilitate the development and deployment of these technologies in the forest sector to improve productive efficiency and reduce costs. These goals are of particular interest and application in improving harvest logistics of beetle-killed timber harvests for biomass when product values are low, potential profit margins are narrow, and cost-minimization, even by small margins, can have large impacts on the financial viability of supply chains.

2. Materials and Methods

We used Garmin Alpha multi-transmitter GNSS-RF units to record the swing movements of a stationary log loader, also called a "shovel", in order to characterize the swinging and slew of the boom. This study was conducted on one research site with four replicated trials, with each trial consisting of 18 cycles. The full swing extent of the machine (360 degrees) was broken into 18 equal arc segments of 20 degrees each (Figure 1). Separation of the swing extent into smaller components made it possible to time the machine movements at varying swing angles to determine if GNSS-RF transponders were able to capture the machine movements accurately at different distances of motion. This was done for three transmission frequency refresh intervals: 2.5, 5.0 and 10.0 s. These three refresh intervals translate into transmission intervals of 24, 12 and 6 transmissions per minute for the 2.5, 5.0 and 10.0 intervals respectively. GNSS-RF transponders were placed in three locations on the machine: one in the right-rear of the cab (close to the center of the machine), one at the full boom extent near the grapple, and one at the heel rack, which is halfway down the most forward boom segment (Figure 1).

To lay out the experiment, a circle with a radius of 12.2 m was delineated in a flat, open area. The open area was chosen purposefully to avoid attenuated GNSS signals and resulting multipath error. The center point was chosen and a fiberglass tape was extended to the chosen radius or 12.2 m. As the individual at the center point rotated around the axis, a field technician on the end of the tape marked the outside edge with paint. Wooden stakes were placed along the circumference of the circle to mark the 18 different angle segments of 20 degrees each. To determine the location of these points, we determined the side length of an octdecagon (18 sided regular polygon) using the circumradius of 12.2 m and the formula side = $2r \cdot \sin(180/n)$, where "r" is the circumradius, and "n" is the number of sides. Each side of the polygon measured 4.23 m. A starting point stake labeled "0" was placed on the circumference of the circle. The location of stake 1 was determined by measuring 4.2 m from stake 18 and finding where along this circumference of the circle this landed. This process was repeated for the remaining stakes (2–17) until there were 18 unique 20 degree angle segments within the full swing extent as shown in Figure 1.

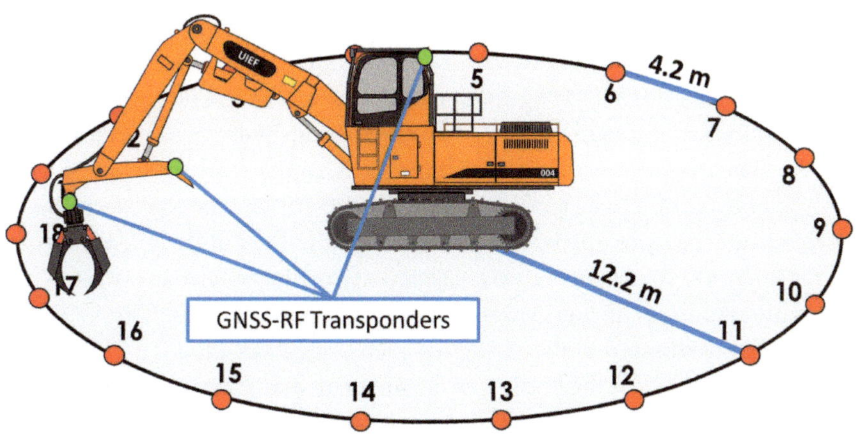

Figure 1. Study layout, consisting of eighteen, twenty-degree shovel swing angle segments.

The machine was then driven to the center of the plot and operated in this location for the duration of the study. Keeping the machine in a stationary position removed any possible error or variation that lateral movements would introduce to the GNSS data. The signal sent to the hand held receiver relies on satellite communication to transmit GNSS locations. The number of satellites with which the receiver can communicate at any point in time alters the accuracy of the measurement at that time, meaning accuracy is variable in time depending on how many satellites are in range. We were not able to record the number of satellites in range for each measurement to quantify this source of error. Therefore, a completely randomized design was used in the data collection process, which is described below. A total of seven (7) Alpha multi-transmitter GNSS units were used for the experiment and two (2) handheld receivers were used in data collection. Three Alpha transponders were utilized in measuring the boom movements at the grapple head and were labeled Boom 1, Boom 2 and Boom 3. Boom 1, 2 and 3 were assigned transmission intervals 2.5, 5.0 and 10.0 s, respectively. Three additional Alpha transponders were used in the data collection for the heel rack and labeled Heel 1, Heel 2 and Heel 3 with transmission intervals 2.5, 5.0 and 10.0 s, respectively. One Alpha transponder was labeled as Cab and was used for both the boom and heel trials. The main hypothesis being tested is that shorter transmission intervals result in better accuracy when classifying cycle elements. Additionally, we hypothesized that shorter intervals would also result in more accurate estimates of the angle of machine swing.

The three Alpha transponders attached to grapple and the cab transponder were synched to one handheld receiver. The additional three transponders for the heel and the cab transponder were synched to the second handheld receiver. Two handheld receivers (full and heel) were used as opposed to one receiver to ensure data would not overlap and be compromised. During the swinging experiment, the boom was extended to 9.5 m from the center axis of the log loader and the heel rack was kept a constant 6.7 m from the center axis of the log loader. After each replicated trial, these distances were checked to ensure consistency throughout the experiment. The transponders were easily attached to the heel rack and grapple of the loader with zip-ties, meaning no modifications had to be done to the machinery and the operator could perform tasks normally, with no changes in operation attributable to the addition of the transponders or any other experimental condition.

In each of the four trials, the shovel covered all 18 of the angle segments in separate cycles in random order predetermined by a random number generator. Through each of the 18 cycles per trial, the start time, stop time at the randomly selected angle, and the return time to the start position was recorded both manually and through the data packets received by the GNSS-RF receiver. Manual timing was conducted on a laptop computer using the Sys.time() command in the R statistical programming environment. The internal clock of the laptop used to conduct the manual timing was synched to the time on the GNSS-RF receiver by synchronizing with the nist.gov time server [26]. Synching the time of the GNSS-RF receivers and the laptop assisted in matching the GNSS-RF data with the manually recorded cycle times ensuring error when merging manually and remotely collected data was minimized.

Three 4.9 m, beetle-killed lodgepole pine logs were used in the experiment to simulate real world use of the loader on a harvest site. For each of the 18 cycles for each of the four replications, the same sequence was followed. This resulted in 72 unique cycles for both the grapple and heel rack GNSS placements. The logs and log loader boom began at the 0 degree marker to start each cycle. Once the operator swung to the required angle, the logs were dropped and the loaded swing time was recorded. The operator would then swing back to the starting position empty, following the same path. The ending time was recorded when the grapple was placed on the ground back at position 0. At this time the operator would then gather the logs and return them to the 0 degree marker to reset for the next cycle, which always started at position 0. Two cycle elements were defined to represent the movements of the log loader: "Swing/Unload" and "Return". "Swing/Unload" is described as the time from the start of the swing for each cycle with a loaded grapple starting at angle 0 until the loader drops the logs at the ending angle measure and starts the swing back to the starting point. At this point, the time from the start of the return swing to the moment the grapple touches the ground back at angle 0 is defined as "Return". The loading element of the cycle, which consists of an unloaded grapple leaving the stop position and collecting logs before the start of the loaded swing element was not included in measurement because it was expected to be identical for each cycle, was not required to test transponder accuracy, and would have introduced additional variation into the timing. Motorola two-way radios were used to communicate with the operator during the experiment, including directions about the selected angle for each cycle.

The resulting data from the field observation data sheets and the data received from the Garmin GNSS-RF receivers were entered into a spreadsheet and imported into the statistical analysis environment "R". The data were then processed, analyzed, and interpreted following the flow diagram represented in Figure 2.

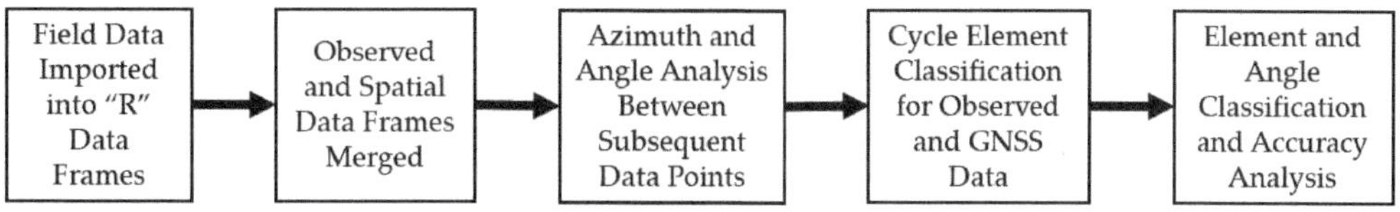

Figure 2. Flow diagram showing the progression of analysis of the spatial and observed data.

A chi-squared ANOVA test was performed to determine the significance of each individual parameter on the proportion of correct classifications. The parameters tested included transmission interval, swing angle, transponder location, cycle element, and the interaction of transmission interval and swing angle. Additionally, a binomial logistic regression was used to determine the influence of the predictor variables on cycle element classification of the GNSS data when compared to the field observations. A regression based two-one sided *t*-test (TOST) equivalence test was performed to analyze whether the predicted angles of the model derived from GNSS-RF data were statistically similar to the observed angle measures. To further analyze the relationship between observed and predicted angle measures derived from the GNSS-RF data, a linear regression based adapted two-one sided *t*-test (TOST) was performed with bootstrapping in order to determine whether the null hypothesis of dissimilarity could be rejected for the observed angle measures and predicted values. Unlike traditional hypothesis testing where a failure to reject null results in a conclusion of indifference, an equivalence test starts with the assumption of dissimilarity, meaning that a rejection of null indicates similarity. This analysis process then shifts the burden of proof onto the model's ability to derive accurate predictions. Equivalence testing was originally derived from bioequivalence testing used in the development of pharmaceutical drugs, and has been successfully adapted and utilized in tree physiology and biometrics research [27,28]. While TOST tests for population-wide agreement, the regression based adaptation discussed by Robinson et al. also addressed point to point agreement between observed and predicted values [28]. This adapted analysis is able to test how well the distributions of the observed angles match the distribution of the predicted angles. Additionally, the ability to use bootstrap resampling makes this statistical approach favorable. Resampling was not done to provide an estimate of the distribution of the predicted values. The data retrieved from the GNSS-RF data packets and subsequent derivations provided us with the predicted value distribution. Rather, the bootstrap resampling is included to provide an estimate of the sampling distribution within the range of predicted angle measures; providing a larger sample size than could feasibly be obtained in fieldwork alone. The ability to increase sample size can strengthen the evidence for similarity in equivalence-based tests [28].

All statistical analysis was performed in R [29]. The equiv.boot function in the "equivalence" package was used to evaluate equivalence tests [30]. A total of 12 equivalence tests were conducted using 95% confidence intervals. These represented all combinations of the three transmission intervals (3), GNSS-RF transponder locations (2), and cycle elements (2). The equiv.boot function analyzes whether the slope and intercept of the regression of predicted angles fall within the desired region of equivalence, determining similarity or dissimilarity to observed angles. The region of equivalence for the regression was tested at $\pm 10\%$ for both the slope and the intercept and 10,000 bootstrap replications were performed [28]. Equivalence testing shifts the burden of proof to difference.

3. Results

3.1. Summary

Transmission interval ($p < 0.001$), transponder location ($p < 0.001$), and swing angle ($p < 0.001$) all affected the correct classification of cycle elements (Table 1). At the grapple, the proportion of verified correct classifications was greater for the 2.5 and 5.0 s transmission intervals than at the heel (Table 2). However, the proportion of correct classification for the 10.0 s interval was slightly higher at the heel than the associated grapple proportion. In all instances, the 2.5 s interval returned higher proportions of success than the other intervals, apart from the 5.0 s interval return element at the heel location. The proportion of correct classifications was least reliable overall for the 10.0 s interval, as expected, followed by the 5.0 s interval.

Table 1. Chi-squared ANOVA table for parameter significance on correct classification proportion.

Variable	*Df* [1]	Deviance	Resid. *Df* [2]	Resid. Dev. [3]	*Pr* (>Chi) [4]
NULL			2680	2054	
Interval	2	74.18	2678	1980	<0.001
Swing Angle	1	42.38	2677	1937	<0.001
Location	1	37.62	2676	1900	<0.001
Cycle Element	1	0.04	2675	1900	0.8421
Interval: Swing Angle	2	1.00	2673	1899	0.6043

[1] Degrees of freedom; [2] Degrees of freedom of the residuals; [3] Deviance of the residuals; [4] *p*-value for the level of significance on correct classification.

Only the 2.5 s interval was able to achieve greater than 90% classification success for a complete cycle and achieved 99% classification success for the Swing Unload element at the grapple location. The 5.0 s interval achieved 90% successful classification at the grapple location. The total swing angles for the duration of the trials captured by the transponders were underestimated. In this instance, the 2.5 s transponder returned the total angle for all 72 trials, and thereby best represented the actual swing totals.

Table 2. Summary data table.

Location	Interval	Element	Data Points	Total Angle (Degrees) Observed (GPS)	Proportion Correct
Grapple			1339		0.84 [1]
	2.5		769	27,360 (23,882)	0.94
		Swing Unload	451	13,680 (12,950)	0.99
		Return	318	13,680 (10,932)	0.88
	5.0		382	27,360 (21,442)	0.90
		Swing Unload	219	13,680 (11,698)	0.89
		Return	163	13,680 (9744)	0.91
	10.0		188	27,360 (14,918)	0.68
		Swing Unload	113	13,680 (8751)	0.71
		Return	75	13,680 (6167)	0.66
Heel			1342		0.79 [1]
	2.5		773	27,360 (24,455)	0.82
		Swing Unload	443	13,680 (12,495)	0.84
		Return	330	13,680 (11,960)	0.80
	5.0		376	27,360 (22,740)	0.86
		Swing Unload	227	13,680 (12,609)	0.84
		Return	149	13,680 (10,131)	0.87
	10.0		193	27,360 (15,558)	0.69
		Swing Unload	110	13,680 (9195)	0.71
		Return	83	13,680 (6363)	0.68

[1] Mean proportion of correctly classified elements across all treatments for Heel or Grapple location.

3.2. Binomial Logistic Regression-Element Characterization

The odds of correctly classifying an element at the 10.0 s transmission interval was decreased 0.262 times when compared to the 2.5 s transmission interval, and was significant at a 95% confidence interval (CI) (Table 3).

Table 3. Logistic regression coefficients associated with model describing variable impact on whether GPS returned correct element classification as represented by field observations.

Variable	Estimate	SE [1]	*p*-Value	Odds Ratio	95% Confidence
Intercept	1.748	0.212	<0.001	5.741	3.813–8.772
Rate 5	0.144	0.320	0.6525	1.155	0.621–2.182
Rate 10	−1.339	0.328	<0.001	0.262	0.138–0.499
Swing Angle	0.004	0.001	<0.001	1.004	1.003–1.006
Location Heel	−0.744	0.124	<0.001	0.476	0.372–0.605
Cycle Element	0.025	0.122	0.8374	1.025	0.807–1.300
Rate 5: Swing Angle	−0.001	0.002	0.3789	0.999	0.996–1.002
Rate 10: Swing Angle	0.000	0.002	0.8781	1.000	0.999–1.003

[1] Standard error of the variable estimate

The 5.0 s interval odds of correct classification was not significantly different than the 2.5 s interval. At each of the 18 observed angle intervals between 20 and 360, the odds of correct classification increased 1.004 times for each increasing interval with a CI exceeding 95%. In practice, this means that larger arcs have a higher incidence of correct classification. The location of the GPS transponders was also found to be a significant predictor (95% CI) of correct classification, with the odds of correct classification decreasing 0.476 times when moving from the grapple to the heel location. In practice, connected to arc distance, for the same arc in degrees, the heel moves a shorter distance than the grapple, resulting in more frequent misclassification. Observed element (Swing/Unload or Return) was not found to create differences in the odds of correct classification of cycle element. The relationships between the transponder location, swing angle, GNSS interval, element and proportion of correctly classified elements are further represented in Figures 3 and 4.

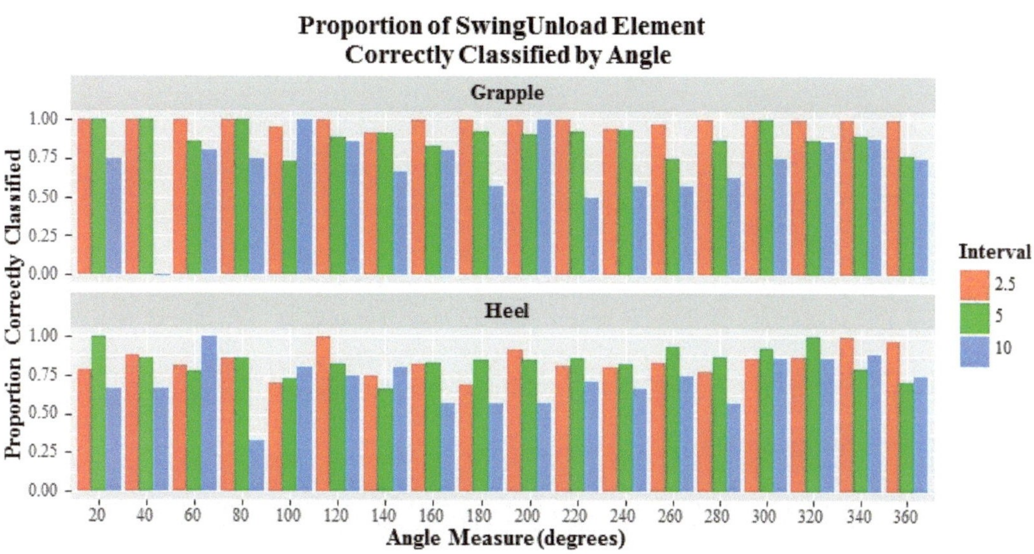

Figure 3. Bar chart representing the proportion of correct classifications of the Swing Unload cycle element at the grapple and heel locations at 18 angle intervals.

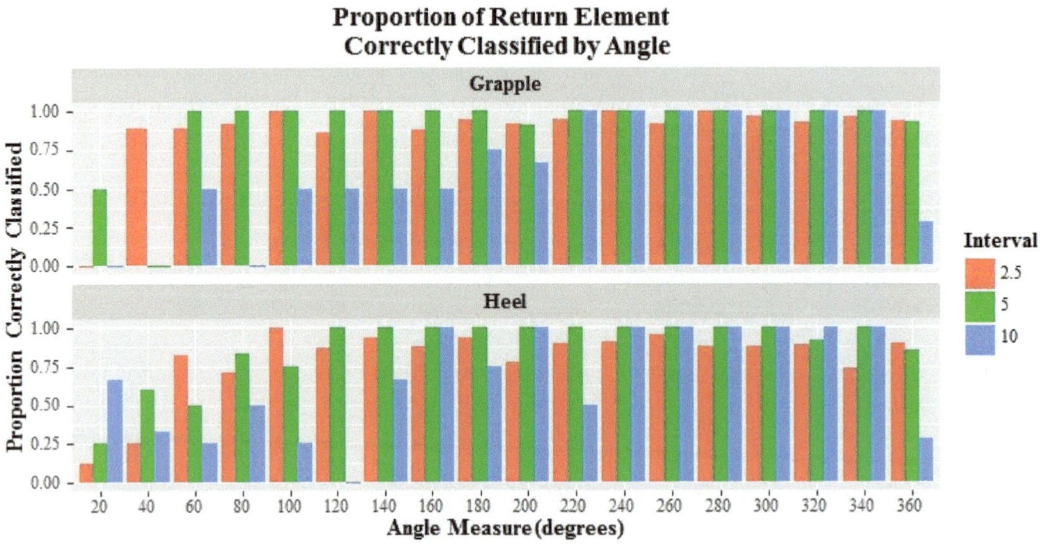

Figure 4. Bar chart representing the proportion of correct classifications of the Return cycle elements at the grapple and heel locations at 18 angle intervals.

3.3. *Swing Angle Analysis*

Figure 5 shows a graphical comparison between the observed angle interval for each trial and the angles derived from GNSS-RF data packets for each transmission interval at both locations and for both elements. It is evident in the figure that the 2.5 and 5.0 s intervals most accurately capture overall angle measures at the grapple location, and that the heel location introduces greater error at both intervals. The solid black line in the figures represents a 1:1 relationship between the observed and the predicted values derived from the GNSS-RF recorded data packets. A clear trend of under prediction of swing angle can be seen in the figure for all transmission intervals, transponder locations, and machine elements. The most severe under-predictions are associated with the 10.0 s transmission interval, however. Both 2.5 and 5.0 s transmission intervals appeared to follow similar trends of underestimation in the predicted angles.

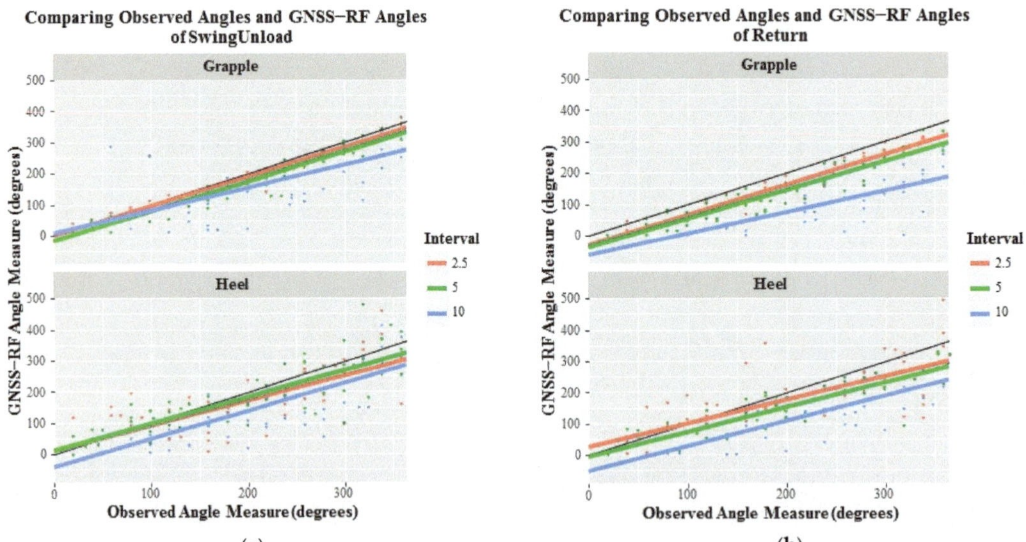

Figure 5. These figures represent the relationship between GNSS-RF derived angle measurements and observed angle intervals at the Grapple and Heel locations for the (**a**) Swing Unload and (**b**) Return elements.

Equivalence-based regression analysis showed that the only combination of GNSS-RF derived angles for which both the intercept and slope were statistically similar to the observed values was the transponder located at the grapple with a 2.5 s interval capturing the swing unload cycle element (Table 4). In all other instances the intercept was dissimilar for the predicted angle values when compared to the observed values. Only the 5.0 s interval swing unload element at the grapple and the 2.5 s interval at the grapple for the return element were found to be similar in regards to the slope of the regression line.

Table 4. Summary of equivalence-based regression results. The model represents the various combinations of transmission interval (2.5,5,10); transponder location, grapple (G) or heel (H); and cycle element, swing unload (SU), and return (R). Sample size is denoted by n, and the approximate joint two one-sided 95% confidence intervals for the slope and intercept are: $(C^{-}{}_{\beta_1}, C^{+}{}_{\beta_1})$ and $(C^{-}{}_{\beta_0}, C^{+}{}_{\beta_0})$, respectively. The former should be contained by the intercept interval of equivalence, $(I^{-}{}_{\beta_0}, I^{+}{}_{\beta_0}) = \overline{y} \pm 25\%$, and the latter by the slope interval of equivalence $(I^{-}{}_{\beta_1}, I^{+}{}_{\beta_1}) = 1 \pm 0.25$.

Model	n	$C^{-}{}_{\mathrm{fi}_0}$	$C^{+}{}_{\mathrm{fi}_0}$	$I^{-}{}_{\mathrm{fi}_0}$	$I^{+}{}_{\mathrm{fi}_0}$	**fi$_0$ Result**	$C^{-}{}_{\mathrm{fi}_1}$	$C^{+}{}_{\mathrm{fi}_1}$	$I^{-}{}_{\mathrm{fi}_1}$	$I^{+}{}_{\mathrm{fi}_1}$	**fi$_1$ Result**
G.SU.2.5	72	186.29	194.13	161.88	197.85	Reject	0.988	1.057	0.9	1.1	Reject
G.SU.5	72	184.46	195.91	146.23	178.72	Fail	0.934	1.042	0.9	1.1	Reject
G.SU.10	65	188.92	221.63	141.74	173.23	Fail	0.539	0.876	0.9	1.1	Fail
H.SU.2.5	72	175.16	205.88	156.19	190.90	Fail	0.533	0.779	0.9	1.1	Fail
H.SU.5	72	177.62	203.34	157.62	192.64	Fail	0.717	0.972	0.9	1.1	Fail
H.SU.10	64	192.70	220.46	129.30	158.04	Fail	0.625	0.893	0.9	1.1	Fail
G.R.2.5	68	195.02	205.03	144.68	176.83	Fail	0.946	1.035	0.9	1.1	Reject
G.R.5	66	196.55	212.46	132.87	162.40	Fail	0.886	1.044	0.9	1.1	Fail
G.R.10	44	229.89	263.50	95.75	117.03	Fail	0.468	0.902	0.9	1.1	Fail
H.R.2.5	67	187.15	218.16	160.66	196.36	Fail	0.603	0.920	0.9	1.1	Fail
H.R.5	63	199.18	221.60	144.72	176.88	Fail	0.864	1.097	0.9	1.1	Fail
H.R.10	48	216.13	245.22	119.30	145.81	Fail	0.699	1.022	0.9	1.1	Fail

The extent to which swing movements associated with each GNSS-RF transponder could be discerned visually varied with transmission interval (Figure 6). Visual interpretation of the tracks indicated that the 2.5 s transmission interval more accurately represented swing movements compared to 5.0 and 10.0 s transmission intervals. The data support this interpretation in that the angle totals best represented the actual swing angle totals for the 2.5 s interval. Higher frequency of data point collection creates a comparably smoother swing track than the 5.0 and 10.0 s transmission intervals. The movements of the cab during the experiment were also captured by an additional transponder located on the rear corner of the cab. The transponders located at the grapple and heel remained equidistance from the cab transponder throughout the duration of the experiment.

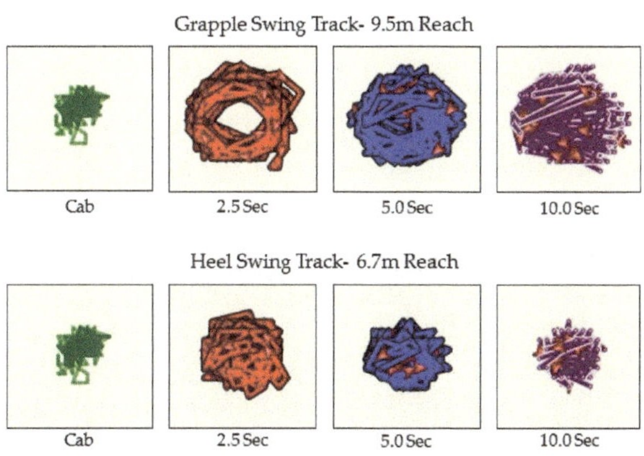

Figure 6. The GNSS-RF transponders create a movement progression track between data points through the duration of the trials which are shown for the two (2) transponder locations and three (3) transmission intervals used in the experiment in addition to the cab location.

4. Discussion

Analysis of the field data returned both expected and surprising results. As anticipated, the 2.5 s transmission interval data returned the highest proportion of correct element classifications. However, the 5.0 s transmission interval returned comparable correctly classified proportions. With 2.5 and 5.0 s transmission intervals capturing 94.4% and 91.7% of the respective overall cycles with at least one correct element for both Swing Unload and Return at the Grapple, accuracy exceeding that described by McDonald and Fulton and similar to that described by Hejazian et al. was obtained when classifying machine cycles using GNSS data [20,24]. The ability to correctly classify at least one observation of each element within each trial was reduced when analyzing the Heel data. The associated percentages of success for the 2.5, 5.0 and 10.0 s intervals at the Heel were 91.7%, 88.9% and 58.3%, respectively. Even so, if, for some operational reason, the transponders were placed at the heel rather than the grapple, a transmission interval of 2.5 or even 5.0 s would provide useful data, depending on accuracy requirements.

When addressing the potential for GNSS-RF transponders to accurately capture angle measurements, a consistent trend of angle underestimation was found across all angles exceeding 180 degrees, though at the smaller angle measures this underestimation is decreased and in some cases there is predicted angle overestimation (Figure 5). The reason for this underestimation at the larger observed angle measures is likely due to transponder data packets representing the machine swing not capturing the full extent of the swing arc. For the GNSS location to represent the true arc angle, the transponder has to record a data packet the exact moment the full swing extent is reached or during any pause at the full extent. The time of a pause at full extent is a greater proportion of the overall time at shorter arcs, resulting in a higher inclusion probability of a data packet representing full swing extent. Therefore, the full extent would be represented more often for smaller swing angles and would result in lower underestimation of swing angles. As seen in Figure 5, any intersection of the 1:1 line and the time interval trend line represents a location where the predicted angles derived from the GNSS-RF data match observed angles. Additionally, Figure 5 and Table 4 indicate overall that the 2.5 s interval is most successful in capturing data to derive accurate angle measures.

In our analysis, there was a trend of decreased GNSS measurement accuracy at the Heel location when compared to the Grapple location. The movement of the transponders around a smaller swing circumference means the transponders are less likely to be able to accurately and concisely plot varying points because the range of the movements is shorter. This is due to the inherent accuracy of GNSS. For example, GNSS accuracy of ± 1 m will have a greater impact on accurately capturing movements when the overall distance traveled 5 m as opposed to 15 m or 20 m. One possible solution to help account for this error when the swing arc is smaller is to incorporate additional sensors and mechanisms to capture grapple and other intricate movements to assist in defining elements [8]. These would augment GNSS data in ways that get around constraints related to GNSS accuracy. At the longer GNSS-RF transmission intervals, minute machine boom movements are not as accurately captured. This is seen by observing both the positional data itself, as well as the successful classification proportions. This is especially clear for the 10.0 s interval data. As suggested by Devlin and McDonnell, the 10.0 s interval may be beneficial for analysis of general machine movements across the landscape [10]. However, precise and intricate machine movements are best captured and analyzed using higher frequency data transmission of locations. Importantly, transmission intervals should be tested and tailored to meet the needs of particular equipment movements rather than relying on a general rule of thumb.

From a visual representation standpoint, it is evident in Figure 3 that the 2.5 s transmission interval results in a plotted machine movement track that has smoother curves and higher accuracy representation of precise machine movements and machine component locations to meet operational analysis objectives. Correct classification of elements and overall cycles exceeding 90% for the 2.5 s interval further support this observation, as does the results of the regression based equivalence test. However, higher frequency transponder transmission rates are also more data intensive, requiring twice the data flow and storage capacity. This intensity could be difficult to manage for large projects with many pieces of equipment, both in terms of data collection and storage, but also real-time analysis, validation, and model optimization. When addressing the ability of the transponder to accurately characterize machine cycles, cycle elements, and angles of swing, the 5.0 s interval was shown to return comparable accuracies to those obtained by the 2.5 s interval transponders, at half the data intensity.

In this particular study, the overall sampled time represented only a small portion of what a full productive work day would entail in a commercial logging operation in beetle-killed timber. With the 2.5 s interval, the available storage space needed for each data file will be utilized twice as fast as the 5.0 s interval. Each file saved can consist of up to 9999 data points, after which the device overwrites existing data by default. Therefore, a new file needs to be saved on the handheld receiver every 6.9 h when working with the 2.5 s transmission interval to avoid overwriting and every 13.8 h when working with the 5.0 s transmission interval with three transponders. If additional transponders are used on multiple pieces of equipment, then available storage space will be used up more quickly. This can present problems with data being overwritten during average work days if the file is not saved, e.g., mid-way through the day. Additionally, it was found that the large data pools associated with the 2.5 s interval made analysis and interpretation of the data more cumbersome and time intensive than the longer intervals, an observation also made by de Hoop and Duprè [25]. Depending on the specific application, machines, and desired level of accuracy, incorporating the 5 s interval could return acceptable, though slightly lower, accuracies than the 2.5 s interval. In turn, longer work cycles could be sampled without concern for overwriting data and analysis and interpretation of those data would prove quicker and easier than with the 2.5 s interval if available memory storage and analytical capacity is a concern. When interest is focused on positioning and analysis of precise machine movements where rapid machine movements may be missed by long transmission intervals, the 2.5 s interval likely provides the best option for transmission frequency, or perhaps the study requires more traditional work study methods that relies on direct rather than passive observation.

An important consideration with our study is that the Garmin Alpha receiver and positioning transponders were located in open line-of-sight conditions, which may not be typical of many forest stands. No transmissions were obstructed by vegetation, topography, or inter-machine positioning. One hundred percent of the time-stamped positional coordinates were received. Prior experience with GNSS-RF data in operational forestry has shown that GNSS position and radio signal propagation quality (RSSI) can interact in complex ways. In operational forestry, there are many situations in which positioning transponders may receive a GNSS signal [14,15], but radio propagation of coordinates to other devices elsewhere on the jobsite is blocked topographically by dense vegetation or interference from other radio systems [14,15]. If a portion of data packets are missing due to radio signal interference, the classification of machine elements may be affected in more complex ways not evident in our controlled experiment. With the quantities of data collected in this way, post-hoc manual and visual inspections of outlier points is almost impossible, so automated data quality control procedures are especially critical in such an environment.

In order to further develop real-time modeling of machine movements on active logging operations, further studies exploring the impacts of forest canopy on GNSS-RF accuracy in particular applications should be explored. For example, in stands with high mortality due to beetle-kill with many trees that have dropped most of their foliage, this may not be a problem, but that remains to be evaluated. Past studies exploring canopy impact in GNSS accuracy have shown large decreases in accuracy depending on canopy cover [10,16–20,31]. However, these studies were for GNSS only and did not study dual effects of canopy on GNSS and RF, or the interaction of these two signals. Because canopy density tends to be low in beetle impacted stands due to high levels of mortality, for example, canopy impacts on GNSS multipath error may be less of a concern than in healthy stands. Additionally, developing a methodology for similar element and cycle classification when the machine is traversing the landscape on steep slopes in beetle impacted stands in biomass utilization operations as opposed to sitting in a fixed location on flat ground will be necessary. Introducing machine movements on slopes into the analysis will add an additional level of complexity but is important for development of subsequent applications to improve the efficiency of harvesting in beetle impacted and unaffected forests alike.

5. Conclusions

Both 2.5 and 5.0 s transmission intervals correctly characterized cycles and cycle elements in the rotation movements of log loaders. However, the 2.5 s interval was the most successful in allowing for the prediction of accurate swing angle measures. Additional studies should be conducted to further refine the methodology and analysis techniques to foster use of this approach in real-time analysis of equipment movements on active beetle-killed harvesting operations. Accurate characterization of machine movements from spatially explicit data through the use of simple, non-intrusive multi-transmitter GNSS-RF allows for an increased level of situational awareness for improved streamlined operational production in the woods where obstructed views from topography or other site features may limit the ability to visually identify the actions of forest machinery at all times. The consumer-grade transponders incorporated into this study returned promising results in this preliminary study as to their accuracy and use in real-time production analysis and support models. The designed and controlled experiment performed in this study provided valuable information which will be used to further develop the real-time analysis model in subsequent studies. Executing these focused studies will allow various machine processes to be analyzed individually, resulting in strong model components representing the various cycle elements of associated harvest machinery. As previously mentioned, once activity profiles for harvest systems are completed, the model will be applied at a landscape scale, providing valuable information to operations foresters and contractors regarding harvest site selection and real-time decision support during harvest.

Further, connecting machine movements to the onboard computers of multiple machines interacting with one another opens new possibilities for real-time decision-support and logistics analysis in beetle-killed harvests and forest operations in general. For example, the movements of a shovel can be paired with subsequent processing activities of additional machinery and personnel working within the harvest boundaries. Incorporation of real-time machine production analysis and model-assisted decision-making will prove a valuable development in the continued advancement of precision forestry. These technologies will allow for streamlined real-time production analysis and feedstock logistics in not only beetle-killed harvests, but more broadly in a wide range of forest operations.

Acknowledgments: This project was funded by the Agriculture and Food Research Initiative Competitive Grant No. 2013-68005-21298 from the USDA National Institute of Food and Agriculture under the Bioenergy Alliance Network of the Rockies (BANR) project. Special thanks to Steve Normington and Tom Dean Logging for their assistance in the field work portion of this study and graciously making time and machinery available, and Eloise Zimbelman for her help with equipment setup and sampling.

Author Contributions: Ryer Becker and Robert Keefe conceived and designed the experiments; Ryer Becker performed the experiments; Ryer Becker, Robert Keefe, and Nathaniel Anderson analyzed the data; Ryer Becker and Robert Keefe wrote the paper with edits and additional text from Nathaniel Anderson.

Conflicts of Interest: The authors declare no conflict of interest. The founding sponsors had no role in the design of the study; in the collection, analyses, or interpretation of data; in the writing of the manuscript, and in the decision to publish the results.

References

1. Wells, L.; Chung, W.; Anderson, N.; Hogland, J. Estimating bioenergy feedstock supply and delivered costs. In Proceedings of the 38th Annual COFE Meeting, Lexington, KY, USA, 19–22 July 2015.
2. Rummer, B.; Prestemon, J.; May, D.; Miles, P.; Vissage, J.; McRoberts, R.; Liknes, G.; Shepperd, W.D.; Ferguson, D.; Elliot, W.; et al. *A Strategic Assessment of Forest Biomass and Fuel Reduction Treatments in Western States*. U.S. Department of Agriculture, Forest Service, Research and Development: Washington, DC, USA, 2003; Volume iii, pp. 1–18.

3. Brown, R.T.; Agee, J.; Franklin, J. Forest restoration and fire: Principles in the context of place. *Conserv. Biol.* **2004**, *18*, 903–912. [CrossRef]
4. Anderson, N.; Chung, W.; Loeffler, D.; Jones, J.G. A productivity and cost comparison of two systems for producing biomass fuel from roadside forest treatment residues. *For. Prod. J.* **2012**, *62*, 222–233. [CrossRef]
5. Jacobson, R.A.; Keefe, R.F.; Smith, A.M.S.; Metlen, S.; Saul, D.; Newman, S.M.; Laninga, T.J.; Inman, D. Multi-spatial analysis of forest residue utilization for bioenergy. *Biofuels Bioprod. Bioref.* **2016**, *10*, 560–575. [CrossRef]
6. Anderson, N.; Mitchell, D. Forest operations and woody biomass logistics to improve efficiency, value and sustainability. *Bioenerg. Res.* **2016**, *9*, 518–533. [CrossRef]
7. Keefe, R.; Anderson, N.; Hogland, J.; Muhlenfeld, K. Woody Biomass Logistics. In *Cellulosic Energy Cropping Systems*, 1st ed.; Karlen, D.L., Ed.; John Wiley & Sons: Hoboken, NJ, USA, 2014; Chapter 14; pp. 251–278.
8. Palander, T.; Nuutinen, Y.; Kariniemi, A.; Vaatainen, K. Automatic time study method for recording work phase times of timber harvesting. *For. Sci.* **2013**, *59*, 472–483. [CrossRef]
9. Strandgard, M.; Walsh, D.; Acuna, M. Estimating harvester productivity in *Pinus radiata* plantations using StanForD stem files. *Scand. J. For. Res.* **2013**, *28*, 73–80. [CrossRef]
10. Devlin, G.J.; McDonnell, K. Performance accuracy of real-time GPS asset tracking systems for timber haulage trucks travelling on both internal forest road and public road networks. *Int. J. For. Eng.* **2009**, *20*, 45–49.
11. Carter, E.A.; McDonald, T.P.; Torbert, J.L. Application of GPS technology to monitor traffic intensity and soil impacts in a forest harvest operation. In Proceedings of the Tenth Biennial Southern Silvicultural Research Conference, Shreveport, LA, USA, 16–18 February 1999.
12. Danskin, S.D.; Bettinger, P.; Jordan, T.R.; Cieszewski, C. A comparison of GPS performance in a southern hardwood forest: Exploring low-cost solutions for forestry applciations. *South. J. Appl. For.* **2009**, *33*, 9–16.
13. Hamzah, K.A. Remote sensing, GIS and GPS as a tool to support precision forestry practices in Malaysia. In Proceedings of the 22nd Asian Conference on Remote Sensing, Singapore, 5–9 November 2001.
14. Keefe, R.F.; Eitel, J.U.H.; Smith, A.M.S.; Tinkham, W.T. Applications of multi-transmitter GPS-VHF in forest operations. In Proceedings of the 47th International Symposium on Forestry Mechanization and 5th International Forest Engineering Conference, Gerardmer, France, 23–26 September 2014.
15. Grayson, L.M.; Keefe, R.F.; Tinkman, W.T.; Eitel, J.U.H.; Saralecos, J.D.; Smith, A.M.S.; Zimbelman, E.G. Accuracy of WAAS-Enabled GPS-RF warning signals when crossing a terrestrial geofence. *Sensors* **2016**, *16*, 912. [CrossRef] [PubMed]
16. Sigrist, P.; Coppin, P.; Hermy, M. Impact of forest canopy on quality and accuracy of GPS measurements. *Int. J. Remote Sens.* **1999**, *20*, 3595–3610. [CrossRef]
17. Hasegawa, H.; Yoshimura, T. Estimation of GPS positional accuracy under different forest conditions using signal interruption probability. *J. For. Res.* **2007**, *12*, 1–7. [CrossRef]
18. Taylor, S.E.; McDonald, T.P.; Veal, M.W.; Grift, T.E. Using GPS to evaluate productivity and performance of forest machine systems. In Proceedings of the First International Precision Forestry Symposium, Seattle, WA, USA, 17–20 June 2001.
19. Veal, M.W.; Taylor, S.E.; McDonald, T.P.; McLemore, D.K.; Dunn, M.R. Accuracy of tracking forest machines with GPS. *Trans. Am. Soc. Agric. Eng.* **2001**, *44*, 1903–1911.
20. Yoshimura, T.; Hasegawa, H. Comparing the precision and accuracy of GPS positioning in forested areas. *J. For. Res.* **2003**, *8*, 147–152. [CrossRef]
21. Wing, M.G.; Eklund, A.; Kellogg, L.D. Consumer-grade global positioning system (GPS) accuracy and reliability. *J. For.* **2005**, *103*, 169–173.
22. McDonald, T.P.; Fulton, J.P. Automated time study of skidders using global positioning system data. *Comput. Electron. Agric.* **2005**, *48*, 19–37. [CrossRef]
23. Strandgard, M.; Mitchell, R. Automated time study of forwarders using GPS and a vibration sensor. *Croat. J. For. Eng.* **2015**, *36*, 175–184.
24. Wang, J.; McNeel, J.; Baumgras, J. A computer-based time study system for timber harvesting operations. *For. Prod. J.* **2003**, *53*, 47–53.
25. De Hoop, C.F.; Duprè, R.H. Using GPS to document skidder motions—A comparison with manual data collection. In Proceedings of the Council on Forest Engineering (COFE) Conference, Coeur d'Alene, ID, USA, 22 July–2 August 2006.
26. NIST Internet Time Servers. Available online: http://tf.nist.gov/tf-cgi/servers.cgi (accessed on 15 May 2016).

27. Robinson, A.P.; Froese, R.E. Model validation using equivalence tests. *Ecol. Model.* **2004**, *176*, 349–358. [CrossRef]
28. Robinson, A.P.; Duursma, R.A.; Marshall, J.D. A regression-based equivalence test for model validation: Shifting the burden of proof. *Tree Physiol.* **2005**, *25*, 903–913. [CrossRef] [PubMed]
29. R Core Team. *R: A Language and Environment for Statistical Computing. R Foundation for Statistical Computing*; R Core Team: Vienna, Austria, 2016; Available online: https://www.R-project.org/ (accessed on 10 September 2016).
30. Robinson, A.P. Equivalence: Provides Tests and Graphics for Assessing Tests of Equivalence. 2016. Available online: https://CRAN.R-project.org/package=equivalence (accessed on 17 October 2016).
31. Bolstad, P.; Jenks, A.; Berkin, J.; Horne, K. A comparison of autonomous, WAAS, real-time, and post-processed global positioning systems (GPS) accuracies in northern forests. *North J. Appl. For.* **2005**, *22*, 5–11.

16

Adaptive Variation and Introgression of a *CONSTANS*-Like Gene in North American Red Oaks

Jennifer F. Lind-Riehl [1] and Oliver Gailing [2,*]

[1] Department of Entomology, University of Wisconsin-Madison, Madison, WI 53706-1598, USA; jflind@mtu.edu

[2] School of Forest Resources and Environmental Science, Michigan Technological University, Houghton, MI 49931-1295, USA

* Correspondence: ogailing@mtu.edu

Academic Editors: Stephen P. DiFazio and John MacKay

Abstract: Oaks provide a model system to study maintenance of species identity by divergent selection since they maintain morphological differences and ecological adaptations despite interspecific hybridization. The genome of closely related interfertile oak species was shown to be largely homogeneous, with a few genomic areas exhibiting high interspecific differentiation possibly as result of strong divergent selection. Previously, a genic microsatellite was identified as under strong divergent selection, being nearly fixed on alternative alleles in the two interfertile North American red oak species: *Quercus rubra* L. and *Quercus ellipsoidalis* E.J. Hill. Further genotyping in two other red oak species—*Quercus velutina* Lam. and *Quercus coccinea* Münchh.—revealed a similar bias for the *Q. ellipsoidalis*-specific allele. To further elucidate the basis of this differentiation, we sequenced the microsatellite in individuals from all four red oak species. Sequence variability was observed in the microsatellite motif which encodes a poly-Q repeat in a *COL* gene involved in phenology and growth. Furthermore, in neighboring (parapatric) *Q. rubra*/*Q. ellipsoidalis* populations, introgression of the *Q. ellipsoidalis*-specific allele into *Q. rubra* occurred at a lower rate than introgression of the *Q. rubra*-specific allele into *Q. ellipsoidalis* despite symmetric interspecific gene flow, indicating potential adaptive introgression. Introgression of adaptive alleles can be an important mechanism for rapid adaptation to new environmental conditions (e.g., climate change).

Keywords: adaptive introgression; Expressed Sequence Tag-Simple Sequence Repeats (EST-SSRs); outlier genes; *Quercus*; *Lobatae*

---

1. Introduction

The transfer of adaptive alleles and traits by hybridization might be an important mechanism of rapid evolution and adaptation to changing environments, e.g., in the face of climate change, and evidence for adaptive trait transfer has been reported in both plant and animal species (reviewed in [1,2]). The availability of genomic resources and analytical methods for the identification of loci under strong divergent selection (outlier loci) [3] allows us to trace the introgression of potentially adaptive alleles in interspecific hybrid zones. Oaks generally reveal porous species boundaries, but morphological species' identity and ecological adaptations (e.g., soil moisture) are generally maintained despite recurrent interspecific gene flow [4–6]. Thus, oaks provide a model for the identification of outlier loci under divergent selection with annotated functions and potential roles in stress tolerance and reproductive isolation between species (e.g., [7]).

The four interfertile North American red oak species—*Quercus rubra* L., *Quercus velutina* Lam., *Quercus coccinea* Münchh., and *Quercus ellipsoidalis* E.J. Hill—exhibit porous species boundaries and recent studies found strong evidence for contemporary interspecific gene flow in sympatric and parapatric stands based on genetic assignment and parentage analyses [5,8–10]. Most recently, Owusu et al. used both genic and nuclear microsatellite markers to better resolve the taxonomic relationships between these four species [10]. The genetic assignment analysis showed a clear separation of all four species, with *Q. rubra* being the most differentiated. After excluding genetically intermediate individuals identified in the genetic assignment analysis, a phylogenetic tree based on population distances at nuclear microsatellite markers revealed *Q. ellipsoidalis* and *Q. coccinea* as most closely related and a clear separation of *Q. rubra* from the *Q. velutina*/*Q. ellipsoidalis*/*Q. coccinea* clade [10]. This result is also supported by previous studies and a recent restriction site associated sequencing (RAD-seq) phylogeny of the genus ([9,11–13]). Genetically intermediate forms were found in contact zones between *Q. ellipsoidalis*, *Q. rubra* and *Q. velutina*, in sympatric stands indicating gene flow among these three species [8,10,14]. High levels of interspecific gene flow were detected especially between *Q. ellipsoidalis* and *Q. velutina*. Thus about 20% of *Q. velutina* individuals had a recent *Q. ellipsoidalis* ancestor and about 30% of *Q. ellipsoidalis* individuals had a recent *Q. velutina* ancestor [8]. This supports previous indications of recent and ongoing hybridization between *Q. ellipsoidalis* and *Q. velutina* [9,11]. Recent studies have shown gene flow between *Q. rubra* and the other three species to be more restricted, yet generally symmetric [9,10,15]. These four species also represent a drought tolerance gradient, with *Q. rubra* as the least drought tolerant followed by *Q. velutina*, *Q. coccinea* and lastly, *Q. ellipsoidalis* as the most drought tolerant [16,17].

Previously, we discovered among 36 genic and eight non-genic microsatellites one genic microsatellite (FIR013), located in the coding region of a *CONSTANS*-like (*COL*) gene, which was nearly fixed on alternative alleles in multiple population pairs of two of these species (*Q. rubra* and *Q. ellipsoidalis*; [7]). This genic microsatellite was originally developed for *Quercus robur* L. and the trinucleotide microsatellite encodes a poly-Q-repeat (poly-glutamine repeat) [18]. *Quercus ellipsoidalis* is characterized by the lack of one repeat unit (138 base pair (bp) allele) as compared to *Q. rubra* (141 bp allele, [7]). The locus has a putative function as a *COL* gene, which is thought to be involved in flowering time and growth [7], both of which can be impacted by water availability [19]. In the European white oak species *Quercus petraea* (Matt.) Liebl., a single nucleotide polymorphism (SNP) located in the same *COL* gene, was identified as significantly associated with bud burst along an altitudinal gradient [20]. Latitudinal and altitudinal gradients, as well as local environmental conditions such as water availability, have been shown to impact bud burst timing in other oak species [21,22]. *Quercus rubra* and *Q. ellipsoidalis* seedlings grown in a common garden exhibited differences in bud burst timing and leaf fall over two consecutive years [23] although this did not hold in natural populations of the same provenance [24]. Thus, differences in flowering time could still contribute to limit gene flow between these two species [24,25] and highly divergent markers, such as FIR013, located in the first exon of a *COL* gene, might be involved in their adaptive divergence and/or partial reproductive isolation. *COL* genes have been linked to adaptive divergence in other species, including the European white oak *Q. petraea* and *Populus tremula* L. [20,26,27]. For example, in *P. tremula*, an allele of the coding poly-E repeat (poly-glutamic acid) microsatellite in *COL2B*, was associated with growth cessation across a latitudinal gradient [26]. Interestingly, *Q. ellipsoidalis* exhibited later leaf fall than *Q. rubra* in a common garden, indicating genetic differences between species in leaf fall [22], a trait which is positively associated with growth cessation [28]. The biological function of single amino acid repeats (e.g., poly-E and poly-Q) has been studied mostly in animal species, with variation in these repeats often associated with genetic disorders [29,30]. However, variation in poly-glutamine repeats, such as the poly-Q repeat found in the *COL* gene, have been shown to be under selection in various species including fish, birds, plants and fungi, making this particular marker an excellent candidate gene, potentially contributing to reproductive isolation and adaptive divergence between the two oak species [31].

In this study, we asked two questions: (1) is the microsatellite size variation due to poly-Q repeat variation in these four red oak species; and (2) is there introgression of potentially adaptive alleles between drought averse *Q. rubra* and drought tolerant *Q. ellipsoidalis*, *Q. velutina* and *Q. coccinea*?

To answer the first question, we have sequenced part of the *COL* gene including the poly-Q repeat in a total of 46 samples representing each species and allele at FIR013. In reference to the second question, we assessed genetic variation in 16 populations (Table S1, Figure 1) at 12 microsatellite markers (including FIR013) and examined relative allelic and genotypic frequencies in each genetically assigned species at FIR013.

2. Materials and Methods

DNA samples were obtained from 16 populations of four red oak species (*Q. rubra*, *Q. ellipsoidalis*, *Q. velutina*, and *Q. coccinea*) from the Great Lakes region (Wisconsin, Michigan, Illinois and Indiana; see Table S1 and Figure 1) [8,10,25]. The samples consisted of eight population pairs of *Q. rubra* and *Q. ellipsoidalis* (parapatric populations), six mixed stands of *Q. ellipsoidalis* and *Q. velutina* in proximity to unsampled *Q. rubra* populations (parapatric populations) and one mixed stand of *Q. ellipsoidalis*, *Q. velutina*, and *Q. rubra* (sympatric population). Finally, one *Q. coccinea* population within the distribution range of *Q. rubra* and *Q. velutina* was also included. Leaf material was stored at −20 °C until DNA extraction using the DNeasy96 Plant Kit (Qiagen, Valencia, CA, USA) following the manufacturer's instructions.

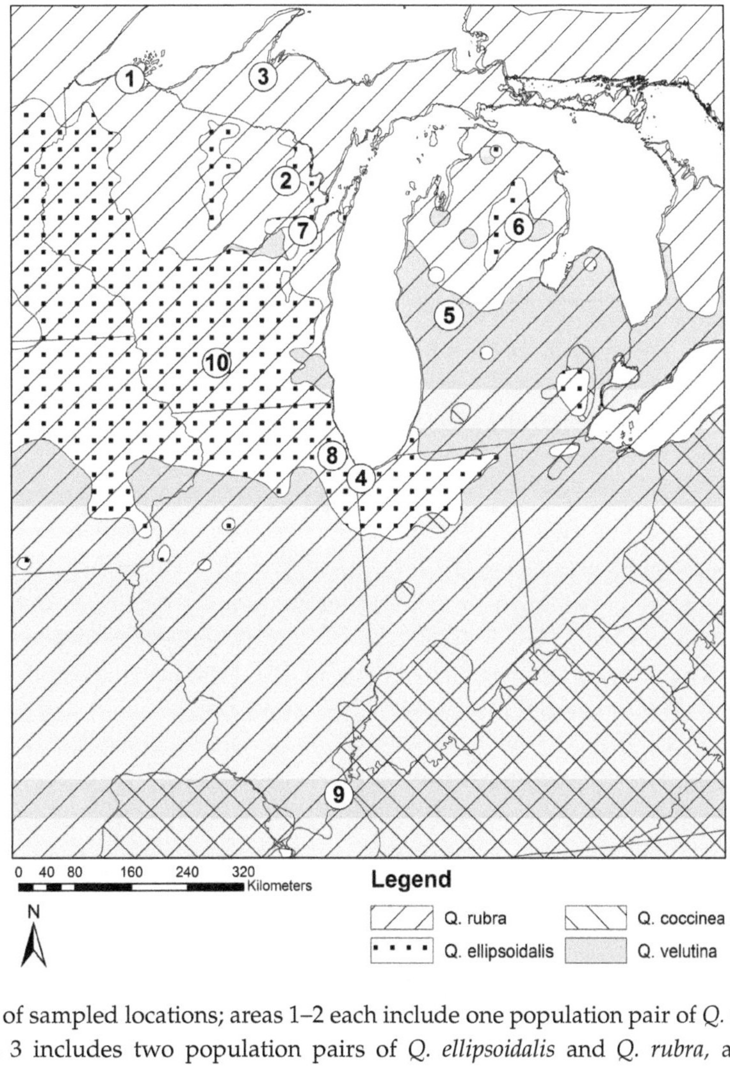

Figure 1. Map of sampled locations; areas 1–2 each include one population pair of *Q. ellipsoidalis* and *Q. rubra*, area 3 includes two population pairs of *Q. ellipsoidalis* and *Q. rubra*, area 4 includes two mixed *Q. ellipsoidalis* and *Q. velutina* populations, areas 5–8 each include one mixed *Q. ellipsoidalis* and *Q. velutina* population each, area 9 includes the single *Q. coccinea* population, and area 10 includes the single sympatric population of *Q. rubra*/*Q. ellipsoidalis*/*Q. velutina*; area number associated with individual populations is shown in Table S1; approximate species ranges are from Little (1971).

Using the primer pair for microsatellite FIR013 (Table S2), we sequenced the microsatellite and flanking regions in 46 samples (all homozygous for either the 138 bp, 141 bp or 144 bp allele at FIR013) from sampled populations of the four red oak species: *Q. ellipsoidalis* (11), *Q. rubra* (9), *Q. velutina* (18), and *Q. coccinea* (8) (see Table S3). Sequencing reactions were run on an ABI3730 DNA Analyzer using the ABI BigDye Terminator Cycle Sequencing Ready Reaction Kit v.3.1 (Applied Biosystems, Foster City, CA, USA) at the Nevada Genomics Center at the University of Nevada in Reno. Sequence data were aligned using the BioEdit Sequence Alignment Editor [32]. Alignment of forward and reverse sequences was completed by using pairwise alignment with the option to allow the two ends to slide. Multiple fragment alignments were completed using the Muscle algorithm (with default settings) in MEGA (Molecular Evolutionary Genetics Analysis) 6.06 [33] and final alignments were generated after careful visual inspection and manual re-editing. Sequence data were translated into amino acid sequences in MEGA using the standard genetic code option.

All individuals from the 16 sampled populations were previously genotyped at 11 microsatellite markers (six expressed sequence tag-simple sequence repeats (EST-SSRs) and five nuclear simple sequence repeats (nSSRs); Table S2). Genotypic data on EST-SSR FIR013 were available for the eight *Q. rubra* and *Q. ellipsoidalis* populations, and we genotyped the remaining seven populations (six sympatric *Q. ellipsoidalis*/*Q. velutina* populations and the sympatric *Q. rubra*/*Q. ellipsoidalis*/ *Q. velutina* population; Table S2) at FIR013 [8,10,25]. A single population of *Q. coccinea* was included for the sequencing effort and for the amplification of FIR013, where eight samples were sequenced and all 20 samples were genotyped at FIR013.

All genetic variation analyses were conducted both with and without the outlier FIR013. Genetic variation parameters were calculated in GeneAlEx 6.41 [34] including the number of alleles per locus (N_a), Nei's unbiased gene diversity (H_e), observed heterozygosity (H_o) and Wright's inbreeding coefficient (F) by population and species. Using only pure individuals from each species (based on previous genetic assignment analyses exluding outlier loci [8,10,25]), F_{ST} and pairwise F_{ST} between species with corresponding significances were calculated in GenePop 4.1 [35]. F_{ST}-based outlier screens were conducted in the program LOSITAN [36] between *Q. rubra* and *Q. ellipsoidalis*, *Q. rubra* and *Q. velutina*, and *Q. velutina* and *Q. ellipsoidalis* using the same settings as detailed in [7]. Specifically, we applied the stepwise mutation model with 50,000 simulations at the 99% confidence level and false discovery rate of 0.10.

Relative frequencies of introgressive forms and hybrids between drought averse *Q. rubra* and the more drought tolerant *Q. ellipsoidalis*, *Q. velutina*, and *Q. coccinea* were calculated based on results from two previous genetic assignment analyses [10,25]. Although slightly different marker sets were used in the genetic assignments, we have shown previously that very similar results are obtained using marker sets of 10, 16 or 44 markers [7,24]. The same method and criteria for genetic assignment analysis in the program STRUCTURE [37] were used as described in [25] and did not include the marker FIR013 or any other outlier marker. Pure species, introgressive forms and hybrids were classified as having >0.90, 0.61–0.89, and 0.4–0.6 proportion of their ancestry in one genetic cluster, respectively. Introgression of the species characteristic FIR013 alleles 138 bp (*Q. ellipsoidalis*) and 141 bp (*Q. rubra*) [7] between species with different drought tolerance was determined by relative frequencies of the 141 bp allele in genetically assigned *Q. ellipsoidalis*, *Q. velutina* and *Q. coccinea* individuals and as the relative frequency of the 138 bp allele in genetically assigned *Q. rubra* individuals.

3. Results

3.1. Nucleotide Sequence Analysis

Sequence variation confirmed that allele size differences were due to variation in a poly-Q repeat (Figure 2). *Quercus velutina* and *Q. coccinea* share the same common allele (138 bp) with *Q. ellipsoidalis* (Figure 3) which is reflected in the sequence variation between these three species and *Q. rubra* (Figure 2). In addition, there was a non-synonymous single nucleotide polymorphism (SNP) within the nucleotide repeat region present in all four species at position 9 (Figure 2). This SNP displayed different allelic patterns in *Q. rubra* as compared to the other three more drought tolerant species, with almost all *Q. rubra* individuals homozygous for the nucleotide "C" (88.9%), while heterozygotes (C/G) are only present in *Q. velutina* (44.4%), *Q. ellipsoidalis* (18.2%) and *Q. coccinea* (50.0%). The frequency of homozygotes (G/G) was similar (between 10%–20%) for all but *Q. ellipsoidalis* (45.5%) (Table S4). Sample sizes were small for all four species and may bias these results.

Sample	Population	Species	Allele	1	2	3	4	5	6	7	8	9	10	11	12	13	14	15	16	17	18	19	20	21	22	23	24	25	26	27	28	29	30	31	32	33	34	35	36
			Protein Sequence	Y			S			H/Q/Y			H/Q			Q			H/Q			Q			Q			Q			Q			Q			H		
2583	SNF-IL-GC	*Q. coccinea*	138	T	A	T	A	G	C	C	A	S	-	-	-	-	-	-	C	A	G	C	A	G	C	A	G	C	A	G	C	A	G	C	A	A	C	A	T
2584	SNF-IL-GC	*Q. coccinea*	138	T	A	T	A	G	C	C	A	C	-	-	-	-	-	-	C	A	G	C	A	G	C	A	G	C	A	G	C	A	G	C	A	A	C	A	T
2585	SNF-IL-GC	*Q. coccinea*	138	T	A	T	A	G	C	C	A	S	-	-	-	-	-	-	C	A	G	C	A	G	C	A	G	C	A	G	C	A	G	C	A	A	C	A	T
2587	SNF-IL-GC	*Q. coccinea*	138	T	A	T	A	G	C	C	A	S	-	-	-	-	-	-	C	A	G	C	A	G	C	A	G	C	A	G	C	A	G	C	A	A	C	A	T
2588	SNF-IL-GC	*Q. coccinea*	138	T	A	T	A	G	C	C	A	G	-	-	-	-	-	-	C	A	G	C	A	G	C	A	G	C	A	G	C	A	G	C	A	A	C	A	T
2592	SNF-IL-GC	*Q. coccinea*	138	T	A	T	A	G	C	C	A	S	-	-	-	-	-	-	C	A	G	C	A	G	C	A	G	C	A	G	C	A	G	C	A	A	C	A	T
2593	SNF-IL-GC	*Q. coccinea*	138	T	A	T	A	G	C	C	A	C	-	-	-	-	-	-	C	A	G	C	A	G	C	A	G	C	A	G	C	A	G	C	A	A	C	A	T
2594	SNF-IL-GC	*Q. coccinea*	138	T	A	T	A	G	C	C	A	C	-	-	-	-	-	-	C	A	G	C	A	G	C	A	G	C	A	G	C	A	G	C	A	A	C	A	T
F052	FC-E	*Q. ellipsoidalis*	138	T	A	T	A	G	C	C	A	G	-	-	-	-	-	-	C	A	G	C	A	G	C	A	G	C	A	G	C	A	G	C	A	A	C	A	T
F002	FC-C	*Q. ellipsoidalis*	138	T	A	T	A	G	C	C	A	G	-	-	-	-	-	-	C	A	G	C	A	G	C	A	G	C	A	G	C	A	G	C	A	A	C	A	T
H002	PV-DC	*Q. ellipsoidalis*	138	T	A	T	A	G	C	C	A	S	-	-	-	-	-	-	C	A	G	C	A	G	C	A	G	C	A	G	C	A	G	C	A	A	C	A	T
H003	PV-DC	*Q. ellipsoidalis*	138	T	A	T	A	G	C	C	A	C	-	-	-	-	-	-	C	A	G	C	A	G	C	A	G	C	A	G	C	A	G	C	A	A	C	A	T
CX35	CNF	*Q. ellipsoidalis*	138	T	A	T	A	G	C	C	A	S	-	-	-	-	-	-	C	A	G	C	A	G	C	A	G	C	A	G	C	A	G	C	A	A	C	A	T
NX20	NNF	*Q. ellipsoidalis*	138	T	A	T	A	G	C	C	A	G	-	-	-	-	-	-	C	A	G	C	A	G	C	A	G	C	A	G	C	A	G	C	A	A	C	A	T
3090	MI-NC	*Q. ellipsoidalis*	138	T	A	T	A	G	C	C	A	G	-	-	-	-	-	-	C	A	G	C	A	G	C	A	G	C	A	G	C	A	G	C	A	A	C	A	T
2898	MI-NC	*Q. ellipsoidalis*	138	T	A	T	A	G	C	C	A	G	-	-	-	-	-	-	C	A	G	C	A	G	C	A	G	C	A	G	C	A	G	C	A	A	C	A	T
28	FC-B	*Q. rubra*	138	T	A	T	A	G	C	C	A	G	-	-	-	-	-	-	C	A	G	C	A	G	C	A	G	C	A	G	C	A	G	C	A	A	C	A	T
2897	MI-NC	*Q. velutina*	138	T	A	T	A	G	C	C	A	S	-	-	-	-	-	-	C	A	G	C	A	G	C	A	G	C	A	G	C	A	G	C	A	A	C	A	T
2909	MI-NC	*Q. velutina*	138	T	A	T	A	G	C	C	A	C	-	-	-	-	-	-	C	A	G	C	A	G	C	A	G	C	A	G	C	A	G	C	A	A	C	A	T
2910	MI-NC	*Q. velutina*	138	T	A	T	A	G	C	C	A	C	-	-	-	-	-	-	C	A	G	C	A	G	C	A	G	C	A	G	C	A	G	C	A	A	C	A	T
2911	MI-NC	*Q. velutina*	138	T	A	T	A	G	C	C	A	C	-	-	-	-	-	-	C	A	G	C	A	G	C	A	G	C	A	G	C	A	G	C	A	A	C	A	T
2915	MI-NC	*Q. velutina*	138	T	A	T	A	G	C	C	A	S	-	-	-	-	-	-	C	A	G	C	A	G	C	A	G	C	A	G	C	A	G	C	A	A	C	A	T
2916	MI-NC	*Q. velutina*	138	T	A	T	A	G	C	C	A	C	-	-	-	-	-	-	C	A	G	C	A	G	C	A	G	C	A	G	C	A	G	C	A	A	C	A	T
2917	MI-NC	*Q. velutina*	138	T	A	T	A	G	C	C	A	S	-	-	-	-	-	-	C	A	G	C	A	G	C	A	G	C	A	G	C	A	G	C	A	A	C	A	T
3052	LS-IN-LC	*Q. velutina*	138	T	A	T	A	G	C	C	A	S	-	-	-	-	-	-	C	A	S	C	A	G	C	A	G	C	A	G	C	A	G	C	A	A	C	A	T
3067	LS-IN-LC	*Q. velutina*	138	T	A	T	A	G	C	C	A	C	-	-	-	-	-	-	C	A	G	C	A	G	C	A	G	C	A	G	C	A	G	C	A	A	C	A	T
3086	HPSNP-IN-LC	*Q. velutina*	138	T	A	T	A	G	C	C	A	S	-	-	-	-	-	-	C	A	G	C	A	G	C	A	G	C	A	G	C	A	G	C	A	A	C	A	T
B004	PV-DC	*Q. velutina*	138	T	A	T	A	G	C	C	A	G	-	-	-	-	-	-	C	A	G	C	A	G	C	A	G	C	A	G	C	A	G	C	A	A	C	A	T
B005	PV-DC	*Q. velutina*	138	T	A	T	A	G	C	C	A	S	-	-	-	-	-	-	C	A	G	C	A	G	C	A	G	C	A	G	C	A	G	C	A	A	C	A	T
B022	PV-DC	*Q. velutina*	138	T	A	T	A	G	C	C	A	S	-	-	-	-	-	-	C	A	G	C	A	G	C	A	G	C	A	G	C	A	G	C	A	A	C	A	T
B023	PV-DC	*Q. velutina*	138	T	A	T	A	G	C	C	A	G	-	-	-	-	-	-	C	A	G	C	A	G	C	A	G	C	A	G	C	A	G	C	A	A	C	A	T
3049	LS-IN-LC	*Q. velutina*	138	T	A	T	A	G	C	C	A	C	-	-	-	-	-	-	C	A	G	C	A	G	C	A	G	C	A	G	C	A	G	C	A	A	C	A	T
JS720	MI-OGC	*Q. velutina*	138	T	A	T	A	G	C	C	A	S	-	-	-	-	-	-	C	A	G	C	A	G	C	A	G	C	A	G	C	A	G	C	A	A	C	A	T
2543	MI-NC	*Q. velutina*	138	T	A	T	A	G	C	C	A	G	-	-	-	-	-	-	C	A	G	C	A	G	C	A	G	C	A	G	C	A	G	C	A	A	C	A	T
F010	FC-C	*Q. ellipsoidalis*	141	T	A	T	A	G	C	C	A	C	C	A	G	-	-	-	C	A	G	C	A	G	C	A	G	C	A	G	C	A	G	C	A	A	C	A	T
F001	FC-C	*Q. ellipsoidalis*	141	T	A	T	A	G	C	C	A	C	C	A	G	-	-	-	C	A	G	C	A	G	C	A	G	C	A	G	C	A	G	C	A	A	C	A	T
F124	FC-A	*Q. rubra*	141	T	A	T	A	G	C	C	A	C	C	A	G	-	-	-	C	A	G	C	A	G	C	A	G	C	A	G	C	A	G	C	A	A	C	A	T
R002	PV-DC	*Q. rubra*	141	T	A	T	A	G	C	C	A	C	C	A	S	-	-	-	C	A	G	C	A	G	C	A	G	C	A	G	C	A	G	C	A	A	C	A	T
R024	PV-DC	*Q. rubra*	141	T	A	T	A	G	C	C	A	C	C	A	G	-	-	-	C	A	G	C	A	G	C	A	G	C	A	G	C	A	G	C	A	A	C	A	T
F149	FC-B	*Q. rubra*	141	T	A	T	A	G	C	C	A	C	C	A	G	-	-	-	C	A	G	C	A	G	C	A	G	C	A	G	C	A	G	C	A	A	C	A	T
NM22	NNF	*Q. rubra*	141	T	A	T	A	G	C	Y	A	C	C	A	G	-	-	-	C	A	G	C	A	G	C	A	G	C	A	G	C	A	G	C	A	A	C	A	T
CM40	CNF	*Q. rubra*	141	T	A	T	A	G	C	C	A	C	C	A	G	-	-	-	C	A	G	C	A	G	C	A	G	C	A	G	C	A	G	C	A	A	C	A	T
115	AL-MC	*Q. rubra*	141	T	A	T	A	G	C	C	A	C	C	A	G	-	-	-	C	A	G	C	A	G	C	A	G	C	A	G	C	A	G	C	A	A	C	A	T
116	TN-HC	*Q. rubra*	141	T	A	T	A	G	C	C	A	C	C	A	G	-	-	-	C	A	G	C	A	G	C	A	G	C	A	G	C	A	G	C	A	A	C	A	T
3048	LS-IN-LC	*Q. velutina*	141	T	A	T	A	G	C	C	A	C	C	A	G	-	-	-	C	A	G	C	A	G	C	A	G	C	A	G	C	A	G	C	A	A	C	A	T
CX23	CNF	*Q. ellipsoidalis*	144	T	A	T	A	G	C	C	A	C	C	A	G	C	A	G	C	A	G	C	A	G	C	A	G	C	A	G	C	A	G	C	A	A	C	A	T

Figure 2. Sequence variation at the poly-Q repeat in the *COL* gene in four red oak species (S = G or C; Y = C or T); codons alternately highlighted in medium grey and white.

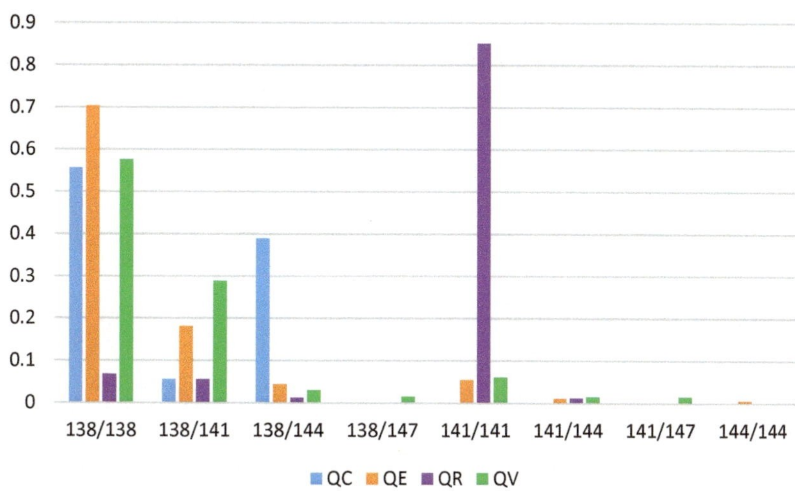

Figure 3. Relative genotypic frequencies in four oak species at the FIR013 locus.

3.2. Genetic Diversity and Differentiation Analysis

Levels of genetic diversity were similar across populations and species (Table S6). *Quercus rubra* showed significant, but low differentiation from the other three species at the 11 non-outlier microsatellites (3% to 4.3%) (Table 1), and very high and significant differentiation at the FIR013 locus (62.5% to 67.2%, Table 1). Genetic diversity and differentiation at all 12 microsatellite markers showed similar patterns (Tables S5 and S7). Differentiation at FIR013 between the drought tolerant species was low at 0.9% (Table 1). All outlier screens between *Q. rubra* and other more drought tolerant species identified FIR013 as an outlier, consistent with the allelic patterns observed for all four species (Table S8, Figures S2–S4).

Table 1. Pairwise F_{ST} values by species across 11 non-outlier microsatellite markers (upper triangle) and at FIR013 (lower triangle) using GeneAlEx [34].

	Q. ellipsoidalis	*Q. rubra*	*Q. velutina*
Q. ellipsoidalis	-	0.043 ***	0.030 ***
Q. rubra	0.672 ***	-	0.032 ***
Q. velutina	0.009 *	0.625 ***	-

* significant at α = 0.05; ** significant at α = 0.01; *** significant at α = 0.001.

In the only *Q. coccinea* population, the 141 bp (*Q. rubra*) allele was rare (3%) and no homozygotes for this allele were found (Table 2; Figure 3). The frequency of the 141 bp allele in *Q. velutina* and *Q. ellipsoidalis* was higher (22% and 15%, respectively) and homozygotes for the 141 bp allele were observed in both species (6% in *Q. velutina*, 5% in *Q. ellipsoidalis*) (Figure 3).

For sympatric *Q. velutina*/*Q. ellipsoidalis* populations that were in close proximity to *Q. rubra* populations, introgression of the 141 bp (*Q. rubra*) allele was considerably higher into *Q. velutina* than into *Q. ellipsoidalis* for three out of the five populations (Figure 4; Table 2: populations MI-NC, MI-OGC, WRPNP-IL-CC). The level of introgression of the 141 bp allele was similar for *Q. velutina* and *Q. ellipsoidalis* in population LS-IN-LC and higher in *Q. ellipsoidalis* than in *Q. velutina* in population HPNSP-IN-LC (Figure 4; Table 2).

Previous genetic assignment analyses [10,25] displayed similar levels of hybrids and introgressive forms between parapatric *Q. rubra* and *Q. ellipsoidalis* populations (p-value = 0.84), with *Q. rubra* populations exhibiting introgression rates (percentage of hybrids and introgressive forms) of 0%–23% ($\overline{x}$ = 14%; M = 17%) and *Q. ellipsoidalis* populations exhibiting introgression rates of 8%–23% ($\overline{x}$ = 16%; M = 16%) (Table 3). While the percentage of hybrids and introgressive forms was very similar in neighboring (parapatric) *Q. rubra* and *Q. ellipsoidalis* populations (last column, Table 3), introgression of the 141 bp allele into *Q. ellipsoidalis* was consistently higher than introgression of the 138 bp allele into *Q. rubra* (populations C-QE/C-QR: 29%/18%, populations N-QE/NQR: 24%/2%, populations FC-QE/FC-QR: 8%–18%/5%–6%) (Table 2). By contrast, in the sympatric population, the introgression of the 141 bp allele into *Q. ellipsoidalis* was lower (5%) than introgression of the 138 bp into *Q. rubra* (19%). Overall, the highest level of introgression of outlier alleles between species was found in C-QR (18%) and C-QE (29%) (Table 2).

Table 2. Relative frequency of FIR013 species-specific alleles in pure *Q. ellipsoidalis* (QE), *Q. rubra* (QR), *Q. velutina* (QV) and *Q. coccinea* (QC) individuals from 15 parapatric populations and one sympatric population.

Population ***	Species	Population Type	Sample Size (N)	Allele 141 *	Allele 138 *
C-QE	QE	parapatric	31	0.29	0.65
C-QR	QR	parapatric	31	0.82	0.18
N-QE	QE	parapatric	31	0.24	0.73
N-QR	QR	parapatric	32	0.98	0.02
FC-A	QR	parapatric	36	0.94	0.06
FC-C	QE	parapatric	40	0.18	0.82

Table 2. *Cont.*

Population ***	Species	Population Type	Sample Size (*N*)	Allele 141 *	Allele 138 *
FC-B	QR	parapatric	36	0.95	0.05
FC-E	QE	parapatric	37	0.08	0.91
HPNSP-IN-LC	QE QV	parapatric	15 8	0.13 0.06	0.77 0.94
LS-IN-LC	QE QV	parapatric	4 4	0.25 0.25	0.63 0.75
MI-NC	QE QV	parapatric	8 17	0 0.24	0.93 0.74
MI-OGC	QE QV	parapatric	6 7	0.08 0.14	0.92 0.79
WI-BC	QE	parapatric	8	0.13	0.88
WRPNP-IL-CC **	QE QV	parapatric	7 8	0.07 0.31	0.93 0.69
	QR		1	0	1.00
SNF-IL-GC	QC	parapatric	20	0.03	0.78
PV-DC	QE QV	sympatric	11 17	0.05 0.21	0.82 0.65
	QR		16	0.63	0.19

* Other minor alleles are not considered here; see Figure 3 and Figure S3; ** All trees in this population were morphologically identified as QE or QV. Genetic assignment analysis identified one individual as QR, ***Full names of populations listed in Table S1.

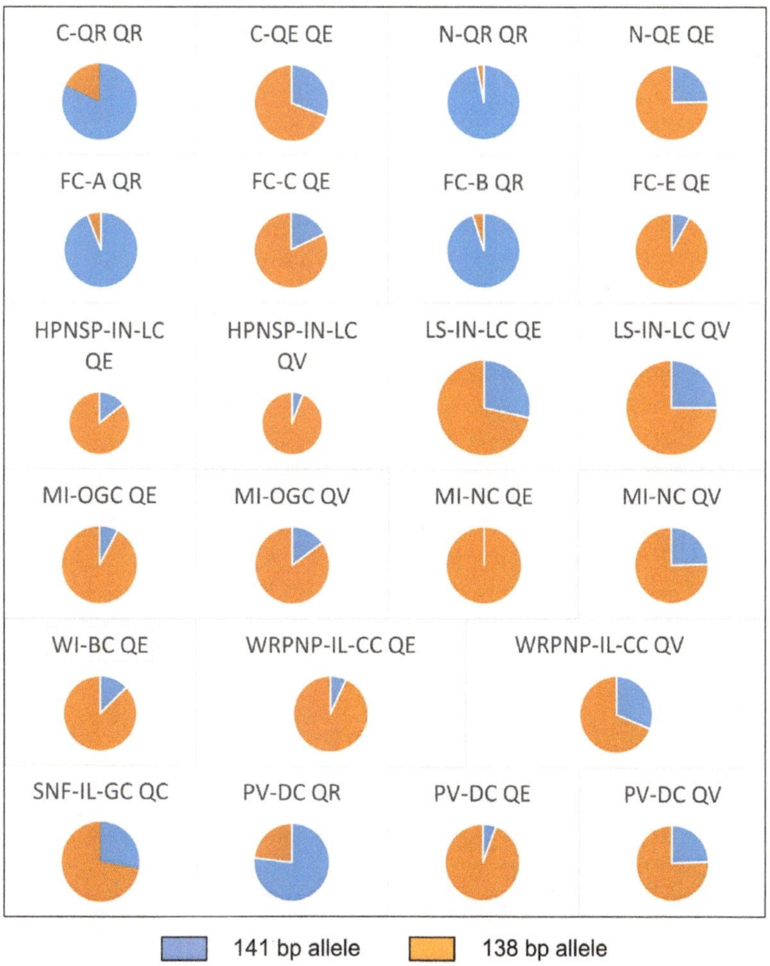

Figure 4. FIR013 allele frequencies in each population separated by species. Orange represents the 138 bp allele and blue represents the 141 bp allele. Full names of populations listed in Table S1.

Table 3. Relative frequency of hybrids and introgressive forms *.

	Q. rubra with Introgression from *Q. ellipsoidalis* or *Q. velutina*	*Q. ellipsoidalis*/ *Q. velutina* with Introgression from *Q. rubra*	F_1 Hybrids of *Q. rubra* with *Q. ellipsoidalis* or *Q. velutina*	Total Introgressive Forms and F_1 Hybrids between *Q. rubra* and *Q. ellipsoidalis* or *Q. velutina*
Parapatric populations of *Q. ellipsoidalis* and *Q. rubra***				
C-QE	-	0.20	0.03	0.23
C-QR	0.23	-	-	0.23
N-QE	-	0.18	0.03	0.21
N-QR	0.20	-	-	0.20
FC-A	0.13	-	-	0.13
FC-C	-	0.10	0	0.10
FC-B	0.00	-	-	0.00
FC-E	-	0.08	0	0.08
Sympatric populations of *Q. ellipsoidalis* and *Q. velutina* in parapatry with *Q. rubra***				
HPNSP-IN-LC	-	0.04	0.00	0.04
LS-IN-LC	-	0.00	0.00	0
MI-NC	-	0.03	0.00	0.03
MI-OGC	-	0.18	0.03	0.21
WI-BC	-	0.04	0.04	0.08
WRPNP-IL-CC	-	0.00	0.00	0.00
Sympatric populations of *Q. ellipsoidalis, Q. velutina* and *Q. rubra***				
PV-DC	0.24	0.11	0.10	0.45
Q. coccinea* population*				
SNF-IL-GC	-	0.00	0.00	0.00

* Based on previously published assignment analyses [10,25]. ** Full names of populations listed in Table S1.

4. Discussion

Quercus rubra is the only species nearly fixed on the 141 bp allele at outlier locus FIR013. The other three species are largely fixed on the 138 bp allele, and this difference is reflected in the sequence variation of the microsatellite itself. Previous work on these four species has also indicated that *Q. rubra* is the most diverged of the four species and that there is contemporary gene flow occurring between all four of these species [8,10]. The lack of one glutamine residue in the poly-Q repeat is a shared character for the three more drought tolerant species, with interspecific gene flow likely responsible for incomplete differentiation. The phylogeny of section Lobatae is in progress [13] and can be used to evaluate whether the character is a synapomorphy for the three more drought tolerant species. The four species represent a drought tolerance gradient with *Q. ellipsoidalis* being the most drought tolerant, followed by *Q. coccinea* and then *Q. velutina*, while *Q. rubra* is considered the most mesophilic red oak species [16,38]. Despite morphological and ecophysiological differences [8,16,38,39] between the four oak species, genetic differentiation between them is low at most nuclear genetic markers [7,8,10,25] as well as at chloroplast markers [14]. This is consistent with the tendency for closely related oak species to frequently hybridize with one other [40]. Thus, genomic regions that display high levels of genetic differentiation are of great interest to elucidate how oaks maintain their species identity despite interspecific gene flow.

Poly-Q repeat function and variation have been studied extensively in regards to human disease, but there is scant information about their function in plant and animal species [29–31]. Thus, further investigation of their functional role and variation is warranted. In addition to the repeat number variation, a non-synonymous SNP within the poly-Q repeat (Figure 2) was found in all four species (C/G) and there was a bias towards both the heterozygotes (C/G) and homozygotes (G/G) in the three more drought tolerant species. The observed SNP change from C to G results in a change of the protein sequence from a glutamine (uncharged polar) to a histidine (positive polar). These sequence differences between species with different drought tolerance might be related to functional differences. Association studies in full-sib families derived from controlled intra- and interspecific crosses, and gene expression analyses could be used to associate sequence variation in the *COL* gene with phenotypic variation.

Genetic markers that exhibit extreme differentiation between species as signatures of divergent selection can be used as diagnostic markers to assess introgression of potentially adaptive alleles between interfertile species in parapatric and sympatric populations. FIR013 provides such a marker and our examination of the relative allelic frequencies at the FIR013 locus in parapatric *Q. rubra*/*Q. ellipsoidalis* populations has shown that introgression of the 141 bp allele is consistently higher into the drought tolerant *Q. ellipsoidalis* than introgression of the 138 bp allele into the mesophilic species, *Q. rubra* (Table 2, Figure 4; [7]). This pattern of asymmetric introgression of outlier alleles suggests different strength of selection against the 138 bp and 141 bp alleles in non-parental environments if interspecific gene flow is symmetric [41–43] as observed in the present study between *Q. rubra* and *Q. ellipsoidalis*. In contrast, in the only sympatric population, introgression of the 141 bp into *Q. ellipsoidalis* is much lower (5%) than introgression of the 138 bp into *Q. rubra* (19%), suggesting that competition between species might affect introgression at FIR013 outlier alleles. Also, the extent of interspecific gene flow between *Q. rubra*/*Q. ellipsoidalis* population pairs was not related to the introgression of alleles 138 bp and 141 bp, again suggesting that introgression of these alleles was also affected by environment-dependent selection. For example, the introgression of allele 138 bp into *Q. rubra* was 2% for N-QR and 18% for C-QR, while interspecific gene flow was very similar (Table 3). In future studies, a systematic assessment of introgression rates and soil and climate variables is needed in both sympatric and parapatric *Q. rubra*/*Q. ellipsoidalis* populations to analyze the effect of interspecific competition on introgression rates at outlier alleles and to test for an association between introgression rates at outlier alleles and environmental variables (e.g., climate, soil, habitat characteristics).

In a seedling common garden experiment, *Q. ellipsoidalis* had a later bud burst and leaf fall, a much slower growth rate and higher mortality than *Q. rubra* over two growing seasons [23]. A slower growth strategy may be an adaptation to a xeric environment [44] and *Q. ellipsoidalis* is generally restricted to dry, sandy pine barrens [7,25] and is considered to be the most drought tolerant red oak species [16,45]. In most mixed *Q. ellipsoidalis*/*Q. velutina* populations and the sympatric population (PV-DC), introgression of the 141 bp allele was higher into *Q. velutina* than into *Q. ellipsoidalis*. *Quercus velutina* is less water efficient than *Q. ellipsoidalis* and generally found in savannas [16,46,47], while *Q. ellipsoidalis* is generally found on very dry sandy sites, possibly creating differences in introgression patterns between the three species in their sympatric range.

Introgression of adaptive alleles results in adaptive trait transfer between species and may be crucial for the adaptation to rapidly changing environmental conditions (e.g., [2]). In addition to heterozygous genotypes (138 bp/141 bp) at outlier locus FIR013, we found rare homozygotes for the 141 bp (*Q. rubra*) allele in both *Q. velutina* and *Q. ellipsoidalis* as well as homozygotes for the 138 bp allele in *Q. rubra* (see Figure 3). Common garden studies with different drought treatments could reveal whether introgression of outlier alleles is associated with differences in drought tolerance. For example, assessments of phenotypic traits, such as growth and water use efficiency for different genotypes at FIR013 (138 bp/138 bp, 138 bp/141 bp, 141 bp/141 bp) in each species, could allow quantification of relative fitness related to variation in the poly-Q repeat of *COL*.

In conclusion, we have identified a genic microsatellite marker with repeat number variation resulting in two major alleles differentiating between the drought averse *Q. rubra* and the drought tolerant oak species *Q. ellipsoidalis*, *Q. velutina* and *Q. coccinea*. This candidate gene may be involved in adaptive differences between the species. Common garden experiments of seedlings with all possible genotype combinations at FIR013 in each species can elucidate the allelic and genotypic effects at FIR013 on survival and other fitness related traits and thus the relation between introgression of FIR013 alleles between species and adaptive trait introgression.

Supplementary Materials: Figure S1:
Relative genotypic frequencies in pure species of *Q. rubra*, *Q. ellipsoidalis* and *Q. velutina* in parapatric populations and one sympatric population of *Q. rubra*, *Q. ellipsoidalis* and *Q. velutina* (starred) at the FIR013 locus, Figure S2: Outlier loci analysis of 12 markers between *Q. rubra* and *Q. ellipsoidalis* as calculated in the program LOSITAN [7] with a 99% confidence interval; FIR013 was found to be an outlier even after the application of 10% false discovery rate (FDR), Figure S3: Outlier loci analysis of 12 markers between *Q. rubra* and *Q. velutina* as calculated in LOSITAN [7] with a 99% confidence interval; 3A05 was found not to be an outlier after the application of 10% FDR, Figure S4: Outlier loci analysis of 12 markers between *Q. velutina* and *Q. ellipsoidalis* as calculated in LOSITAN [7] with a 99% confidence interval; 3A05 was found not to be an outlier after the application of 10%

FDR, Table S1: Location of collection sites for the 16 *Quercus* spp. populations, Table S2: Microsatellite marker characteristics, Table S3: Characteristics of samples used in the sequencing effort, Table S4: Relative frequency of a single nucleotide polymorphism in the coding region of a *COL* gene (Position 9, see Figure 2), Table S5: Genetic variation parameters by population and species across the 12 microsatellite markers, Table S6: Genetic variation parameters by population and species across 11 non-outlier microsatellite markers, Table S7: Pairwise F_{ST} value by species across all 12 microsatellite markers using GeneAlEx [6], Table S8: F_{ST}-based outlier screening results across the 12 microsatellite markers using LOSITAN [7].

Acknowledgments: This work was made possible through funding from the Hanes Trust and the McIntire Stennis fund (USDA-NIFA # 1001471). Additional funding was provided by the Life Science and Technology Institute (LSTI) and Ecosystem Science Center (ESC) at Michigan Technological University. Samples from certain populations were provided by Andrew Hipp (PV-DC, WI-BC, MI-NC, MI-OGC, WRPNP-IL-CC, HPNSP-IN-LC, LS-IN-LC) and Alexis Sullivan (N-QE, N-QR, C-QE, C-QR). Published data from Sullivan et al. (2016) and Owusu et al. (2015) were utilized for genetic variation assessment and genetic assignment information.

Author Contributions: O.G. and J.F.L.-R. conceived and designed the experiments; J.F.L.-R. performed the experiments; O.G. and J.F.L.-R. analyzed the data and wrote the paper.

Conflicts of Interest: The authors declare no conflict of interest.

References

1. Arnold, M.L. *Evolution through Genetic Exchange*; Oxford University Press: Oxford, UK, 2006.
2. Arnold, M.L. *Divergence with Genetic Exchange*; Oxford University Press: Oxford, UK, 2016.
3. Scotti-Saintagne, C.; Mariette, S.; Porth, I.; Goicoechea, P.G.; Barreneche, T.; Bodénès, C.; Burg, K.; Kremer, A. Genome scanning for interspecific differentiation between two closely related oak species (*Quercus robur* L. and *Q. petraea* (Matt.) Liebl.). *Genetics* **2004**, *168*, 1615–1626. [CrossRef] [PubMed]
4. Gailing, O.; Curtu, A.L. Interspecific gene flow and maintenance of species integrity in oaks. *Ann. For. Res.* **2014**, *57*, 5–18. [CrossRef]
5. Moran, E.V.; Willis, J.; Clark, J.S. Genetic evidence for hybridization in red oaks (*Quercus* sect. *Lobatae*, Fagaceae). *Am. J. Bot.* **2012**, *99*, 92–100. [CrossRef] [PubMed]
6. Lepais, O.; Gerber, S. Reproductive patterns shape introgression dynamics and species sucession within the European white oak species complex. *Evolution* **2011**, *65*, 156–170. [CrossRef] [PubMed]
7. Lind-Riehl, J.F.; Sullivan, A.R.; Gailing, O. Evidence for selection on a *CONSTANS*-like gene between two red oak species. *Ann. Bot. Lond.* **2014**, *113*, 967–975. [CrossRef] [PubMed]
8. Sullivan, A.R.; Owusu, S.A.; Weber, J.A.; Hipp, A.L.; Gailing, O. Hybridization and divergence in multi-species oak (*Quercus*) communities. *Bot. J. Linn. Soc.* **2016**, *181*, 99–114. [CrossRef]
9. Hipp, A.L.; Weber, J.A. Taxonomy of Hill's oak (*Quercus ellipsoidalis*: Fagaceae): Evidence from AFLP data. *Syst. Bot.* **2008**, *33*, 148–158. [CrossRef]
10. Owusu, S.A.; Sullivan, A.R.; Weber, J.A.; Hipp, A.L.; Gailing, O. Taxonomic Relationships and Gene Flow in Four North American *Quercus* Species (*Quercus* section *Lobatae*). *Syst. Bot.* **2015**, *40*, 510–521. [CrossRef]
11. Hipp, A.L.; Weber, J.A.; Srivastava, A. Who am I this time? The affinities and misbehaviors of Hill's oak (*Quercus ellipsoidalis*). *Int. Oak J.* **2010**, *21*, 27–36.
12. Hipp, A.L. Hill's oak: The taxonomy and dynamics of a Western Great Lake endemic. *Arnoldia* **2010**, *67*, 2–14.
13. Hipp, A.L. What do we mean by oak species and the oak phylogeny? Oak species coherence and parallel ecological diversification in the red and white oaks. In Presented at the IUFRO Genomics and forest tree genetics, Arcachon, France, 30 May–3 June 2016.
14. Zhang, R.; Hipp, A.L.; Gailing, O. Sharing of chloroplast haplotypes among red oak species suggests interspecific gene flow between neighboring populations. *Botany* **2015**, *93*, 691–700. [CrossRef]
15. Sullivan, A.R. Speciation with Gene-Flow in eastern North American Red Oaks (*Quercus* section *Lobatae*). Master's Thesis, School of Forest Resources and Environmental Science, Houghton, MI, USA, 2013.
16. Abrams, M.D. Adaptations and responses to drought in *Quercus* species of North America. *Tree Physiol.* **1990**, *7*, 227–238. [CrossRef] [PubMed]
17. Thompson, R.S.; Anderson, K.H.; Bartlein, P.J. *Atlas of Relations between Climatic Parameters and Distributions of Important Trees and Shrubs in North America-Hardwoods*; U.S.G. Survey: Denver, CO, USA, 1999.

18. Durand, J.; Bodenes, C.; Chancerel, E.; Frigerio, J.-M.; Vendramin, G.; Sebastiani, F.; Buonamici, A.; Gailing, O.; Koelewijn, H.-P.; Villani, F.; et al. A fast and cost-effective approach to develop and map EST-SSR markers: Oak as a case study. *BMC Genom.* **2010**, *11*, 570. [CrossRef] [PubMed]
19. Schmalenbach, I.; Zhang, L.; Reymond, M.; Jimenez-Gomez, J.M. The relationship between flowering time and growth responses to drought in the Arabidopsis Landsberg *erecta* x Antwerp-1 population. *Front. Plant Sci.* **2014**, *5*, 609. [CrossRef] [PubMed]
20. Alberto, F.J.; Derory, J.; Boury, C.; Frigerio, J.M.; Zimmerman, N.E.; Kremer, A. Imprints of natural selection along environmental gradients in phenology-related genes of *Quercus petraea*. *Genetics* **2013**, *195*, 495–512. [CrossRef] [PubMed]
21. Cavender-Bares, J.; Pahlich, A. Molecular, morphological, and ecological niche differentiation of sympatric sister oak species, *Quercus virginiana* and *Q. geminata* (Fagaceae). *Am. J. Bot.* **2009**, *96*, 1690–1702. [CrossRef] [PubMed]
22. Ducousso, A.; Guyon, J.; Kremer, A. Latitudinal and altitudinal variation of bud burst in western populations of sessile oak (*Quercus petraea* (Matt) Liebl). *Ann. Sci. For.* **1996**, *53*, 775–782. [CrossRef]
23. Gailing, O. Difference in growth, survival and phenology in *Quercus rubra* and *Q. ellipsoidalis* seedlings. *Dendrobiology* **2013**, *70*, 71–79. [CrossRef]
24. Collins, E.; Sullivan, A.R.; Gailing, O. Limited effective gene flow between two interfertile red oak species. *Trees* **2015**, *29*, 1135–1148. [CrossRef]
25. Lind, J.F.; Gailing, O. Genetic structure of *Quercus rubra* L. and *Quercus ellipsoidalis* E.J. Hill populations at gene-based EST-SSR and nuclear SSR markers. *Tree Genet. Genomes* **2013**, *9*, 707–722. [CrossRef]
26. Ma, X.-F.; Hall, D.; Onge, K.R.S.; Jansson, S.; Ingvarsson, P.K. Genetic differentiation, clinal variation and phenotypic associations with growth cessation across the *Populus tremula* photoperiodic pathway. *Genetics* **2010**, *186*, 1033–1044. [CrossRef] [PubMed]
27. O'Malley, K.G.; Banks, M.A. A latitudinal cline in the Chinook salmon (*Oncorhynchus tshawytscha*) *Clock* gene: Evidence for selection on PolyQ length variants. *Proc. R. Soc. B Biol. Sci.* **2008**, *275*, 2813–2821. [CrossRef] [PubMed]
28. Weih, M. Genetic and environmental variation in spring and autumn phenology of biomass willows (*Salix* spp.): Effects on shoot growth and nitrogen economy. *Tree Physiol.* **2009**, *29*, 1479–1490. [CrossRef] [PubMed]
29. Faux, N.G.; Bottomley, S.P.; Lesk, A.M.; Irving, J.A.; Morrison, J.R.; Garcia de la Banda, M.; Whisstock, J.C. Functional insights from the distribution and role of homopeptide repeat-containing proteins. *Genome Res.* **2005**, *15*, 537–551. [CrossRef] [PubMed]
30. Lobanov, M.Y.; Klus, P.; Sokolovsky, I.V.; Tartaglia, G.G.; Galzitskaya, O.V. Non-random distribution of homo-repeats: Links with biological functions and human disease. *Sci. Rep.* **2016**, *6*, 26941. [CrossRef] [PubMed]
31. Haerty, W.; Golding, G.B. Low-complexity sequences and single amino acid repeats: Not just "junk" peptide sequences. *Genome* **2010**, *53*, 753–762. [PubMed]
32. Hall, T.A. Bioedit: A user-friendly biological sequence alignment editor and analysis program for Windows 95/98/NT. *Nucleic Acids Symp. Ser.* **1999**, *41*, 95–98.
33. Tamura, K.; Stecher, G.; Peterson, D.; Filipski, A.; Kumar, S. MEGA6: Molecular Evolutionary Genetics Analysis Version 6.0. *Mol. Biol. Evol.* **2013**, *30*, 2725–2729. [CrossRef] [PubMed]
34. Peakall, R.O.D.; Smouse, P.E. GeneAlEx 6: Genetic analysis in Excel. Population genetic software for teaching and research. *Mol. Ecol. Notes* **2006**, *6*, 288–295. [CrossRef]
35. Raymond, M.; Rousset, F. GenePop (Version-1.2)—Population genetics software for exact tests and ecumenicism. *J. Hered.* **1995**, *86*, 248–249.
36. Antao, T.; Lopes, A.; Lopes, R.; Beja-Pereira, A.; Luikart, G. LOSITAN: A workbench to detect molecular adaptation based on a Fst-outlier method. *BMC Bioinform.* **2008**, *9*, 323. [CrossRef] [PubMed]
37. Pritchard, J.K.; Stephens, M.; Donnelly, P. Inference of population structure using multilocus genotype data. *Genetics* **2000**, *155*, 945–959. [PubMed]
38. Barton, A.M.; Gleeson, S.K. Ecophysiology of seedlings of oaks and red maple across a topographic gradient in eastern Kentucky. *For. Sci.* **1996**, *42*, 335–342.
39. Gailing, O.; Lind, J.F.; Lilleskov, E. Leaf morphological and genetic differentiation between *Quercus rubra* L. and *Q. ellipsoidalis* E.J. Hill populations in contrasting environments. *Plant Syst. Evol.* **2012**, *298*, 1533–1545. [CrossRef]

40. Van Valen, L. Ecological species, multispecies, and oaks. *Taxon* **1976**, *25*, 233–239.
41. Minder, A.M.; Widmer, A. A population genomic analysis of species boundaries: Neutral processes, adaptive divergence and introgression between two hybridizing plant species. *Mol. Ecol.* **2008**, *17*, 1552–1563. [CrossRef] [PubMed]
42. Evans, P.D.; Mekel-Bobrov, N.; Vallender, E.J.; Hudson, R.R.; Lahn, B.T. Evidence that the adaptive allele of the brain size gene microcephalin introgressed into *Homo sapiens* from an archaic *Homo* lineage. *Proc. Natl. Acad. Sci. USA* **2006**, *103*, 18178–18183. [CrossRef] [PubMed]
43. Whitney, K.; Randell, R.A.; Rieseberg, L.H. Adaptive introgression of herbivore resistance traits in the weedy sunflower *Helianthus annuus*. *Am. Nat.* **2006**, *167*, 794–807. [CrossRef] [PubMed]
44. Motzkin, G.; Ciccarello, S.C.; Foster, D.R. Frost pockets on a level sand plain: Does variation in microclimate help maintain persistent vegetation patterns? *J. Torrey Bot. Soc.* **2002**, *129*, 154–163. [CrossRef]
45. Abrams, M.D. Comparative water relations of three successional hardwood species in central Wisconsin. *Tree Physiol.* **1988**, *4*, 263–273. [CrossRef] [PubMed]
46. Abrams, M. Distribution, historical development and ecophysiological attributes of oak species in the eastern United States. *Ann. Sci. For.* **1996**, *53*, 487–512. [CrossRef]
47. Jensen, R.J. *Quercus* Linnaeus sect. *Lobatae* Loudon, the red or black oaks. In *Flora of North America North of Mexico*; Flora of North America Editoral Committee; Oxford University Press on Demand: Oxford, UK, 1997; Volume 3, pp. 447–468.

PERMISSIONS

All chapters in this book were first published in FORESTS, by MDPI; hereby published with permission under the Creative Commons Attribution License or equivalent. Every chapter published in this book has been scrutinized by our experts. Their significance has been extensively debated. The topics covered herein carry significant findings which will fuel the growth of the discipline. They may even be implemented as practical applications or may be referred to as a beginning point for another development.

The contributors of this book come from diverse backgrounds, making this book a truly international effort. This book will bring forth new frontiers with its revolutionizing research information and detailed analysis of the nascent developments around the world.

We would like to thank all the contributing authors for lending their expertise to make the book truly unique. They have played a crucial role in the development of this book. Without their invaluable contributions this book wouldn't have been possible. They have made vital efforts to compile up to date information on the varied aspects of this subject to make this book a valuable addition to the collection of many professionals and students.

This book was conceptualized with the vision of imparting up-to-date information and advanced data in this field. To ensure the same, a matchless editorial board was set up. Every individual on the board went through rigorous rounds of assessment to prove their worth. After which they invested a large part of their time researching and compiling the most relevant data for our readers.

The editorial board has been involved in producing this book since its inception. They have spent rigorous hours researching and exploring the diverse topics which have resulted in the successful publishing of this book. They have passed on their knowledge of decades through this book. To expedite this challenging task, the publisher supported the team at every step. A small team of assistant editors was also appointed to further simplify the editing procedure and attain best results for the readers.

Apart from the editorial board, the designing team has also invested a significant amount of their time in understanding the subject and creating the most relevant covers. They scrutinized every image to scout for the most suitable representation of the subject and create an appropriate cover for the book.

The publishing team has been an ardent support to the editorial, designing and production team. Their endless efforts to recruit the best for this project, has resulted in the accomplishment of this book. They are a veteran in the field of academics and their pool of knowledge is as vast as their experience in printing. Their expertise and guidance has proved useful at every step. Their uncompromising quality standards have made this book an exceptional effort. Their encouragement from time to time has been an inspiration for everyone.

The publisher and the editorial board hope that this book will prove to be a valuable piece of knowledge for researchers, students, practitioners and scholars across the globe.

LIST OF CONTRIBUTORS

Ladislav Dzurenda
Department of Woodworking, Faculty of Wood Sciences and Technology, Technical University in Zvolen,
T.G. Masaryka 24, 960 53 Zvolen, Slovakia

Emília Hroncová and Juraj Ladomerský
Department of Environmental Management, Faculty of Natural Sciences, Matej Bel University,
Tajovského 40, 974 01 Banská Bystrica, Slovakia

Carla Iglesias
Department of Natural Resources and Environ-mental Engineering, University of Vigo, Vigo 36310, Spain

António José Alves Santos and Helena Pereira
Centro de Estudos Florestais, Instituto Superior de Agronomia, Universidade de Lisboa, Tapada da Ajuda, Lisboa 1349-017, Portugal

Javier Martínez
Department of Mathematics and Statistics, Centro Universitario de la Defensa, Escuela Naval Militar, Marín 36920, Spain

Ofélia Anjos
Centro de Estudos Florestais, Instituto Superior de Agronomia, Universidade de Lisboa, Tapada da Ajuda, Lisboa 349-017, Portugal
Instituto Politécnico de Castelo Branco, Castelo Branco 6001-909, Portugal

Aurélie Terrier and Mathieu Paquette
Chaire Industrielle en Aménagement Forestier Durable (NSERC-UQAT-UQAM), Université du Québec à Montréal; P.O. Box 8888, Stn. Centre-ville, Montréal, QC H3C 3P8, Canada

Sylvie Gauthier and Martin P. Girardin
Natural Resources Canada, Canadian Forest Service, Laurentian Forestry Centre, 1055 du PEPS, P.O. Box 10380, Stn. Sainte-Foy, Québec, QC G1V 4C7, Canada

Sylvain Pelletier-Bergeron
Département des Sciences du Bois et de la Forêt, Université Laval, 2405 de la Terrasse, Québec, QC G1V 0A6, Canada

Yves Bergeron
Chaire Industrielle en Aménagement Forestier Durable (NSERC-UQAT-UQAM), Université du Québec à Montréal; P.O. Box 8888, Stn. Centre-ville, Montréal, QC H3C 3P8, Canada
Forest Research Institute, Université du Québec en Abitibi-Temiscamingue, 445 blvd de l'Université, Rouyn-Noranda, QC J9X 5E4, Canada

Thomas D. Lee
Department of Natural Resources and the Environment, University of New Hampshire, 56 College Road, Durham, NH 03824, USA

Stephen E. Eisenhaure
Office of Woodlands and Natural Areas, University of New Hampshire, Kingman Farm, Route 155, Madbury, NH 03823, USA

Ian P. Gaudreau
Department of Civil and Environmental Engineering, University of New Hampshire, 33 Academic Way, Durham, NH 03824, USA

Matthew J. Reilly
Department of Forest Ecosystems and Society, College of Forestry, Oregon State University, Corvallis, OR 97330, USA
USDA Forest Service, Southern Research Station, Athens, GA 30602, USA

Kenneth Outcalt, Joseph J. O'Brien and DaleWade
USDA Forest Service, Southern Research Station, Athens, GA 30602, USA

Craig A. Carr
Department of Animal and Range Sciences, Montana State University, PO Box 172900, Bozeman, MT 59717, USA

William C. Krueger
Department of Animal and Rangeland Sciences, Oregon State University, 112 Withycombe Hall, Corvallis, OR 97331, USA

Ville Luoma, Ninni Saarinen, Mikko Vastaranta and Markus Holopainen
Department of Forest Sciences, University of Helsinki, P.O.Box 27 (Latokartanonkaari 7), 00014 Helsinki, Finland
Centre of Excellence in Laser Scanning Research, Finnish Geospatial Research Institute FGI, National Land Survey, 04310 Masala, Finland

Michael A. Wulder and Joanne C. White
Canadian Forest Service (Pacific Forestry Centre), Natural Resources Canada, 506 West Burnside Road, Victoria, BC V8Z 1M5, Canada

Juha Hyyppä
Department of Remote Sensing and Photogrammetry, Finnish Geospatial Research Institute FGI, National Land Survey, Geodeetinrinne 2, 04310 Masala, Finland
Centre of Excellence in Laser Scanning Research, Finnish Geospatial Research Institute FGI, National Land Survey, 04310 Masala, Finland

Ed-Haun Chang
Mackay Junior College of Medicine, Nursing and Management, Beitou, Taipei 112, Taiwan

Guanglong Tian
Environmental Monitoring and Research Division, Monitoring and Research Department, Metropolitan Water Reclamation District of Greater Chicago, 6001 W. Pershing Road, Cicero, IL 60804, USA

Chih-Yu Chiu
Biodiversity Research Center, Academia Sinica, Nangang, Taipei 11529, Taiwan

Fermín J. Alcasena
Agriculture and Forest Engineering Department (EAGROF), University of Lleida, Alcalde Rovira Roure 191, Lleida 25198, Spain

Michele Salis
Euro-Mediterranean Center on Climate Change (CMCC), IAFES Division, Via Enrico De Nicola 9, Sassari 07100, Italy

Alan A. Ager
USDA Forest Service, Rocky Mountain Research Station, 72510 Coyote Road, Pendleton, OR 97801, USA

Rafael Castell
Bomberos de Navarra, Agencia Navarra de Emergencias (ANE), Calle Aoiz 35 bis 3, Pamplona 31004, Spain

Cristina Vega-García
Agriculture and Forest Engineering Department (EAGROF), University of Lleida, Alcalde Rovira Roure 191, Lleida 25198, Spain
Forest Sciences Centre of Catalonia, Carretera de Sant Llorenç de Morunys km 2, Solsona 25280, Spain

Huimei Wang
Key Laboratory of Forest Plant Ecology, Northeast Forestry University, Harbin 150040, China

WenjieWang
Key Laboratory of Forest Plant Ecology, Northeast Forestry University, Harbin 150040, China
Northeast Institute of Geography and Agricultural Ecology, Chinese Academy of Sciences, Changchun 130102, China

Scott X. Chang
Department of Renewable Resources, University of Alberta, 442 Earth Science Building, Edmonton, AB T6G 2E3, Canada

Anne-Christine Mupepele and Carsten F. Dormann
Biometry and Environmental System Analysis, University of Freiburg, 79106 Freiburg, Germany

Eugenia Iturritxa
Production and Plant Protection, Neiker Tecnalia, Apartado 46, Vitoria Gasteiz 01080, Spain

Tyler Trask and Cheryl L. Patten
Department of Biology, University of New Brunswick, P.O. Box 4400, Fredericton, NB E3B 5A3, Canada

Nebai Mesanza
Production and Plant Protection, Neiker Tecnalia, Apartado 46, Vitoria Gasteiz 01080, Spain
Department of Biology, University of New Brunswick, P.O. Box 4400, Fredericton, NB E3B 5A3, Canada

Rosa Raposo
INIA-CIFOR, Ctra. La Coruña Km 7.5, Madrid 28040, Spain
Sustainable Forest Management Research Institute, University of Valladolid-INIA, Avda. Madrid 44, Palencia 34004, Spain

Margarita Elvira-Recuenco
INIA-CIFOR, Ctra. La Coruña Km 7.5, Madrid 28040, Spain

Radu-Daniel Pintilii, Ion Andronache, Daniel Constantin Diaconu, Daniel Peptenatu, Cristian-Constantin Drăghici and Ana-Maria Ciobotaru
University of Bucharest—Research Center for Integrated Analysis and Territorial Management; 4-12, Regina Elisabeta Avenue; 030018 Bucharest, Romania

Răzvan Cătălin Dobrea
Bucharest University of Economic Studies—Faculty of Management; 6, Piața Romană Square; 010374 Bucharest, Romania

Martina Zeleňáková
Institute of Environmental Engineering, Technical University of Kosice, Vysokoskolska 4, Kosice 04200, Slovakia

Rasmus Fensholt
University of Copenhagen, Department of Geosciences and Natural Resource Management, Oster Voldgade 10, DK-1350 Copenhagen, Denmark

Huiyan Qin and Yingli Huang
Department of Forestry Economics, College of Economics & Management, Northeast Forestry University, Harbin 150040, China

Lingbo Dong
Department of Forest Management, College of Forestry, Northeast Forestry University, Harbin 150040, China

Ryer M. Becker and Robert F. Keefe
Forest Operations Research Lab, College of Natural Resources, University of Idaho, 875 Perimeter Drive, Moscow, ID 83844-3322, USA

Nathaniel M. Anderson
Rocky Mountain Research Station, U.S. Forest Service, Missoula, MN 59801, USA

Jennifer F. Lind-Riehl
Department of Entomology, University of Wisconsin-Madison, Madison, WI 53706-1598, USA

Oliver Gailing
School of Forest Resources and Environmental Science, Michigan Technological University, Houghton, MI 49931-1295, USA

Index